FOREWORD

This dictionary is a new addition to a series designed for use in schools. It is intended for students of geography up to A-level, but we hope that it will also be helpful to students of related subjects and to anyone interested in geography.

The book contains concise definitions of over 2500 headwords. In a book of this size it is not possible to include all the specialist terms used in geography. We have tried to include the more important terms that the student is likely to encounter at school level, but would welcome any comments regarding serious omissions or indeed about the content of any of the entries.

We would like to thank all the people who have cooperated in producing this book. The team of contributors is listed on the acknowledgements page.

D1824652

ACKNOWLEDGEMENTS

Series Editor
   John Daintith B.Sc. Ph.D.

Consultant Editors
   Laurie Bolwell M.A. Ph.D.
   Clifford Lines M.Sc.

Contributors
   Laurie Bolwell M.A. Ph.D.
   Jennifer Bridgland B.Sc.
   Robin J. Griffith-Jones B.A. M.A.
   Charles V. Hopkins B.A.
   David Hughes M.A.
   J.A.A. Jones B.A. A.K.C. M.Sc. Ph.D.
   Clifford Lines M.Sc.
   Christopher J. Lowe B.A. M.Sc.
   Martin Meggs B.A. M.Sc.
   Michael G. Rowan B.A.
   John Spencer B.Sc. P.G.C.E.

DICTIONARY
of
GEOGRAPHY

Editor
Jackie Smith B.A.

Published by Charles Letts and Co Ltd
London, Edinburgh and New York

Published 1983 by Charles Letts and Co Ltd
Diary House, Borough Road, London SE1 1DW

Copyright © 1983 Charles Letts and Co Ltd

ISBN 0 85097 579 4

Compiled and prepared for typesetting by
Market House Books Ltd, Aylesbury

Printed by Charles Letts (Scotland) Ltd

This book is a companion to

KEY FACTS DICTIONARY OF BIOLOGY
KEY FACTS DICTIONARY OF CHEMISTRY
KEY FACTS DICTIONARY OF ECONOMICS
KEY FACTS DICTIONARY OF MATHEMATICS
KEY FACTS DICTIONARY OF PHYSICS

HOW TO USE THIS BOOK

Headwords The headwords in the dictionary are printed in bold-face type. Synonyms of the term come immediately after the headword in brackets. For example:

pie graph (divided circle diagram) A graph . . .

Here 'divided circle diagram' is simply another term for 'pie graph'. The same style is used for abbreviations. For instance the entry:

Greenwich Mean Time (GMT) The mean local time . . .

'GMT' is a common abbreviation for 'Greenwich Mean Time'.

Level markers Throughout the book we have tried to separate the text into two levels by using the marker †. In reading through an entry information up to the † is considered suitable for lower-level work; information after the † is suitable for a higher level. For instance, the first part of the entry for 'atoll':

A circular or horse-shoe shaped coral reef, enclosing within it a lagoon.

is lower-level information. The remainder of the entry:

† The origin of atolls . . .

is additional higher level information.

In using these level markers there are three points to note:

(1) Certain headwords have two or more separate definitions, numbered 1, 2, etc. Each definition is treated as an entirely separate entry from the point of view of level.

(2) The cross references (see below) are all placed at the end of the definition irrespective of level.

(3) The 'lower' and 'higher' level sections of the text are designed to correspond roughly to O- and A-level work respectively. However, there is a variation in the content of different syllabuses and in the treatment given by different teachers. This book has not been based on the syllabus of any single examining board, but covers all boards. Consequently, the level markers should **not** be treated as definite indication of the difference between O- and A-level work.

Cross references These direct the user to other entries at which additional information may be found. All cross references are placed at the end of the definition. Unless they directly follow higher level text or are specifically marked they apply to both lower and higher levels of content.

A

aa A volcanic lava with a rough blocky structure. *Compare* pahoeĕoe. [Hawaiian]

ablation The process in which snow and ice is lost from the surface of an ice sheet or glacier. This may be through melting, sublimation (the direct transference of water to the atmosphere from the solid to the gaseous state), or abrasion.

Abney level An instrument used in surveying to measure the degree of slope from an observer to an object. It consists of a small telescope linked to a semicircular protractor, and a pointer with a spirit level attached. An object sighted through the telescope that is not on a level with the observer will throw the bubble in the spirit level out of centre. By adjusting a screw the bubble can be brought into view in the telescope. The pointer then shows how much the telescope is tilted from the horizontal.

aborigines The earliest known inhabitants of a land or those found in possession of the land by early colonists. The term may be applied to human inhabitants, animals, and plants. The word is now applied in particular to the Aborigines of Australia.

abrasion The wearing down of a land surface by moving weathered rock debris. This debris may be transported by water (e.g. by rivers, ice, or sea) or by wind. *See also* abrasion platform, scour.

abrasion platform A level rock platform cut into a slope or cliff. It is most commonly a feature of coastal areas, where it is also known as a wave-cut platform, and is generally produced in hard resistant rocks.

absolute drought *See* drought.

absolute humidity The amount of water vapour per unit volume of air, usually expressed in grams per cubic metre. *Compare* relative humidity, specific humidity.

absolute temperature (thermodynamic temperature) A temperature measured on a scale based on the lowest temperature that is physically possible ($-273.15°C$) known as absolute zero or 0 kelvin (K). It was formerly measured in degrees absolute (°A) but is now measured in kelvins (the degree sign is now officially omitted). For meteorological purposes the absolute temperature is taken as the Celsius temperature plus 273°C; for example $-10°C$ is 263 K and 100°C (boiling point) is 373 K.

absorption In meteorology, the capacity of the atmosphere to transform radiation into a different form of energy instead of transmitting or reflecting it. About 15% of the radiation from the Sun that reaches the upper layers of the atmosphere is absorbed, mainly by the water vapour, carbon dioxide, and ozone within the atmosphere, before it reaches the Earth's surface.

abyssal Denoting the deepest parts of the ocean, generally taken to be those areas with a depth of about 4000 m or more. Conditions are fairly uniform at such depths; no sunlight penetrates, the water is cold, and the pressure of the water is very great. No plant life is able to exist but some specialized animals have adapted to the environment. These feed either on each other or on particles of organic matter that drift down from the surface waters.

abyssal plain (deep-sea plain) The generally level area of the ocean floor that lies between about 3500 m and 5500 m below the surface of the ocean. The regularity of the abyssal plain is broken by such features as trenches, mid-ocean ridges, sea mounts, etc. The ocean floor at these depths is covered by ooze or red clay.

accessibility The degree to which a location is capable of being reached or is 'get-at-able'. Accessibility has an economic meaning – the extent to which a resource is within reach of economic exploitation, e.g. a forest. It also has a social meaning – the extent to which different groups are able, irrespective of geographical location, to obtain services or goods, e.g. because of differences in income different groups have differing access to private housing.

accessibility isopleth A line on a map that joins places having equal access to a particular location, e.g. places with equal degrees of accessibility to an out-of-town shopping centre.

access road A road that provides an approach or entry to a given point, e.g. to a motorway entry point, to an industrial estate, or to the warehouse bays of shops in a pedestrian precinct.

acclimatize To adapt or become accustomed to a new environmental condition. Usually it means to adapt to a different climatic regime though sometimes the term is applied when plants adapt to changed soil conditions. In the USA the word acclimate is used in place of acclimatize.

accordant drainage A pattern of drainage that closely reflects the rock type and structure over which the streams flow.

accretion 1. In meteorology, the process by which ice crystals grow through collision with water droplets in clouds.
2. The accumulation of materials.

accumulated temperature †A measure of the variation of temperature from a given norm for a particular place over a specified period, measured in degree-days. This is most commonly used for determining whether an area is suitable for the cultivation of particular crops. Over a given period (e.g. 30 days) the difference between each mean daily temperature and the base value (e.g. in the UK 5.5°C is taken as the temperature below which plant growth is inhibited) are added together, the result indicating the

effectiveness of temperature over the given period.

acid lava A stiff viscous molten rock that flows slowly, has a high melting point (about 850°C), and congeals into thick tongues before having travelled far from the volcanic vent. The resulting landform is a steep-sided dome, typically represented by the Puys of the Auvergne, France. Acid lava is rich in silica (at least 66%), and contains quartz crystals (at least 10%). Examples from the geological record can be found in Cumbria, Fife, and Snowdonia. *See also* basic lava, intermediate lava.

acid rock An igneous rock that is very rich in silica (over 66%) and aluminium, and which has at least 10% of its composition made up of quartz crystals. It is relatively light and forms the main material of the continents. Examples of acid rocks include granite, obsidian, and rhyolite.

acid soil A soil with a pH value below 7.0. †Acidity is associated with abundant hydrogen ions and a lack of exchangeable bases, especially calcium. In the soil this may result from such factors as leaching, acid rainwater, organic matter containing few bases, or acid parent material. In moist climates the prevailing process of leaching acidifies soils; most British soils, for example, are slightly acid. Examples of acid soils include podzols and brown earths.
See pH.

action area In the UK, an area of land indicated in a structure plan where major change is expected. This can be by improvement, development, or redevelopment, or some combination of these, by public or private bodies. Action areas were introduced by the 1968 Town and Country Planning Act. *See* structure plan.

active fault A break in the Earth's crust along which movement is taking place. Earthquakes occur as a result of this movement. An example is the San

2

Andreas Fault, which runs through California, USA; the San Francisco earthquake of 1912, which virtually destroyed the city, was the result of movement along this fault.

active layer †In permafrost regions, the top layer of soil that seasonally thaws; this can vary in depth from a few centimetres to several metres. The active layer is important for the production of a wide range of periglacial landforms. *See also* frost heaving, patterned ground, solifluction, thermokarst.

active volcano A volcano that continually or periodically erupts. *Compare* dormant volcano, extinct volcano.

adiabatic lapse rate †*See* adiabatic process.

adiabatic process †An atmospheric process in which changes take place in the pressure and volume of a body of gas without an actual loss or gain of heat from outside. As a consequence, the temperature of the air mass changes. Adiabatic changes commonly occur in an ascending or descending air mass. As a parcel of air is displaced to an environment of lower pressure (i.e. when it rises) its volume increases and its temperature consequently falls. The rate at which temperature decreases in rising expanding air is called the *adiabatic lapse rate*. If the rising air parcel is unsaturated it cools at the dry adiabatic lapse rate, which has a constant value of 0.98°C per 100 m. If the rising air is saturated it cools at the saturated adiabatic lapse rate. This differs from the dry adiabatic rate because of latent heat released by condensation of vapour. Unlike the dry rate, the saturated rate depends on the air temperature, being higher at low temperatures. *See also* dry adiabatic lapse rate, saturated adiabatic lapse rate, lapse rate.

adit A horizontal or nearly horizontal passage cut into a hillside for mining purposes. The passage is excavated to reach coal seams or veins of mineral ore. Adits are sometimes used to drain away excess water from mines. Examples include some of the coal mines in Alabama, USA, and tin mines in Bolivia.

administrative centre A settlement that is responsible for the control of its region. It may be the capital on a national scale or an important town on a local scale.

adobe Originally, an unburnt brick that was dried in the sun. The term has now been extended to:
(1) the earth or clay from which the brick is made. The word is used particularly in the W USA where it is now used to describe any alluvial or playa clay in arid regions.
(2) a house or building made from the clay or sun-dried bricks.

adret A slope that faces towards the equator, therefore receiving more sunshine than slopes that face away from the Sun. Adret slopes tend to be better drained and warmer, and have an important influence on agriculture, especially in mountainous regions. *Compare* ubac. *See also* Sonnenseite. [French]

adsorption The adhesion of particles of one substance to the surface of another, such as the attraction of liquids or gases to a solid surface. In soils, this includes the concentration of base salts (e.g. calcium and potassium) of microscopic size (colloids) surrounding larger mineral particles. Adsorption also refers to the process by which a thin film of water surrounds individual particles in sediments.

advection The horizontal movement of an air mass across the Earth's surface. Such movement is especially important when the air mass is moving from a different source area and is bringing in different conditions of temperature and humidity. Coastal and frontal areas often experience this and advection fog frequently results.

advection fog Fog formed by the horizontal movement of warm moist air over a cooler surface, which reduces the temperature of the lower layers of the air to give condensation. It occurs in summer over cool seas; for example, near the Grand Banks of Newfoundland, Canada, and off the coast of California.

aeolian Denoting the processes or landforms caused by or related to the wind. These processes or forms are most significant in arid and semiarid regions, but can also be important in some coastal and periglacial environments.

aerial photograph A photograph of the Earth's surface taken from an aircraft. There are two types of aerial photograph. The first is the vertical aerial photograph in which the camera points vertically downwards producing a photograph resembling a plan of the ground. The second type is the oblique aerial photograph where the camera axis points at an angle to the ground producing a view of the ground similar to that obtained from a high building or from an aircraft window. *See also* stereoscope.

age The smallest division of geological time. The rocks laid down during an age are known as a formation. A number of ages together comprise an epoch.

age pyramid (age/sex pyramid) A type of bar graph that illustrates the structure of a population. Adjoined horizontal bars are placed on either side of a central vertical axis, which represents age categories. The columns representing the male and female population are shown on either side of the central axis. The length of each column varies with either the number or percentage of the total population in each group.

agglomerate A rock made up of large fragments of angular material cemented together by finer material. The term is normally applied to rocks

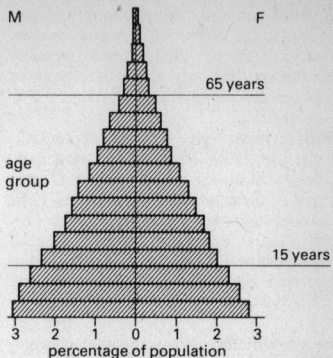

Age pyramid

formed by explosive volcanic activity and in these the cement is of ash or tuff. *See also* breccia.

aggradation The process by which a land surface is built up by the deposition of debris and other solid material, especially by rivers. †It usually results from the river's load-bearing capacity being reduced. This may be as a consequence of an increase in the amount of debris supplied to the river, a loss of speed or volume of flow, or a rise in base level. Material is deposited on the stream bed, which accumulates with time. *Compare* degradation.

aggregate A mass or cluster of particles, most commonly used in reference to soils.

agricultural region An extensive area of land that produces a distinctive pattern of agriculture, different from that to be found in adjacent areas. For example, the wheat-alfalfa-cattle crescent of the Argentine Pampas and the dairy-farming regions of North Island, New Zealand. †A number of models have been developed to explain regional variations in agriculture, the best known are those by von Thünen (1826) and O. Jonasson (1926). O. E. Baker (1925), D. Whittlesey (1927), and others have attempted to identify and delimit agricultural regions.

agriculture The cultivation of the soil in order to grow crops and rear livestock. The essential purpose of agriculture is the production of food from the land for human or animal consumption.

agronomy The science of managing the land. It includes the theoretical background and practical aspects of growing crops, maintaining the soil's fertility, and the rearing and care of farm animals.

A horizon A soil horizon consisting of well mixed organic and inorganic material that lies near the ground surface. †A horizons usually include a large proportion of humus, which gives the layer its dark coloration, and have had many of their soluble minerals removed by leaching (i.e. they are eluvial zones). A horizons are subdivided according to the precise qualities of their humus.
See soil horizon. *See also* soil profile.

air mass A mass of air, with similar properties of temperature and moisture covering a large area of the Earth's surface and bounded by fronts. According to the nature of their source regions, air masses may be classified on a basis of temperature (tropical or polar) and humidity (maritime or continental); they can therefore be polar maritime (mP), polar continental (cP), tropical ∙maritime (mT) or tropical continental (cT).

airport A location with facilities for the arrival and departure of civil aircraft. The facilities available will vary depending on the size and significance of the airport. Although some airports, such as those at Bangkok and Abu Dhabi, are also used by service aircraft, the principal facilities are for the use of passengers and freight and the service facilities are confined to one part of the airport.

air stream Any movement of air that is constant in direction and of reasonably long duration, e.g. the westerly air stream. It is more marked at

higher levels in the atmosphere where surface disturbances do not occur. *See also* jet stream.

ait (eyot) A small island in the middle of a river.

Aitoff's projection †An equal-area map projection in which the Earth's shape is represented by an ellipse. The equator is twice the length of the central meridian and both are straight lines; all other meridians and parallels are curved. The angles of intersection of the meridians and parallels alter only slightly towards the margins. Near the centre of the projection land areas have a fairly accurate shape but at the E and W margins there is considerable distortion.

albedo (reflection coefficient) The proportion of radiation reflected by a surface, often expressed as a percentage. The albedo of cloud tops is 30–80%. For the Earth as a whole it averages about 40% and ranges from about 9% over coniferous forest to over 35% above sandy desert areas.

alfalfa (lucerne) A green clover-like plant grown as a feed crop for cattle and sheep. Deep roots enable the plant to produce rich foliage even in light porous soils, such as those of the middle Pampas in Argentina. High yields over a period of five to eight years after planting make this an ideal forage crop. It is grown extensively in the Midwest and W states of the USA. In W Europe it is called lucerne. [Spanish]

alidade A surveying instrument used for drawing directions in plane table surveys. In its simplest form it consists of a ruler with raised sights, a slot at one end and a hair line at the other. Having lined up the sights with a distant object the edge of the ruler can be used for drawing a direction line (ray). More elaborate alidades include telescopic sights and can also be used for measuring vertical angles. *See* plane table.

alkali flat A plain of sediment with a high proportion of alkali salts, characteristic of arid and semiarid regions. It is produced by the evaporation of a shallow mineral-rich lake, often in the dry season, which leaves behind a residue with a high concentration of sodium and potassium. In the wet season desert streams flowing into these areas produce playa lakes.

alkaline rock An igneous rock that is rich in feldspars composed chiefly of sodium and potassium silicates; it contains 45–52% silica and magnesium is the second most abundant mineral. It is dark and heavy, and forms the foundation of the ocean floor, extending underneath the continents. The term is now becoming obsolete.

allotment A small portion of land rented by an individual as a cultivated garden at a distance from the home. The land may be privately or publicly owned. On old maps such as tithe plans an allotment refers to a portion of land that has been assigned to a particular person or to a specific purpose, e.g. fuel allotments on areas of peat.

alluvial *See* alluvium.

alluvial cone A cone-shaped accumulation of thick coarse alluvial material similar to, but steeper than, an alluvial fan. The landform is more typical of arid and semiarid areas where short-lived torrents carrying large loads are abruptly checked on entering a main valley or plain. The stream drops most of its load, no longer having the capacity to carry it, and steep deposits rapidly build up.

alluvial fan A fan-shaped landform composed of alluvium deposited where a constricted river enters a main valley or emerges from mountains onto a plain. The apex of the fan is located at the point at which the river emerges. The speed of the river is abruptly checked at this point, it divides to form several distributaries and deposits its load. Examples include the alluvial fans of the upper Rhône Valley, Switzerland, where mountain torrents leave the Bernese and Pennine Alps for the main valley.

alluvial flat A horizontal surface composed of alluvium deposited by a river in times of flood. Extensive flat areas of alluvium are known as *alluvial plains*.

alluvial terrace A level area of alluvium that lies above the present river level. Alluvial terraces are formed when a rejuvenated meandering river cuts into alluvial deposits on an old floodplain or lake bed leaving the remains, often as a pair on either side of the river, at a higher level. River terraces are similar in their formation but are composed of coarser material than alluvium.

alluvium Material deposited by a river along its course; this may be as cones, fans, deltas, and floodplains and includes deposits on the stream bed, in estuaries, and in lakes. The term is sometimes restricted to particles of silt size (e.g. by the Geological Survey of the UK) but in other definitions it may be extended to include sands and gravels. Alluvial deposits are agriculturally fertile, although subject to flooding, and can contain important minerals such as gold, platinum, and diamonds.

alp 1. A high mountain.
2. The upland pasture on the side or shoulder of a mountain, especially in Switzerland. Animals are driven up to the pasture in summer.

Alpine Referring to or belonging to the Alps, but also used to refer to the physical features (relief, climate, soils, etc.) of mid-latitude mountain environments.

alpine glacier *See* valley glacier.

altimeter An instrument for measuring altitude, which is used in aircraft and for surveying. The altitude measured may be the height above the ground

6

or the height above mean sea level. The two chief types of altimeter are the *pressure altimeter*, a form of aneroid barometer that measures changes in atmospheric pressure with height, and the *radio altimeter*, which measures the time taken by radio waves from an aircraft to reflect back from the Earth's surface.

altiplano A region of elevated plateaus in the Andes of Latin America, lying at heights of approximately 3500– 4500 m above sea level. It lies mainly within W Bolivia, but extends into Peru in the N and Argentina in the S. Although climatic conditions are harsh it is of significant agricultural value to the Bolivian Indians, producing the staple diet of potatoes and barley. [Spanish]

altitude 1. The height of an object or place above mean sea level.
2. (*Astronomy*) The angle of a heavenly body above the horizon, for example the altitude of the Sun at noon on 21 June at the tropic of Cancer is 90°.
3. (*Surveying*) The angle of elevation between the horizontal plane of the observer and a higher point.

altocumulus (Ac) A fleecy cloud that occurs in bands or waves. Blue sky can usually be seen between the clouds although they sometimes join up. Altocumulus clouds occur at about 2400–6000 m. They are usually a sign of fair weather.

altostratus (As) A cloud formation occurring at medium heights in the atmosphere, with a base level of 2000 –7000 m, and composed of either ice particles or water droplets or a mixture of both. It usually forms a greyish sheet covering much of the sky, through the thinner parts of which the Sun may be dimly seen although it displays no halo. Precipitation is not common but altostratus cloud usually indicates rain to come as it heralds the approach of a warm front.

aluminium A silvery-grey metal. It is the most abundant metallic element and the third most abundant element after oxygen and silica, comprising 8.13% by weight of the Earth's crust. As it is extremely light, relatively strong, resists corrosion, and is a good conductor of electricity it is extensively used in the manufacture of motor cars, aircraft, electrical goods, and kitchen utensils. The metal is extracted chiefly from bauxite ore, which is widely found in the tropics (e.g. in Guinea, Guyana, Jamaica, and W Africa) but the smelting is carried out in the USA, Canada, and N Europe.

anabatic wind A local wind that blows upslope during the afternoon in mountain areas. Air above the valley slopes is heated to a greater extent than air at the same height above the centre of the valley. Air thus rises up from the slopes and feeds an upper return current. *Compare* katabatic wind. *See* valley wind.

anafront †A frontal surface at which warm air is rising over a wedge of cold air. Condensation and precipitation are more extensive with this type of front. *Compare* katafront.

anemometer An instrument that measures the velocity and often direction of the wind. The most common is the *cup anemometer*, in which symmetrically mounted cups rotate on a vertical axis, the speed of rotation indicating the strength of the wind. Others include the *pressure-tube anemometer*, in which the indication is the pressure built up in a tube that is closed at one end; the *pressure-plate anemometer*, which measures the pressure exerted by the wind on a metal plate; and the *hot-wire anemometer*, in which the current required to keep the temperature of a platinum wire constant indicates the wind speed. An *anemograph* is an anemometer that automatically records wind speed and direction on a continuous trace.

7

aneroid barometer An instrument for measuring atmospheric pressure. It consists of a metal box from which the air is virtually exhausted. The sides are flexible and expand and contract with changes in air pressure. These movements are amplified and register on a needle on a calibrated circular dial. The instrument is light and portable making it useful for measuring height in the field; it gives readings accurate to ±1.5 m under favourable conditions.

annular drainage A pattern of drainage in which streams follow roughly circular paths. Such patterns are usually produced on domed structures where the rivers follow the outcrops of the weaker beds of rock. *See also* trellis drainage.

Antarctic 1. That part of the Earth's surface lying within the Antarctic Circle.
2. Denoting the S polar regions within which Antarctica lies (e.g. the Antarctic ice sheet). The term is also applied to features or conditions (e.g. climatic) that resemble those of the S polar regions but occur elsewhere. *Compare* Arctic.

Antarctic Circle The line or parallel of latitude 66°32′S. On about 22 December (midsummer) along this parallel of latitude the Sun does not set below the horizon and on about 21 June (midwinter), it does not rise. The number of such days increases towards the South Pole. At the pole itself there are six months when the Sun never sets and six months when it never rises. This happens because the Earth's axis is inclined at an angle of 66½° relative to the plane of its orbit, so that the South Pole is tilted towards the Sun in December, and away from it in June. *See also* Arctic Circle.

antecedent drainage (antecedence) A form of discordant drainage in which the drainage pattern exists before the onset of a period of tectonic folding and uplift; the rivers downcut through the folds that rise in their path at approximately the same rate, thus maintaining their general pattern and direction. For example, the Brahmaputra River, which has carved deep gorges through the E Himalayas. *See also* discordant drainage.

anthracite A type of hard brittle shiny coal. It is formed when bituminous coal is subjected to great pressure, resulting from deep burial and earth movement, accompanied by a slight increase in temperature, which drives off nearly all the volatile material. This forms a coal high in carbon content (85–98%) and low in oxygen and hydrogen. Anthracite burns at a high temperature with a short smokeless flame. It occurs in areas of tightly folded rocks, e.g. S Wales.

anthropogeography The subsection of geography that focuses upon the study of the products of human social life in relation to the Earth. Some writers have defined anthropogeography as another name for human geography. Others have defined it more narrowly, e.g. geographical studies in social and physical anthropology. As a result of this confusion the term is not much used today.

anticline A rock structure in which the beds of rock are folded into an arch, i.e. the beds dip outwards from the crest, the oldest rocks occurring in the core. Anticlines alternate with synclines and are formed through compression of the rock layers.

anticyclone (high) An area of high pressure in the atmosphere. It is shown on a weather chart by a series of roughly circular closed isobars around the area of highest atmospheric pressure. Winds, which are light, circulate in a clockwise direction around the high-pressure system in the N hemisphere and in an anticlockwise direction in the S hemisphere. An anticyclone generally moves more slowly than a cyclone and is a shallower feature. Two types of anticyclone are recognized. A *cold anticy-*

clone results from radiational cooling by contact with a cold land surface (e.g. over Canada and Siberia in winter). The cold air is confined to the lower levels of the anticyclone. A *warm anticyclone* has warm air in the lower parts of the troposphere over which lies cold air in the stratosphere. Warm anticyclones are generally located in the subtropical high-pressure belts. They occasionally move polewards to form a blocking high in the higher latitudes. Calm weather conditions are generally associated with anticyclones as a result of the subsidence of air. In summer fine conditions are general; in winter the presence of an anticyclone may bring clear and frosty or foggy conditions. *Compare* depression. *See also* blocking high.

antimeridian The half of the great circle on the opposite side of the globe from a meridian. The antimeridian is always 180° from a given meridian. For example, the antimeridian to longitude 0° is 180°, which passes through the Pacific Ocean.

antipodes Points on the surface of the Earth directly opposite to each other if a straight line were drawn through the centre of the Earth. Australia and New Zealand, which lie roughly on the opposite side of the Earth to the British Isles, are often referred to as the Antipodes. [From Greek *antipous*: having the feet opposite]

antitrades Winds blowing in a generally westerly direction in the upper atmosphere (about 1800 m) above the northeast or southeast trade winds. (The term was formerly used to describe surface westerly winds in high latitudes but this usage is now obsolete.)

anvil cloud A flattening out of the top of a cumulonimbus cloud resembling, when seen from the ground, the shape of a blacksmith's anvil. It forms when the top of a cumulonimbus cloud reaches the base of the stratosphere and high-level winds cause the ice crystals at that level to spread out horizontally.

apartheid The policy of racial separation, which has been practised in the Republic of South Africa since 1948. The White and non-White populations are segregated residentially, economically, and socially. [Afrikaans: apartness]

aphotic zone That part of the ocean below the level to which sufficient sunlight is able to penetrate to allow photosynthesis to occur (about 90 m). The development of plant life is therefore prevented in this zone. *Compare* photic zone.

apparent solar day *See* solar day.

apparent time *See* local time.

applied climatology The use of the science of climatology in the prediction (forecasting) of future conditions, both long and short term. This is of particular importance in agricultural planning.

applied geography The application of geographical techniques to the practical investigation and solution of problems at both local and world scale. An example is the use of the cartographic and survey methods and analytical techniques of the geographer in such fields as town and country planning, conservation, and development studies.

aquifer (aquafer) A bed of rock that holds water and also allows water to flow freely through it. *See also* artesian basin.

arable land Farm land that is suitable for ploughing or cultivation or is actually ploughed and cropped. It includes land that is temporarily fallow or cultivated with rotation grasses. The first British Land Use Survey (1931) included market gardens as arable whereas in the Second Land Use Survey (1961) market gardening was given a separate category.

arboriculture The cultivation and study of trees and shrubs. The term is normally used by geographers in connection with the growth and maintenance of forest trees such as larch, spruce, and beech. In the UK the Forestry Commission is responsible for the forests, nearly all of which are owned by the Commission under the supervision of the state.

arc A group of islands or volcanoes that lie in the form of a curved string. An arc usually follows a line of folding and weakness in the Earth's crust, which gives rise to the volcanic activity and earth movements.

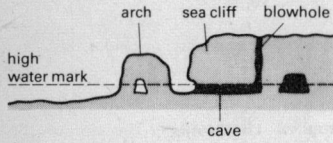

Arch

arch A natural opening or arch through a mass of solid rock. It is most commonly the result of marine erosion. Caves cut into coastal headlands may be further eroded by the action of the sea until an opening is worn right through the headland, forming an arch. With the continuation of erosion the arch will eventually collapse leaving a remnant block known as a stack, e.g. Durdle Door, Dorset.

Archaean A term usually regarded as synonymous with Precambrian. In the USA and Canada it is used to refer to the most ancient Precambrian strata.

archipelago A group of islands that lie in fairly close proximity. Examples include the islands of the Aegean Sea, to which the term was originally applied, and the Tuamotu Archipelago. [From Italian]

arc of the meridian The path taken by a meridian or a part of a meridian. Each meridian (line of longitude) passes from the North to the South Pole and each forms an imaginary arc on the Earth's surface with places along the arc having the same degree of longitude.

Arctic That area of the Earth's surface lying within the Arctic Circle. In a wider sense the term is also used to describe those areas where climatic conditions resemble those of the Arctic proper. *Compare* Antarctic.

Arctic air mass (A) A very cold air mass that has its source region over the Arctic Ocean. In many cases this term is used synonymously with polar air mass.

Arctic Circle The line or parallel of latitude 66°32'N. On about 21 June (midsummer) along this parallel of latitude the Sun does not set below the horizon and on about 22 December (midwinter), it does not rise. The number of such days increases towards the North Pole. At the pole itself there are six months when the Sun never sets and six months when it never rises. This happens because the Earth's axis is inclined at an angle of 66½° relative to the plane of its orbit, so that the North Pole is tilted towards the Sun in June, and away from it in December. *See also* Antarctic Circle.

Arctic front †A semipermanent but fairly inactive frontal zone found to the N of the polar front in latitudes 50°N–60°N in the N hemisphere where cold Arctic air meets cool polar air. *See also* Bjerknes polar front model.

arcuate delta A fan-shaped delta with its rounded margin extending into the sea. The fan shape is a result of the division and subdivision of channels. The classic example of an arcuate delta is the Nile Delta. *See also* delta.

area 1. The extent of a particular expanse of land or surface enclosed within a specified boundary; it is measured in square units; e.g. square kilometres (km²).
2. Any expanse of land forming a region, district, or locality.

Area of Outstanding Natural Beauty
(AONB) In the UK, a tract of land
that, because of its particular landscape value, is considered worthy of
protection above that elsewhere in the
countryside, with the exception of the
national parks. Areas of Outstanding
Natural Beauty are the responsibility
of local planning authorities who have
powers to preserve and enhance the
natural beauty of the area. Designated
areas include the Gower Peninsula,
the Malvern Hills, and the Chiltern
Hills. *See also* national park.

arenaceous Denoting a medium-
grained sedimentary rock with a sandy
texture, such as sandstone and coarse
siltstone. Arenaceous rocks are also
often collectively known as sandstones. They have a grain size (on the
Wentworth scale) of 1/16 mm to
2 mm.

arête A sharp jagged mountain ridge
with steep sides found in glaciated or
formerly glaciated areas. It is formed
when the headwalls of two adjacent
cirques are eroded back until they
meet. *See also* cirque. [French]

argillaceous Denoting a fine-grained
sedimentary rock composed largely of
clay particles, with a grain size (on
the Wentworth scale) of less than 1/16
mm. Argillaceous rocks include siltstones, clays, mudstones, shales, and
marls.

arid Denoting any climate or region in
which the rainfall is insufficient or
barely sufficient to support vegetation.
It is sometimes defined as an area
having an average annual rainfall of
less than 250 mm but this is not a
generally accepted definition.

arroyo The bed of a stream in arid
and semiarid areas. Usually these beds
are dry but following rainfall they
become occupied by short-lived torrential streams, which are strongly
erosive producing steep valley sides
and leaving the valley floors strewn
with debris. This term is usually lim-

ited in its use to North and South
America. *See also* wadi. [Spanish]

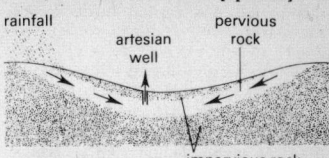

Artesian basin

artesian basin A geological structure
in which the beds dip toward a central point forming a basin with a permeable layer of rock (aquifer) sandwiched between impermeable beds.
Water seeping into the aquifer is
trapped between the impermeable layers where it accumulates and eventually the rock becomes saturated. The
chief example in the UK is the
London Basin, in which permeable
chalk and Lower Eocene sandstones
lie below the London Clay and above
the Gault Clay.

A boring sunk into an artesian basin
is known as an *artesian well*. If the
point at which the well is sunk lies
below the level of the water table in
the aquifer, water will be forced
upwards to the ground surface by
hydrostatic pressure. Artesian wells are
the main source of water in many arid
and semiarid regions, such as parts of
Australia and the USA.

asbestos A fibrous mineral that occurs
in several forms, the most important
commercially being chrysotile. It can
be spun and woven into cloth or compressed into boards but its most
important properties are its resistance
to heat and decay; for these reasons
its main uses are for making fireproof
clothing and in the construction
industry. The major noncommunist
producers are Canada, South Africa,
Zimbabwe, and Italy.

ash Fine-grained particles of lava
thrown out during volcanic eruptions.
The term is usually applied to particles that are larger than volcanic dust,
but occasionally the terms are used
synonymously. Volcanic ash may be
carried great distances from the erup-

tion by the prevailing wind, e.g. following the eruption of Krakatoa (1883) ash was carried twice around the world.

ash cone A cone-shaped hill composed of volcanic ash that is built up around a volcanic vent. Ash produced during eruptions will tend to settle around its point of origin, building up successive layers with each subsequent eruption. Examples include the ash cones in Iceland.

aspect The direction in which a slope faces. The amount of sunshine received by a slope will vary according to its aspect. This has a strong influence on the development of soils, vegetation, and drainage and on the location of settlements. *See also* adret, ubac.

association (*Biogeography*) A group of plant species that share very similar environmental conditions. Associations are divisions of major plant communities, each association having its own distinctive dominant species. For example, bracken and heather associations may contribute to the overall community of dry heathland. *See also* community, stand.

asthenosphere A zone of rocks forming part of the Earth's upper mantle, lying immediately below the lithosphere between about 50 and 240 km below the Earth's surface. The rock in this zone is in a semimolten plastic state and it is thought that this enables the lithosphere to move above it. Within the asthenosphere the velocity of seismic waves is considerably reduced. *See also* lithosphere.

asymmetrical fold An anticline or syncline in which one of the limbs has a steeper dip than the other. It is the stage between a simple fold and an overfold and usually indicates a medium degree of folding pressure.

asymmetrical valley A valley in which the slopes on one side are steeper than those on the other. Many such valleys in the British Isles and Europe

are thought to result from periglacial processes, which tend to exaggerate differences in slopes because of variations in aspect.

Atlantic polar front †The frontal zone of the N Atlantic lying to the S of the Arctic front where cold polar air and warm tropical air converge. Depressions are formed along this front and it greatly affects the weather of Western Europe. *See also* Bjerknes polar front model.

Atlantic type of coastline (discordant coast, transverse coast) A coastline that cuts across the general structural trend of an area. It produces a characteristically irregular outline with drowned valleys and strongly eroded headlands with steep cliffs, such as the coastline of SW Ireland. *Compare* concordant coast.

atlas A collection of maps bound in a volume. [The term was first used by Gerhard Mercator in 1595 in the title of a collection of maps. It was a common practice at that time to ornament the title page of such collections with an illustration of the mythical Atlas, who bore the world on his shoulders]

atmometer *See* evaporimeter.

atmosphere The layer of gases enveloping the Earth. It is composed of nitrogen (78.0%), oxygen (20.95%), argon (0.93%), and carbon dioxide (0.03%), together with small proportions of other gases and varying amounts of water vapour. Some solid particles from such sources as volcanic eruptions, sea salt, and wind-blown dust are also contained in the atmosphere. Nearly 97% of the atmosphere lies within 29 km of the surface of the Earth although its outer edge lies at about 10 000 km. There are several layers in the atmosphere, the lowest of which is the troposphere in which most meteorological phenomena occur. Above this lie the stratosphere, mesosphere, and thermosphere.

atmospheric circulation *See* general circulation of the atmosphere.

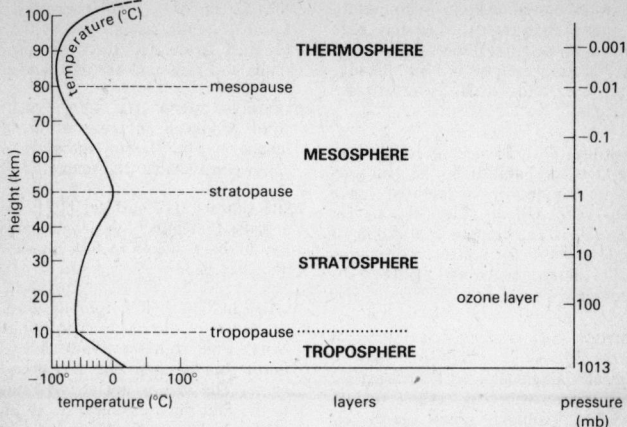

Atmosphere

atmospheric pressure The pressure exerted by the atmosphere as a result of its weight above a unit area of the Earth's surface. It is expressed in millibars (mb) and measured with a mercury barometer. The average atmospheric pressure at sea level is 1013.25 mb. Sea-level values can range from extremes of about 890 mb in hurricanes to 1060 mb in anticyclones. Atmospheric pressure decreases logarithmically with height.

atoll A circular or horseshoe shaped coral reef, enclosing within it a lagoon. †The origin of atolls has been a controversial issue within geomorphology for over 100 years. Charles Darwin suggested that atolls begin as fringing reefs surrounding volcanic islands; as the islands gradually subside, the fringing reefs grow upwards so that the top of the reef remains at sea level. When the original island is completely submerged a circular reef will remain. Some reefs, however, occur in areas where there has been no subsidence. R. A. Daly put forward another theory in which he suggested that a rise in sea level might be responsible. He attributed this to the gradual melting of ice following the Quaternary glaciation, returning water

to the sea and raising its level. Recent research, including borings, has given credibility to Darwin's theories of submergence.

attrition The wearing down of rock particles by other rock particles while being transported by water, wind, or ice. This process results in the particles becoming gradually smaller, smoother, and more rounded, e.g. beach pebbles.

aurora A luminous phenomenon seen in the sky at night in high latitudes. It may be visible as arcs of light or as coloured curtains, streamers, and rays. It is known as the *aurora borealis* (or *northern lights*) in the N hemisphere and as the *aurora australis* (or *southern lights*) in the S hemisphere. †The aurorae are produced by charged particles from the Sun captured by the Earth's magnetic field at heights of about 100 km.

autobahn In Germany, a road with four or more lanes, without intersections and with limited access, designed to enable motor vehicles to maintain continuous high speeds for long distances without stopping. For example, the autobahn from Bonn to Frankfurt. In many countries a word has been created to describe these roads; in the UK it is motorway, in France autoroute, and in Italy autostrada. [German]

autonomy The practice of self-government or the ability to be self-governing. Autonomy is essential for a truly independent state. Spain, for example, is autonomous but within its borders the Basque region is fighting for its independence and therefore its autonomy.

autotroph An organism that obtains its energy directly from inorganic sources, such as sunlight or a rock surface, and from these synthesizes organic materials. For example, green plants are autotrophic. †Autotrophs are primary producers, belonging to the first trophic level, and thus supply energy for other life forms. Most autotrophs convert solar energy by photosynthesis (i.e. are *photoautotrophic*), but others such as some blue-green algae convert chemical energy (i.e. are *chemoautotrophic*).
Compare heterotroph.

autumn The season of the year between summer and winter. In the N hemisphere it is defined as extending from about 21 September, the autumnal equinox, to about 22 December, the winter solstice; in the S hemisphere it extends from about 21 March to about 21 June. More commonly autumn is understood to include the months of September, October, and November in the N hemisphere.

avalanche A slide or fall of a mass of snow and ice in mountainous areas, often carrying with it much rock debris and vegetation. Avalanches occur when snow and ice accumulate to such an extent that the mass is unable to support itself and slides downslope under the influence of gravity. Large-scale avalanches can be very destructive with loss of life and property, particularly in populated areas such as

the Swiss Alps, and measures are taken to control them, such as the use of steel avalanche sheds to protect roads and railways. *See also* landslide.

avalanche wind The fierce rush of wind produced in front of an avalanche by the falling mass. It may cause considerable destruction.

axial plane (axial surface) †In geology, a plane that bisects the angle between two limbs of a fold in rock strata. *See also* axis of fold.

axis, Earth's A line running through the centre of the Earth, connecting the North Pole with the South Pole. The Earth rotates upon this axis every 24 hours. The axis is not at right angles to the Sun but is inclined at an angle of 23½°. This inclination, when combined with the orbit of the Earth around the Sun, results in seasonality, varying lengths of night and day at different times of the year, and changes in the altitude of the Sun at different times of the year.

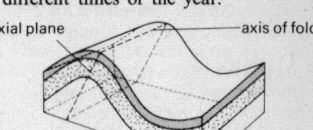

axial plane — axis of fold

Axis of fold

axis of fold †An imaginary line that marks the centre of a fold in rock strata. The strata dip away from the axis in an anticline and rise on both sides of a syncline. The concept is essentially used to describe the direction in which a fold dips (or rises). If the axis is horizontal the fold is upright, if the axis is tilted the fold is described as pitching or plunging. *See also* axial plane, fold.

azimuth 1. (*Surveying*) The bearing or angle between an imaginary line joining the observer and the North or South Pole and a specific point. The bearing may be measured in degrees between 0° and 180° E or W from the pole, or in degrees between 0° and 360° clockwise from true north.

2. (*Astronomy*) The angle between the plane of the meridian of the observer and the vertical plane passing through a heavenly body.

azimuthal equidistant projection *See* zenithal equidistant projection.

azimuthal projection *See* zenithal projection.

azonal soil (skeletal soil) A soil in which the profile is undeveloped and little or no differentiation into separate soil horizons has taken place. Azonal soils are often young, i.e. soil-forming processes have had insufficient time to operate, as on recently deposited alluvium, reclaimed land, wind-blown deposits, or a new lava flow. Bare rocky surfaces such as screes are azonal soils. *Compare* intrazonal soil, zonal soil.

Azores high (Azores anticyclone) The subtropical anticyclone generally centred over the N Atlantic Ocean near the Azores Islands. It moves further S in winter and further N in summer. Occasionally it may extend NE to affect Western Europe and the UK by blocking the westerly circulation.

B

backhaul rates †In transport studies, the charges made for carrying goods along routes where the flow of traffic is lighter in one direction than in the other. Lower charges may be quoted for the direction where the traffic flow is lighter since it costs little more to run a train, boat, or truck with a load instead of empty. For example, coal from the Lake Erie ports is carried much more cheaply to Lake Superior than is iron ore in the opposite direction.

backing An anticlockwise change of direction of a wind, e.g. from W to SW to S. *Compare* veering.

backslope The gentler slope of a cuesta, which dips away from the scarp slope. *Compare* scarp. *See also* dip slope.

backwall *See* headwall.

backwash The flow of sea water back down a beach towards the sea under the influence of gravity after a wave has broken. *See also* swash.

backwash effects †*See* cumulative causation model.

backwater A stretch of water that has become bypassed by the main flow of a stream, although still joined to it, and consequently has a very low rate of flow. It may form when the neck of a meander is cut through by the stream leaving its old channel as a backwater. *See* oxbow.

backwearing †The wearing back of a slope with little change in its steepness. The idea of backwearing was first put forward by the German geomorphologist Albrecht Penck, as an alternative to the downwearing advocated by W. M. Davies. The issue remains unresolved, but it is generally thought that the concept of backwearing has greater application in arid and semiarid environments than in humid temperate environments. *Compare* downwearing. *See also* cycle of erosion.

badlands An elevated arid or semiarid area, with sparse vegetation, dissected by deep gullies leaving sharp ridges and platforms of exposed bare rock. The name is derived from the Bad Lands of South Dakota, USA, but is widely applied to environments with similar characteristics.

baguio A tropical storm experienced in the Philippines. It occurs especially from July to November.

Bai-u season The season of heaviest rain in parts of China and Japan in late spring and summer. As the rains occur during the plum-ripening season they are also known as the *plum rains*. [Japanese]

bajada (bahada) A type of alluvial plain, occurring towards the centre of intermontane basins in arid or semi-arid areas. Bajadas are produced when several alluvial fans join together to form a gently sloping surface lying between upland areas and playa lakes. [Spanish]

balance of nature †*See* ecological regulation.

balance of trade The difference between the value of a country's exports and its imports. It is limited to trade that involves the buying of goods overseas and the export of goods to overseas countries and does not include invisible items such as banking and insurance services. There is rarely a balance between exports and imports and any difference is known as the *trade gap*.

balloon sonde (ballon sonde) A hydrogen-filled balloon carrying meteorological instruments that is released into the atmosphere. As it rises its instruments record changes in temperature and pressure with altitude. The balloon eventually bursts and the instruments return to Earth by parachute. Recordings at very high altitudes can be obtained. *Compare* radiosonde.

bamboo A family of large often long-lived grasses (*Bambuseae*), native to tropical and subtropical areas (e.g. monsoonal Asia). The strong woody hollow stems of bamboo may attain heights of 30 m in a few months' growth. They are used as a common and cheap construction material for housing, boats, and bridges. Young shoots are also eaten as a vegetable. Bamboos thrive in deep fertile soils, but each species fails periodically after a long-delayed flowering and seeding that exhausts the plant's energy.

bank The edge of a river or lake. It is best shown where the river has considerable powers of vertical erosion. *See also* sandbank.

banket A conglomerate of Precambrian age, composed largely of quartz pebbles and containing gold. It is found in the Witwatersrand region of South Africa, where it is the main gold-bearing rock, and Ghana. [Afrikaans]

bankfull †The state of flow in which a river channel is filled to the top of its bank. Although river channels vary considerably in their size and in their environment they show a fairly regular recurrence of reaching bankfull stage, about once in 1.5 years. Once water has overflowed, flood stage is reached.

banner cloud A stationary cloud that resembles a banner streaming out on the lee side of a mountain peak. The banner cloud that appears downwind of the Matterhorn is a notable example. It occurs as a result of condensation in air being lifted through eddying in the lee of the peak.

bar 1. An offshore deposit of mud, sand, and shingle in the form of a ridge running roughly parallel to the coastline. Bars are formed by the process of longshore drift; this orients the sand and shingle with the main currents and can occasionally result in the blocking of bays, estuaries, and river mouths. *See also* barrier beach, bay bar.
2. An alluvial deposit in the channel of a stream.
3. (*Meteorology*) The unit of atmospheric pressure, equal to 10^5 pascals (10^6 dynes per square centimetre). One

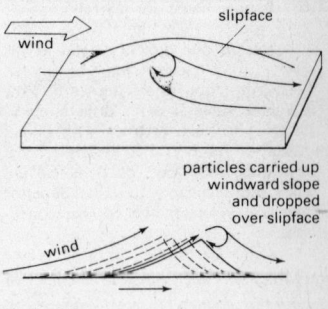

Barchan dune

bar is equal to a column of mercury 750.062 mm high in a barometer, measured at sea level in latitude 45°N at 0°C. In practice the *millibar* (mb), one thousandth of a bar, is commonly used.

barchan (barchane, barkhan) A crescent-shaped sand dune, the horns of which point away from the direction of the dominant wind. In cross section the leeward slope is relatively steep and the windward slope gentle. This asymmetry is due to eddies being set up by the prevailing wind blowing over the crest of the dune. Barchan's 'migrate' as grains of sand are blown up the windward slope and roll down the leeward slope. They can vary in height from a few metres to over 30 metres, and occur singly or in groups. The best examples can be found in the Sahara and in Turkestan (Central Asia).

barley One of the grass cereals; its grain is used as a food, particularly for cattle, and the straw is either burnt or used in cattle stalls. The grain is also allowed to sprout and then dried in a kiln to make malt, which is used in brewing beer and for making spirits such as whisky.

baroclinic †A state of the atmosphere where the surfaces of constant pressure intersect surfaces of constant temperature. This is believed to be partly responsible for the formation of mid-latitude depressions. *Compare* barotropic.

barograph A self-recording barometer (usually an aneroid barometer). A continuous record (*barogram*) of atmospheric pressure is traced on a moving drum.

barometer An instrument for measuring atmospheric pressure. The mercury barometer was invented in 1643 by Torricelli. It measures the height of a column of mercury that the atmosphere is able to support in a vertical glass tube. Adjustments need to be made for variations in gravity (standardized to 45°N), temperature (standardized to 12°C), and altitude. *See also* aneroid barometer.

barometric gradient *See* pressure gradient.

barometric pressure *See* atmospheric pressure.

barotropic †The state of the atmosphere where surfaces of constant pressure are parallel to surfaces of constant temperature. This is a useful theoretical concept but may never occur in reality. *Compare* baroclinic.

barrage A man-made structure to check the flow of a river and hold back the water. In Egypt barrages provide water from the Nile to irrigate the neighbouring desert areas. The barrage on the Thames at Woolwich controls the tidal flow, preventing flooding in the riverside areas of London. *See also* dam.

barrel The standard unit of volumetric measurement of petroleum or crude oil, equal to 159 litres (42 US gallons or 35 Imperial gallons).

barren lands (barrens) Flat open lands with vegetation limited to small trees and sparse shrubs. The name is applied to the tundra plains of N Canada, which are also referred to as the Arctic prairies.

barrier beach An offshore sandy bar that lies above high-tide level, usually separated from the coast by a lagoon. It may form a *barrier island* where it is sufficiently high above high-tide level and has dunes lying on it. Barrier beaches and islands are common on relatively shallow gently sloping offshore areas. Some of the best examples occur on the Atlantic coast of the USA where seaside resorts, such as Palm Beach and Miami Beach in Florida, have been sited on them.

barrier lake A lake formed by the natural damming of a river valley, this can occur as a result of landslides, avalanches, lava flows, or glacial deposition. Barrier lakes are usually temporary phenomena as the material responsible for blocking the valley is generally unresistant and easily eroded.

barrier reef A wide coral reef lying parallel to the coastline but separated from it by a wide deep lagoon. Barrier reefs are believed to have formed with the submergence of a flat platform on which coral was able to grow. The largest example is the Great Barrier Reef off the E coast of Australia, which extends for some 2000 km. Barrier reefs also occur around islands forming a continuous ring of coral. *See also* atoll, coral reef, fringing reef.

barysphere The very dense interior of the Earth, lying beneath the crust, and including the mantle and the core. [From Greek: heavy sphere]

basal sapping †The erosion and removal of debris at the foot of a slope. This may result from a number of different processes:
(1) Wave action at the foot of a sea cliff.
(2) The undermining of laterite-capped tablelands in tropical environments.
(3) The disintegration of rocks along the headwall of a cirque as a result of freeze-thaw processes.
(4) Lateral erosion at the foot of a bluff along a river.

basalt A fine-grained dark igneous rock of basic composition, comprising over 90% of all volcanic rocks. The dark colouring is due to its chemical composition: it consists chiefly of plagioclase feldspar and pyroxene, with magnetite, apatite, and olivine as accessory minerals. Basalt flows readily and often forms extensive sheets, which can be thousands of metres thick and extend over hundreds of square kilometres, as in Iceland and the Deccan of India. Basalts commonly solidify into hexagonal columns, e.g. the Giant's Causeway in Antrim, Northern Ireland, and Fingal's Cave on the Scottish island of Staffa in the Hebrides.

baseflow †That portion of water in a stream channel that is derived from ground-water sources. It is a relatively slow steady transfer of water from within the ground and over short time periods is not greatly affected by heavy rainfall. The amount of baseflow reflects seasonal variations in precipitation, evapotranspiration, and vegetation. *See also* hydrograph.

base level The lowest level to which a stream or river can erode. Sea level is the most common base level for a river, although lakes, waterfalls, and tributary streams will provide temporary local base levels. It is relatively uncommon for streams to erode to absolute base level as this requires a long period of stability without uplift or other earth movements occurring. More common is stream erosion to a series of base levels, producing a highly irregular long profile. *See also* long profile.

base line A line on the Earth's surface that is measured very accurately as part of a triangulation survey. All the points within the triangulation are based upon this line and hence great accuracy is required. It may be measured by a long tape, which is supported above the ground, or by a geodimeter or tellurometer. In the first triangulation of the UK a number of check base lines were used to ensure that errors did not occur some distance from the original base line.

basic demographic equation †*See* demographic equation.

basic lava A lava with a low viscosity and low melting point that flows readily, often for great distances. It is rich in iron, magnesium, and other metallic elements, but poor in silica. It forms shield volcanoes, such as those of the Hawaiian islands, which are widespreading domes with very gentle slopes. Sometimes cracks in the Earth's crust allow vast amounts of lava to flood over the surrounding countryside to form basaltic plateaus thousands of metres thick and covering thousands of square kilometres. Examples are found in Antrim, Northern Ireland, the Columbia–Snake River region of North America, and the Deccan region of India. *Compare* acid lava.

basic rock An igneous rock that is quartz free and contains a low percentage of silica (45–55% by weight); it is composed of minerals such as calcium feldspars. Rock of this type forms the asthenosphere. Examples of basic rocks include basalt, dolerite, and gabbro.

basic slag Waste material obtained from steelworks that make steel by the basic process whereby limestone is mixed with coke and iron in a furnace. The limestone combines with impurities to form liquid slag, which floats on top of the molten metal and can be drawn off. When cool it hardens and, being rich in lime, is used as a fertilizer and for making concrete.

basin 1. A large-scale depression in the surface of the Earth, produced by erosion or by the underlying geological structure.
2. In geology, a structural downfold in the Earth's crust in which the rocks dip towards the same central point.
3. The area drained by a single river system, i.e. a drainage basin.

basin-and-range A region of mountain ridges and intermontane basins produced by the tilting of faulted blocks, e.g. the Great Basin region of the W USA.

basin cultivation The growing of crops in small hollows or basins, often formed by raising low ridges of soil around a plot of land. This type of cultivation is used by subsistence farmers in areas of the tropics, such as S Nigeria where rainfall is high and soil erosion common. The small hollows check rainwash and collect water for the growing crops.

basin irrigation A means of providing water for cultivation that involves the flooding of basin-like hollows sur-

rounded by earth banks. The hollows vary considerably in size. Basin irrigation has been practised in Egypt and the Sudan for many centuries using the summer flood water of the Nile to fill the basins. This method is being replaced by concrete feeder canals supplying stored water.

basket-of-eggs relief *See* drumlin.

batholith (bathylith) A very large dome-shaped intrusion of igneous rock, typically several kilometres in depth and extending over hundreds of square kilometres. It is usually composed of acid rocks, such as granite and diorite, and is always associated with an area where mountain building has taken place. Examples of exposed batholiths include Dartmoor, Devon, and the Mourne Mountains in Northern Ireland.

bathyorographical Denoting maps that show both the relief of the land and depths of the oceans. Whereas relief is shown in a variety of shadings and styles, the depth of the oceans is normally shown in gradations of colour from white to deep blue.

bauxite A clay-like amorphous material, the chief ore of aluminium. It is formed when silica is removed from feldspars by chemical weathering. The remaining aluminium hydroxide is insoluble and collects in rock-like layers at or near the surface as the parent rock weathers away. It forms in tropical and equatorial zones where there is abundant rainfall to leach out the silica and a dry period to allow the precipitation of the aluminium hydroxide. Leading producers of bauxite include Jamaica, Surinam, Guyana, Guinea, Sierra Leone, the USA, and Australia.

bay A wide curving indentation in a coastline, usually lying between two headlands.

bay bar A barrier of sand or shingle that extends across the mouth of a bay from one headland to another, thus being connected to the land at both ends. Bay bars commonly develop from spits, which grow through longshore drift until they reach the next headland. Once closed off by a bay bar a bay will gradually fill up with sediment. *See* longshore drift.

bayou A sluggish backwater or marshy area adjacent to a river, frequently occurring on deltas or floodplains. The term is commonly used in the USA, especially Louisiana, where it refers to oxbow lakes. [From Louisiana French]

beach The accumulations of material along the coast, usually defined as lying between the highest point reached by storm waves and the low-water spring-tide line. The material may consist of a wide range of particles, from large rock fragments, shingle, and sand, to fine mud and silt. It lies on an eroded platform of solid rock – the wave-cut platform.

beach cusp †A curved triangular-shaped ridge of sand or shingle found on beaches between high- and low-water marks. The ridges of the beach cusps point towards the sea and alternate with small rounded depressions, which vary in width from 5 to 60 m and give the beach a scalloped appearance. No satisfactory explanation has yet been put forward for their formation but beach cusps are believed to result from the powerful swash and backwash of waves.

beach ridge (full) A ridge of sand or shingle along a beach, generally running parallel to the shore. Beach ridges tend to be well sorted, with particles increasing in size up the beach. They are formed through the action of constructive waves.

beaded esker An esker that consists of alternate thin ridges and mounds of sand and gravel of fluvioglacial origin resembling beads on a string. The mounds or swellings mark pauses in the retreat of a glacier while the thin

ridges accumulated when the glacier was retreating. *See also* esker.

bearing A horizontal angle measured clockwise from a specific reference line to a point. If the reference line is a meridian (i.e. true north) the angle will be a *true bearing*. If the angle is measured from the magnetic north, as measured on a compass, the angle is a *magnetic bearing*. For example, the true bearing from Gatwick airport to Heathrow is approximately 335°, whereas the magnetic bearing is 343°.

Beaufort wind scale A scale of wind force. It was devised in 1805 by Admiral Sir Francis Beaufort and modified in 1926. The scale, which ranges from 0 to 12, is based on easily observable indicators of wind strength, such as smoke, tree movement, and damage incurred.

beck A small stream in the N of England.

bed 1. In geology, a single layer or stratum of sedimentary rock. The bed of rock is separated from the beds above and below by distinct surfaces known as *bedding planes*.
2. The bottom of a river channel, lake, or sea.

bedding The arrangement of sedimentary rocks in layers. *See also* bedding plane, stratum.

bedding plane The surface or plane of deposition on which sediments were laid down and which separates successive beds of sedimentary rocks in the geological sequence. It is not necessarily horizontal. In sandstones, for example, the bedding plane may be marked by a change in colour or in grain size, or perhaps by a thin layer of clay. Sediments tend to split readily along the bedding planes.

bedload Material transported by rivers that remains in contact with, or close to, the channel bed; it moves by rolling, sliding, or saltation. At higher velocities of river flow bedload may

Beaufort wind scale

Beaufort number	wind	speed (knots)	observed effects
0	calm	< 1	smoke rises vertically
1	light air	1- 3	smoke indicates movement
2	light breeze	4- 6	wind felt on face, leaves rustle
3	gentle breeze	7-10	leaves and twigs in constant motion, flag extended
4	moderate breeze	11-16	small branches move, dust raised
5	fresh breeze	17-21	small trees sway
6	strong breeze	22-27	large branches in motion, whistling in telephone wires
7	near gale	28-33	whole trees in motion
8	gale	34-40	twigs broken off trees
9	strong gale	41-47	slight structural damage to buildings
10	storm	48-55	widespread structural damage to buildings, trees uprooted
11	violent storm	56-63	severe damage (rarely experienced inland)
12	hurricane	>64	severe damage (rarely experienced inland)

become suspended load. *See also* saltation, suspension load.

bedrock The solid rock lying beneath soil and weathered material. It is usually only exposed as an outcrop on steep slopes where no soil is present.

beet, sugar *See* sugar beet.

behavioural geography †An approach to human geography that recognizes that not all decisions made by man are the result of rational economic forces and argues that decisions are greatly influenced by the perceptions people have of their environment. Space is regarded as having social as well as physical attributes and so, for example, perceptions of how desirable part of a city is as a place to live influences how and where different residential areas develop.

beheading *See* capture, river.

belt A broad strip of land characterized by one dominant feature, such as a crop or climatic type. Examples are the Corn Belt and Cotton Belt in the USA.

ben In Scotland, a summit or peak of a mountain, e.g. Ben Nevis.

bench A narrow terrace-like landform, characterized by a flat surface with a steep slope at the back. Benches may be erosional in origin, the most common being produced by lakes or rivers that have experienced a change in base level. They also result from structural processes, e.g. step faulting, and artificial excavations, such as quarrying and opencast mining.

bench mark A mark made by surveyors to record a point of known position and height above mean sea level that can be referred to subsequently when further surveys are carried out. The mark is cut into a permanent feature, such as a rock or the stone facing of a building. It consists of a broad arrow with a horizontal bar through its apex. The arrow points upwards unless the location is below sea level, in which case it is inverted. Bench marks are shown on Ordnance Survey (OS) maps by the letters BM and the height.

benefit cost analysis †*See* cost benefit analysis.

Bergeron–Findeison theory †A theory of precipitation formation. It was put forward by a Swedish meteorologist, Tor Bergeron, in 1933 and modified by Walter Findeison. The theory assumes that water droplets and ice crystals can co-exist in clouds at temperatures of between −15°C and −30°C. The relative humidity of air is greater with respect to an ice surface than to a water surface; this means that ice crystals can grow more rapidly at the expense of supercooled water droplets. Small splinters of ice become detached and increase the number of ice crystals in the cloud. These ice crystals readily join together when they collide and snowflakes form, which eventually become so dense that they fall through the cloud. Depending on the temperatures through which they fall these may reach the ground as snowflakes or raindrops. The Bergeron–Findeison theory fits most observed facts in temperate latitudes but is not relevant in tropical areas where clouds produce much rainfall but have temperatures above freezing point.

bergschrund A deep crevasse in a cirque glacier that frequently occurs near the headwall of a cirque. At this point moving ice is breaking away from ice attached to the headwall. [German]

berg wind A warm dry wind, similar to the föhn wind, that blows in South Africa from the interior plateau down to the coastlands. It is most frequent in winter when the plateau is covered by a strong anticyclone and pressure is low over the ocean. The wind may last for two to three days giving oppressive weather conditions.

berm †A ledge or shelf of shingle or coarse sand found above the high-water mark on a beach. Berms are built up from material thrown up by storm waves.

Bernhard's index of concentration †An index that indicates the degree of dispersion and concentration of settlements. It takes into account the number of settlements, the number of houses, and the area being examined. It is calculated by the formula:
$$C = (H/S)\,(A/S)$$
where C is the index of concentration, H/S is the number of houses in each settlement, and A/S is the number of settlements in a given area.

Bessemer process A method for making steel from pig iron, invented by Sir Henry Bessemer in 1855. The process takes place in a *Bessemer converter*, a large steel vessel. Air is blown through molten pig iron and impurities in the iron (carbon, silicon, and manganese) are oxidized. The impurities form slag, which can be drawn off. Phosphorus is removed from the iron by reaction with the basic refractory lining of the converter.

B horizon A soil horizon occurring in many soil profiles, between the A and C horizons, from which it is usually well distinguished. The B horizon is often known as the subsoil. †The B horizon receives material leached down from the A horizon, e.g. fine clay particles or ferric oxides. This

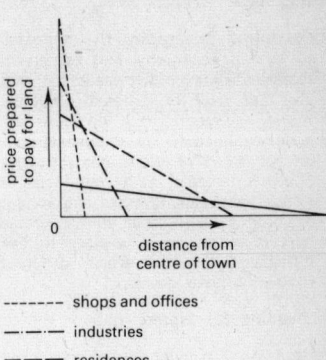

----- shops and offices

—·—·— industries

— — — residences

——— intensive agriculture for local market

Bid rents

material is sometimes concentrated in the form of a pan. B horizons are sometimes called illuviated horizons. *See* soil horizon. *See also* hardpan.

bid rent theory †A theory that attempts to explain the locations of functions within a town or city in terms of economic factors. The basic assumption is that the relative efficiencies of using land in different ways in particular locations is measured by rent-paying ability (i.e. *bid rents*) of a land user. Competition for different locations within the city by different functions produces the most efficient pattern of land use.

In the graph the bid rents of four selected land-use categories – shops, industries, residences, and agriculture – have been plotted against distance from the city centre. The *rent gradient* (i.e. the degree of steepness of the rent curve) is steepest for shops. Shops require central locations with the greatest accessibility to the whole population of the city to maximize profits; they can outbid most other potential users near the city centre but this ability declines rapidly away from it. Industries would benefit from central locations but cannot afford the absolute centrality of the shops and so their bid rent is lower at the city centre but declines less steeply away from it. Residences cannot compete with retail and industrial users and so the rent curve is less steep. Intensive agriculture is outbid by all the urban users and has the least steep rent curve; it thus occupies the margins of the urban area.

The urban land market is thus seen as a land value surface. The market centre is the point of highest site value. As rent declines with distance the value of land falls and as the land value falls, the land use changes. *See also* urban land value surface.

bifurcation ratio †The quantitative relationship between the number of streams in a given order and the number of streams in the next order. To obtain the ratio the number of streams in one order is divided by the number of streams of the next highest order. Bifurcation ratios usually range from 3.0 to 5.0. *See* stream number, stream order.

bight A wide coastal indentation, similar to a bay. Examples include the Great Australian Bight and the Bight of Bonny.

bill A long narrow headland, peninsula, or promontory; e.g. Portland Bill.

bioclimatology The study of climate in relation to organic life, including human beings. This includes the determination of climatic conditions suitable for human habitation.

biogeography The scientific study of the distribution of species of plants and animals. This may be on a global or on a local scale. Biogeography seeks to explain such distributions with reference to environmental factors, particularly climate, soil, and man. It is divided into phytogeography (the distribution of plants) and zoogeography (the distribution of animals). *See also* zoogeography.

biomass †The weight or volume of organic matter per unit area. It may be measured, for example, in grams of dry matter per square metre, so that comparisons between different environments can be made. Biomass may refer to the total organic content of the biosphere, to that of a specific habitat such as a pond, or to that of a medium such as soil.

biosphere †That part of the Earth and its atmosphere where life is able to exist. The surface of the land and the upper layer of the sea contain the greatest density of living organisms.

biotic Denoting living organisms. For example, the *biotic environment* is made up of living components (i.e. plants and animals) as distinct from abiotic components, such as rocks and air.

bird's foot delta (birdfoot delta) A type of delta that extends out into the

Mississippi Delta 1 _5km_

sea in a finger-like pattern with sediment deposited on either side of the distributary streams. It is composed of very fine sediment, largely silt, and develops in conditions where currents and tides are comparatively weak. The river has only a few distributaries, in contrast to arcuate deltas, and these channels are well defined. One of the best examples is the Mississippi Delta. _See also_ delta.

birth rate, crude _See_ crude birth rate.

bitter lake A lake that contains high proportions of mineral salts in solution due to high rates of evaporation. As a result the waters have a bitter taste. Examples include the Great Bitter Lake in Egypt. _See also_ salt lake.

bituminous coal A soft black coal, containing 48–83% carbon. It is the most common type of coal and is often known as household coal. Various types of bituminous coal exist and are named after the uses to which they are most frequently put, e.g. coking coal and gas coal.

Bjerknes polar front model †The theory concerning the life cycle of a depression developed by Norwegian meteorologists and described by J. Bjerknes and others in the early part of the 20th century.
In the model a polar front lies E–W across the N Atlantic separating polar maritime air to the N and tropical maritime air to the S, both air masses moving towards the E. A local disturbance along this front creates a bulge of warm air extending into the cold air and this moves E along the front at approximately the same speed as the warm air stream. This bulge is called a frontal wave, the leading edge being the warm front and the trailing edge the cold front. This may remain a small feature travelling along the polar front, in which case it is known as a stable wave. However the wave often becomes unstable and while moving E, rapidly increases in size, greatly distorting the polar front. Pressure falls at the crest of the wave creating closed isobars and at this stage the feature becomes a depression. The bulge of warm air is now large and is known as the warm sector, while in the cold air around the distortion of the polar front has caused a shift in the wind pattern, with the winds swinging in an anticlockwise direction around the centre of low pressure. Development continues with the warm sector narrowing as the cold front catches up to the warm front progressively from the centre of the low. This forms an occlusion and as the warm air is pinched out and rises, it cools and the depression eventually weakens and dies.

The theory was developed before the upper air flow was understood so that it does not fully explain the origin of depressions. Weather forecasting in temperate areas is based largely on this theory.

black alkali soil †_See_ solonetz.

blackband ironstone A layer of carbonaceous ironstone that occurs in the Coal Measures. It is the most valuable of the iron ores of the Coal Measures. As it has a coal content of 10–20% and virtually no clayey material it can be smelted economically. It occurs as a bedded rock forming the upper part of a coal seam sequence, which suggests that it was formed in shallow lagoons among the coal swamps but shielded from muddy sediment. Examples can be found in S Staffordshire, S Wales, and the Scottish coalfields.

black cotton soil *See* regur.

black earth *See* chernozem.

black lead *See* graphite.

blanket bog A bog that forms under conditions of high rainfall and high humidity. It usually occurs as a continuous cover over the land, broken only by steep slopes and rocky outcrops. *See also* bog, swamp.

blast furnace A furnace for producing pig iron from iron ore. A tall steel container lined with heat-resistant bricks is filled from the top with iron ore, coke, and limestone. A hot blast of air is blown into the base of the furnace and the coke burns melting the ore. This settles at the bottom of the furnace to be drawn off every three to five hours. The furnace is continuously topped up until the lining needs renewal or demand for the pig iron drops.

blight (planning) A condition that occurs when a property is affected by the development proposals of a local authority. Often these plans refer to developments which will take place at some uncertain future date, thus if the property owner wishes to sell his property he may not be able to, except at a substantially lower price than he would have expected.

blind valley In limestone regions, a valley, either dry or still containing a stream, that ends abruptly with a rock wall when its stream disappears underground through a cave or sinkhole.

blizzard A strong bitterly cold wind that is accompanied by dry powdery snow and/or ice crystals. The snow is often whipped up from the ground and visibility is considerably reduced. The term was originally applied to the cold northwesterly gales accompanied by snow that cross the USA in winter,

but is now used more widely. It is particularly prevalent in Antarctica.

block diagram A three-dimensional diagram of a particular area or specific landform. Since the block is imagined to be cut from the Earth's crust it shows the underlying geological structure as well as the Earth's surface so that the relationships between structure and surface landforms can be seen. For example, a block diagram of a limestone region.

block disintegration A type of mechanical weathering, resulting chiefly from frost action. When water present in the pores and joints in rocks freezes ice crystals grow which exert great pressure on the rock. This pressure causes the rock to fragment and crack along lines of weakness, producing a land surface littered with large angular boulders, usually referred to as felsenmeer. *See also* weathering.

block faulting The breaking up of an area of the Earth's crust into blocks, which may be raised or lowered. Raised blocks are called horsts and lowered blocks are called grabens. *See also* fault, fault block.

blocking high An area of atmospheric high pressure (anticyclone), which remains relatively stationary in the zone of the mid-latitude westerlies. It thus blocks the normal routes taken by depressions, which then move to the NE and SE of the anticyclone. Blocking highs occur most frequently over NW Europe and the NW Pacific. They are associated with periods of settled weather.

block lava A lava with a surface that is broken into large rough jagged blocks. This results when the congealing surface of partly crystallized lava is broken up by gases erupting from the lava under pressure. It is often regarded as being synonymous with aa.

block mountain An upland mass created by the uplift of land through earth movements between faults or by the sinking of the land outside the faults. *See* horst.

blossom showers (mango showers) The rain showers that occur from March to May in the monsoon region of SE Asia.

blowhole A near-vertical cleft that provides an outlet from a cave near the sea shore to the cliff top. Waves surging into the cave force spray out of the blowhole. Blowholes form as the result of erosion along a joint or fault. The term *gloup* is sometimes used in Scotland.

blowout A depression or hollow formed in sand or loose earth by wind erosion, frequently as a result of the removal of the protective vegetation cover. Once exposed, the loose soil or sand is easily eroded by the eddying of the wind, enlarging and deepening the hollow. *See also* deflation.

bluff A steep prominent slope, usually occurring in river valleys. It is produced by a river eroding laterally by cutting into the valley sides, for example on the outside of a meander, thus widening the valley floor.

bocage An area of countryside divided into small fields, each of which is surrounded by hedges, often containing fully grown trees. There may also be small patches of woodland, remnants of the forests that were once more extensive. The term is normally used to describe the countryside of Brittany and Normandy in France although similar scenery is found in many parts of the UK, particularly in the West Country. [French]

bog An area of wet spongy ground consisting of decomposing moss and other vegetation. It often forms with the growth of moss, especially sphagnum, on the surface of a shallow pond or lake. Decaying moss and other vegetable matter will gradually accumulate, infilling the pond, to produce a quaking bog. With further compaction peat will ultimately form.

boll weevil A small grey beetle about 5 mm long. The beetle is a pest in Central America, Mexico, and the S states of the USA. It lays its eggs inside young cottonseed pods (bolls) and the eggs hatch into grubs that eat the seed and cotton fibres. Widespread crop destruction in the early 1900s in Mississippi and the other S states of the USA resulted in farming diversity in the affected areas, and the growing of cotton to the N and W where cold winters kill the hibernating weevils.

bolson An inland basin surrounded by mountains and sometimes containing a shallow lake that is found in arid and semiarid areas. It is frequently covered with alluvial deposits. The term is usually limited in application to the inland basins of the SW USA and Mexico.

bomb, volcanic A large mass of volcanic lava; it is originally liquid when thrown out by the volcano but cools while spinning through the air to assume a rounded, often spindle, shape. Volcanic bombs are included in the class of pyroclasts. *See* pyroclast.

bonitative map †A map showing the potentiality of an area. It normally shows land that is favourable or unfavourable for specific kinds of economic development. For example, the potential hydroelectric power resources of Malawi. In addition, it may show the potentiality of an area for improvement, e.g. the mapping of the economic, recreational, and residential potential of the deltas of the SW Netherlands were evidence of the possible benefits of the Delta Plan.

Bonne's projection A simple conical map projection that has been modified to remove exaggerations of scale along the parallels by spacing all meridians their true distance apart along every parallel. The central meridian is straight, while the others are curved. The parallels are concentric circles at their true distance apart. This is an equal-area projection, easy to draw, with reasonable shapes around the central meridian. It is frequently used for atlas maps of large areas. *See* map projection.

bora A cold dry gusty northerly or northeasterly wind that blows down the mountains to the E coast of the Adriatic Sea and N Italy. It is most common in winter and results from high pressure over central Europe and a deep depression to the SW over the Mediterranean. The trend of mountains, bays, and islands has a tunnelling effect on the wind; mean speeds are around 50 km per hour but gusts of up to 209 km per hour have been recorded at Trieste, Italy. The bora is usually accompanied by clear skies and cold dry weather but may bring heavy cloud and rain or snow. [From Italian]

border 1. The line separating one country from another. The administrative power of a country is halted at the border and on transport routes customs posts are set up to monitor traffic.
2. The district lying along the edge of a country, e.g. the Border Country in the UK is the area of land lying either side of the Scottish–English and the Welsh–English borders.

bore A tidal wave moving upstream in a shallow river estuary. It is produced when an incoming high spring tide becomes increasingly constricted as it travels up the estuary and is slowed by friction at its base and by the opposing river current. This causes the water to ride up in the form of a broken wave moving upstream. The best-known example in the UK is the Severn bore, occurring at spring tide with a height of about 1 m. Other examples include the large bore that travels up the Tsing Kiang River in N China and attains heights of about 4 m and that of the River Hooghly in India.

Boreal 1. A climatic zone in W. Köppen's classification of climates. It is characterized by cold snowy winters and short but warm summers. The

zone extends over America, Europe, and Asia between latitudes 40°N and 65°N.
2. The northern coniferous forests.
3. A climatic period from about 7500 to 5500 BC characterized by cold winters and warm summers with a dominant vegetation of pines and hazels.

borough In England and Wales, a town that has a corporation and special privileges, which were granted by royal charter. It is of lower status than a city. In the USA some states have boroughs that correspond to incorporated towns in other states. Greater New York is itself divided into five boroughs. In Australia the term is used to describe a municipal centre of a certain minimum size and population.

boss A small igneous intrusion with a roughly circular outline, a steep angle of contact with the surrounding rock, and a diameter of less than 25 km. *Compare* stock.

bottom 1. In the USA, a low-lying alluvial plain, e.g. the Mississippi River bottoms.
2. A dry valley in chalk or limestone areas.
3. The floor of a lake, sea, or ocean.

boulder A large rounded mass of rock that has been shaped by erosion and transported, for example by ice, from its source. A boulder is defined as having a diameter of over 200 mm in the UK and over 256 mm in the USA (on the Wentworth scale).

boulder clay *See* till.

boundary The line demarcating the recognized limits of an established political unit, administrative region, or geographical region, e.g. a state, county, or district council.

bourne A temporary stream found in chalk country. It customarily only flows during the winter when the water table rises above the level of the valley floor following heavy rainfall.

braided river A river in which the main channel has divided into a complex network of shallow diverging and converging streams separated by bars. Braided rivers occur when the river is unable to carry all the load supplied to it. It thus deposits some of the material within the main channel and the stream is forced to flow around the deposits. Braided rivers are common in periglacial and ice-margin environments, where the load of streams is greatly increased by glacially derived material. *See also* aggradation.

brave west winds The westerly and northwesterly winds of the S hemisphere that blow between latitudes 40°S and 65°S. The winds are characterized by their great force and regularity. The latitudes in which they blow are known as the Roaring Forties.

breaker A mass of turbulent water travelling towards land. It is produced when a wave passes from deep to shallow water. The resultant friction at the base of the wave causes it to increase in height and the crest to fall forward and break. Breakers can be either constructive or destructive depending largely on how they break. *See also* constructive wave, destructive wave.

break-of-bulk point The location at which a cargo is transferred from one form of transport to another. At this point transport costs will increase because of the extra handling incurred. The entrepôt port of Rotterdam in the Netherlands is a break-of-bulk point for Swedish iron ore, which is off-loaded from ocean-going ships to Rhine barges for conveyance to the steelworks of the Ruhr.

break of slope A point on a hillslope where the angle of steepness of the slope changes noticeably. It can be a result of structural or erosional factors.

breakwater A barrier built into the sea to absorb the impact of waves and protect the coast.

breccia A rock consisting of angular fragments embedded in a natural cement. It may be formed in a variety of processes, e.g. it can be the product of material ejected from volcanic vents, in which case it is found close to the vent, or it can be a sedimentary rock occurring as thin lens-shaped beds laid down on the sides of submarine ridges in a geosyncline. A *fault breccia* can occur along a fault plane where the rock has been crushed and broken. [Italian]

breckland Heathland or woodland that has been cleared, ploughed for cultivation, and later abandoned to revert to its natural state. A region known as Breckland covers part of S Norfolk and N Suffolk near Thetford. The land is sandy and marginally productive, providing heathland habitats of great ecological interest.

breeze A wind between force 2 and force 6 on the Beaufort wind scale. The term is generally applied to winds due to convection that occur regularly during the day or night. *See* land breeze, sea breeze.

brickearth A relatively recent fine-grained deposit, found overlying the gravel of some river terraces, e.g. those of the River Thames. It is thought to be derived from loess, a fine wind-blown deposit, which has been redeposited by the river. The name originated from its use as a raw material for brick manufacture.

brickfielder A hot dry dusty wind experienced in Victoria, Australia, especially in summer. It blows in front of a depression from the continental interior in a southerly direction. The wind may blow for several days at a time with temperatures exceeding 38°C.

bridge A structure carrying a road, path, or railway over a gap to facilitate communications. The gap may take the form of an estuary, river, valley, or a similar obstacle. The point on a river where a bridge can be built may be a significant factor in locating a settlement, e.g. London developed at the lowest bridging point of the Thames.

bridging point (bridge point) The point at which a river is or can be bridged. The *lowest bridging point* (the point nearest the river mouth where a bridge can be built) is especially significant as it is often the site of an important route centre, e.g. Newcastle upon Tyne.

bridle path A path that is fit for horse riders but not for motor vehicles. Bridle paths serve as well-marked routeways for ramblers and youth-hostellers.

brine A solution of water with a high concentration of salt. It occurs, for example, in salt marshes, mud flats, and some lakes with high rates of evaporation.

broad A wide stretch of slow-moving fresh water, either a broadening of a river or an area of water linked to the river near its estuary. The Norfolk Broads in East Anglia are thought to be partially the result of the removal of peat in the past, and flooding when sea levels rose. They are widely used for recreation purposes (e.g. fishing and sailing) and for ecological studies.

brook A small stream.

brown coal A type of coal with a high carbon content which is economically very important. Several varieties are found, from a soft brown coal, which appears woody and has a carbon content of 60–69%, to a hard brown coal, which has a carbon content of 71–77%. It is found in West and East Germany and other parts of Europe.

brown earth (brown forest soil) A widely occurring zonal soil that is associated with those areas originally covered by deciduous woodland and having moist climates with moderate

rainfall. Brown earths are found equatorwards of the main podzol soil zone and cover large areas of the mid-latitudes. The soils do not have the distinct layering of a podzol and generally consist of a dark humus-rich A horizon overlying a weakly developed B horizon. Humus is rapidly incorporated into the soil by the soil fauna. The soils are free draining but not usually strongly leached. There is no downward movement of sesquioxides. Generally brown earths are neutral or moderately acid. They are extensively used for agriculture.

brown forest soil *See* brown earth.

brown steppe soil A zonal soil of temperate grasslands, occurring in drier parts than the neighbouring and similar chestnut soils. †Steppe soils are a lighter brown in colour than the chestnut soils as the humus content is less, and the accumulation of calcium is both more marked and more shallowly positioned in the soil profile. *Compare* chestnut soil.

brunizem *See* prairie soil.

buckwheat An annual herbaceous plant, native to Asia, the seeds of which are used in Europe for feeding poultry, horses, and cattle. In North America buckwheat flour is the basis for small cakes, which are fried and eaten for breakfast.

buffer state A relatively small and weak state lying between two larger and more powerful ones that may be in conflict. The weaker country acts as a buffer to reduce the likelihood of conflict, e.g. the Mongolian Peoples Republic, which lies between the USSR and China. A buffer state is often in danger of being overrun itself; for example, Poland was created as a buffer state between Germany and Russia and was invaded by Germany at the start of World War II (1939).

bund An artificial quay, dam, or embankment, or a small ridge between rice fields. The word was adopted by

29

the British in India and is used there extensively. In Shanghai the term was used for the main waterfront quay along which large foreign-owned offices were built. [Hindustani]

buran A strong northeasterly or northerly wind experienced in central Asia. It occurs throughout the year but is particularly frequent in winter as a fierce cold blizzard. [Russian]

burgh A town in Scotland that has a charter, e.g. Melrose. *Royal burghs* are those that have had their charters granted by a ruling monarch. *See also* borough.

bush The more or less untouched country beyond that cleared by man, i.e. where native vegetation still dominates. Bush varies from relatively well-wooded areas to sparsely vegetated desert. The term is used particularly in Africa, Australia, and New Zealand.

bush veld A type of grassland with scattered trees and spiny scrub that occurs in tropical and subtropical Africa. It is adapted to a climate with a dry season. [From Afrikaans *bosveld*: tree field]

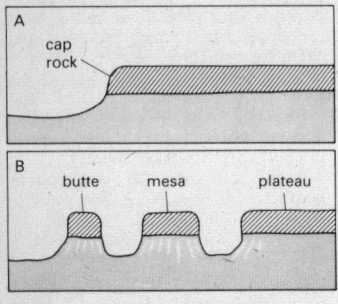

Butte

butte A small flat-topped hill with steep sides, found usually in arid and semiarid areas. Buttes are usually capped by a layer of hard resistant rock, which protects the underlying rocks from erosion. [French: mound]

Buys Ballot's law †A law postulated by the Dutch meteorologist Christoph Buys Ballot in 1857, which states that if an observer in the N hemisphere stands with his back to the wind atmospheric pressure will be lower on his left hand than on his right; in the S hemisphere pressure is lower on his right. This is the result of the Coriolis force on the Earth. Expressed in a different way the law states that in the N hemisphere winds move in an anticlockwise direction around centres of low pressure and clockwise around centres of high pressure; in the S hemisphere the reverse is true. *See also* Coriolis force.

bypass A road that has been built to divert traffic around a settlement to relieve traffic congestion. One of the most famous in the UK is the Exeter bypass, which carries holidaymakers to and from the West Country, a notorious black spot for traffic jams in peak holiday periods.

C

caatinga (catinga) Low open scrub forest of the E corner of Brazil. Vegetation includes cacti, thorn bushes, and mimosa, adapted to survive the frequent severe droughts. [Tupi Indian]

cacao A tree of equatorial and tropical regions that bears large oval seed pods containing beans from which cocoa and chocolate are made. Since chocolate is consumed mainly in temperate lands cacao is a valuable export for some Third World countries. In some areas it is grown on plantations developed with European or US capital. Production is centred in Ghana, Nigeria, E Brazil, and the Caribbean region. [Mexican]

Cainozoic *See* Cenozoic.

cairn A pile of rough stones, originally of use as a memorial, but commonly used today as a landmark on footpaths in mountain and moorland areas. [Gaelic]

calcareous Composed of or containing a high proportion of calcium carbonate ($CaCO_3$); the term is usually applied to rocks and soils with such a composition.

calcicole †A plant that requires a soil rich in lime. Chalk grassland, for example, is an assemblage of calcicoles. Many crops do best in neutral to alkaline conditions. *Compare* calcifuge.

calcification The process by which calcium carbonate accumulates within soils to form a concentration. †The depth of the redeposited material varies according to climate: in arid conditions, where soil water moves upwards by capillarity, calcification takes place at the surface, but in humid areas leaching removes calcium to lower horizons. Subsequent cementation produces a calcareous pan.

calcifuge †A plant that grows in acidic environments. Examples include heath and bog plants such as heather and cotton grass. *Compare* calcicole.

calcimorphic soils †A group of intrazonal soils that result from the weathering of limestones and are therefore markedly alkaline. The calcareous parent material usually prevents any development of acidity regardless of how much leaching takes place. Calcimorphic soils are usually dark, organic rich, and contain abundant soil fauna. Rendzinas are calcimorphic soils.

caldera A large basin-shaped volcanic crater. †In acid lava regions a caldera may form when a gigantic explosion blows away the original cone. In basic lava regions melting of the cone base results in the collapse of the cone into the underlying magma chamber. Examples include Crater Lake, Oregon (USA), and Lake Toba, Indonesia, which has an area of 1800 km². [Spanish]

Caledonian orogeny A period of mountain building in NW Europe that began in the late Silurian period and lasted until the early Devonian period. The Caledonian orogeny was responsible for producing the dominant NE–SW trend that extends from Ireland across Scotland and through Scandinavia. Folds dating from the Caledonian orogeny are present in much of upland Scotland, the Lake District, and Wales.

caliche 1. An evaporite deposit, containing sodium nitrate, sodium chloride, and other minerals, that occurs in arid areas. It may result from lakes drying out or from the leaching of bird guano. Caliche deposits vary in thickness; in the Atacama Desert of Chile and Peru they are extensive enough for economic exploitation. 2. In parts of the USA, a hard resistant crust of calcium carbonate that forms in the surface layer of soil. [Spanish]

calina A warm dust-laden haze that occurs along the Mediterranean coast of Spain in summer. [Spanish]

calving The breaking off of the front, or snout, of a glacier on reaching the sea, forming an iceberg. The term can also describe the further break-up of icebergs themselves.

Cambrian The earliest geological period of the Palaeozoic era and the system of marine rocks that were laid down during this period. Throughout the British Isles there is a marked unconformity between the Precambrian and the Cambrian. Deposition of these rocks started about 570 million years ago and continued for about 70 million years. In Britain the rocks of the Cambrian are largely sandstones, grits, and slates, occurring most extensively in Wales. The Cambrian rocks contain the first shelled fossil remains and are the earliest rocks for which fossils can be used for dating and correlation; trilobites, brachiopods, and gastropods were abundant. [Named by the British geologist Adam Sedgwick, who first identified (1836) the system of rocks in N Wales, after Cambria, the Roman name for Wales]

Campbell–Stokes recorder An instrument for measuring and recording the duration of bright sunshine. As the Sun moves round during the day a spherical lens focuses its rays to burn a trace on a recording card. The length of the trace indicates the sunshine duration. Allowances are made for seasonal changes in the position of the Sun. The instrument has limitations in that it only records bright sunshine – early morning and late evening sunshine may not be detected.

campo The tropical grassland (savanna) that covers the interior plateau of Brazil, particularly Matto Grosso. The equatorial rainforest of the Amazon Basin separates the campos from the similar llanos further N. As with other grasslands, the proportion of trees to open plain varies. Development of the campos for mineral extraction and farming is taking place. *See* savanna. [Portuguese: grassy plain]

canal A man-made waterway used to transport goods or irrigation water. Canals may be built to link inland towns or cities to the coast, e.g. the Manchester Ship Canal; to provide a shorter journey for shipping, e.g. the Corinth Canal, Greece; or to improve existing waterways, e.g. the St Lawrence Seaway, North America. Canals are particularly useful for transporting bulky goods when the time taken in transit is not highly significant.

Cancer, tropic of The parallel of latitude at 23°32′N. It is the most northerly limit along which the Sun appears directly overhead; this occurs at noon on about 21 June, the summer solstice in the N hemisphere. *See also* Capricorn, tropic of.

cane, sugar *See* sugar cane.

canopy The uppermost leafy layer in a woodland or forest (e.g. equatorial rainforest), comprising the crowns of trees of similar height growing closely together. The continuous cover formed by the canopy intercepts rain and casts dense shade, which influences the environment of lower plants. *See also* stratification.

canyon A deep valley with steep near-vertical sides. It generally forms in arid or semiarid areas, where a river is fed with water from a distant source and is capable of rapid downcutting. The lack of rainfall prevents erosion of the valley walls and these remain steep. The most notable example of a canyon is the Grand Canyon of Arizona, USA, which is over 1.5 km deep in some stretches. [From Spanish *cañon*]

cape A headland or promontory projecting into the sea, e.g. the Cape of Good Hope, South Africa, and Cape Wrath, Scotland.

Cape doctor A strong southeasterly wind that blows chiefly in summer in Cape Town, South Africa. Its name derives from the fresh conditions produced by the wind and the belief that it blows germs out to sea.

capillarity In soils, the property resulting from surface tension by which water is held as a thin film around soil particles and in minute spaces (capillary pores) between them. The water (known as *capillary water*) may be drawn up through the soil as a result of evaporation and transpiration. The soil kept moist in this way is known as the *capillary fringe*. †Unlike gravitational water, capillary water cannot freely drain away and remains available for plant roots unless evaporation is intense. Fine-textured clays retain a greater proportion of moisture by capillarity than do coarser soils. *See also* field capacity.

capital 1. A town or city that is the chief town of a country, province, or state and contains the seat of government. The term is also used to describe a particular function for which a town or city is especially important, e.g. São Paulo is the commercial capital of Brazil, Hollywood the film capital of the USA.

2. (*Economics*) The stock of money and goods used for promoting and conducting a business. Capital is one of the factors of production, together with land, labour, and entrepreneurial skills. It may be fixed or circulating: *fixed capital* consists of durable goods such as plant, machinery, and buildings, whereas *circulating capital* consists of stocks of raw materials, semi-finished goods, and components, which are used in the manufacturing process.

capital goods Goods that will be used in the manufacture of other goods. For example, looms are capital goods that are used to manufacture textiles. Raw materials such as wood and china clay are also capital goods since they are used to produce such things as paper and porcelain.

capitalism An economic system that is based on the private ownership of capital and the belief in profit as an incentive for investment. Capitalism also involves the belief in a free market economy in which prices are the result of the unhampered interplay of the forces of supply and demand.

Capricorn, tropic of The parallel of latitude at 23°32'S. It is the most southerly limit along which the Sun appears directly overhead; this occurs at noon on about 21 December, the summer solstice in the S hemisphere. *See also* Cancer, tropic of.

cap rock 1. A layer of comparatively hard resistant rock lying over weaker rocks; it thus protects the underlying rocks from erosion. Cap rocks are important in arid and semiarid areas in the formation of mesas and buttes. **2.** An impermeable layer of rock overlying a salt dome or an oil-bearing rock.

capture, river (beheading) A process by which a stream erodes headwards to such an extent that it intercepts the course of a neighbouring stream. The main stream is then said to have captured or beheaded the second stream, the headwaters of which are diverted

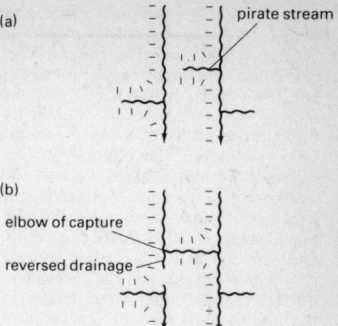

River capture

into the main channel. The point of interception is referred to as the *elbow of capture* as the course of the stream alters abruptly, commonly at right angles. The captured stream undergoes a reduction in volume and is usually referred to as a misfit stream because it is too small for the valley through which it flows. †At the present time this explanation of river capture is subject to some criticism; it is thought increasingly that many examples are a result of glacial interference.

carbonaceous Denoting sedimentary deposits formed principally from the remains of plants and other organic materials, e.g. coal, peat, and petroleum.

carbonation (carbonation-solution) A process of chemical weathering of rocks. It is produced by rainwater containing small amounts of carbon dioxide, from the atmosphere or soil, in solution (weak carbonic acid) and is especially effective on limestones. When exposed to the solution the limestone, which is composed largely of calcium carbonate, is converted to calcium bicarbonate and removed in solution:
$$CaCO_3 + H_2O + CO_2 \Rightarrow Ca(HCO_3)_2$$
The process is the chief erosion agent in limestone country.

carbon cycle The circulation of carbon through the atmosphere, the oceans, and the Earth's surface. Carbon is stored in living organisms, such as vegetation, and also in the form of carbonate minerals (e.g. limestone) and fossil fuels. It is released into the atmosphere, as carbon dioxide, chiefly through the action of living organisms on land and in the oceans. Smaller amounts are also released through the weathering of carbonate minerals, the decay of organic elements in the soil, the burning of fossil fuels (e.g. coal), and volcanic activity. Carbon dioxide can be returned to the biosphere through photosynthesis by plants; in the oceans carbonate of lime is ultimately formed from the shells and skeletons of marine creatures.

carbon dating †*See* radiocarbon dating.

Carboniferous A geological period in the Palaeozoic era extending from about 345 million years ago to about 280 million years ago. It was named after the extensive deposits of coal that were formed during this period. It is divided into two parts: the Upper and Lower Carboniferous. The Lower Carboniferous is composed of thick extensive deposits of Carboniferous Limestone, formed in warm tropical seas. This was followed by the Millstone Grit, formed under deltaic conditions, and the Coal Measures of the Upper Carboniferous, which were derived from decayed and compressed swamp vegetation. The Carboniferous rocks are some of the most important in economic terms, providing the bulk of the world's coal, and also containing significant reserves of iron ore, oil, and oil shale. In the USA the period is often divided into the Mississippian, beginning about 345 million years ago, and the Pennsylvanian, beginning about 320 million years ago.

cardinal points The four main directions of the compass: north (N), south (S), east (E), and west (W).

carr †A type of vegetation that grows in wet places. Carr develops only in base-rich lowland environments such as are found in E Norfolk. The dominant plants are trees, particularly alder and willows. In hydroseres, carr succeeds reed swamp. *See* hydrosere.

carrying capacity of land The ability of an area of land to support people or animals. The capacity can be measured in terms of the unit of land that is required to support one person or one animal. For example, it has been calculated that the alluvial deltas of SE Asia, which are used for growing rice, have a carrying capacity of 500 people per square kilometre. By comparison, the semiarid regions of W Africa may only have a capacity of one person to 10 square kilometres.

carse In Scotland, a lowland area near the estuary of a river. These areas are usually rich agriculturally, e.g. the Carse of Gowrie and Carse of Forth. [Scottish]

cartography The science of constructing maps and charts. It includes the making of original surveys, the selection of suitable map projections, and decisions on colours, layer tinting, and other visual representations.

cascade A small waterfall or a series of small waterfalls down rock steps. It is often an artificial feature created for ornament.

cascading system †A system consisting of a series of individual components that are dynamically linked by the transfer of energy from one component to another. The energy output from one component becomes the input for another. In assessing the behaviour of cascading systems scientists are concerned with the rate of throughput. An example is the basic hydrological cycle cascade in which some energy output from rain becomes an input for water moving down hillsides and along drainage systems. *See* system.

cash crop A crop that is primarily grown for sale and not for use by the grower and his family. For example, a farmer may grow some barley as feed

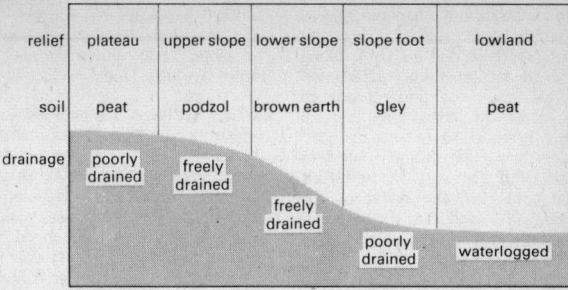

relief	plateau	upper slope	lower slope	slope foot	lowland
soil	peat	podzol	brown earth	gley	peat
drainage	poorly drained	freely drained	freely drained	poorly drained	waterlogged

Soil catena

for his cattle and some as a cash crop, selling it to brewers for the production of malt.

caste A class system in Hindu society based on the hereditary principle. It is an important feature of Indian society and is the basis of its division. Within each caste the members are socially equal and united by their religion. Often members of the same caste follow the same trade. Each caste has little social contact with other castes.

castellanus cloud A form of cloud resembling the turrets of a castle when viewed from the side. *Altocumulus castellanus* (Ac cas) are medium-level clouds with a well-defined base and *cirrus castellanus* (Ci cas) are very similar but found at higher levels.

castle kopje *See* kopje.

cataract A large waterfall, succession of falls, or a series of rapids, as on the River Nile.

catastrophism A concept, now generally disregarded, that the features of the Earth result from sudden catastrophic events, such as earthquakes, floods, and volcanic eruptions, rather than evolving slowly with time through continuous processes. *Compare* uniformitarianism.

catch crop A crop that is either grown when the ground is fallow between the harvesting of one crop and the planting of the next, or grown in between the rows of an existing crop. For example, in W Europe mustard is sometimes planted as a forage crop after an early cereal crop such as barley has been harvested. Crops of vegetables are sometimes grown between rows of olive trees in Mediterranean lands.

catchment The area from which a single river system collects its waters. The boundary of a catchment area is defined by the watershed. *See also* drainage basin, †watershed.

catena †A sequence of soil types usually developed from similar parent material that occurs from the top to the bottom of a hillslope in an area of similar relief. It is therefore an interaction of topography and soil-forming processes. Catenary change is normally continuous rather than abrupt. Soil characteristics reflect differences in drainage, slope angle, and position. The idea that such a sequence exists was first proposed by the British soil scientist G. Milne (1936).

causse (causses) A region of limestone plateaus in the SW of the Massif Central in France (the Grand Causses). These plateaus are characterized by features typical of limestone uplands, e.g. a lack of surface drainage, bare rock pavements, thin soils, and an uneven surface topography. The term is sometimes applied to limestone regions elsewhere. *See also* karst. [French]

35

cave An underground chamber that is accessible from the surface. Caves are most frequently found in cliffs along coasts and in limestone areas. On coasts, caves are commonly produced by marine erosion expanding lines of weakness in rocks to form a cylindrical tunnel in a cliff; this narrows with distance from the sea. In limestone regions, caves are the result of the rock being dissolved through carbonation by underground streams. This expands joint planes and bedding planes in the rock to form caves, occasionally of great size. *Vadose caves* are those formed above the water table by water percolating down; *phreatic caves* are found below the water table and formed by ground water. When caves collapse in limestone areas they can form gorges (e.g. Cheddar Gorge, Somerset) or depressions known as dolines. The term *cavern* is often used synonymously with cave but may be used to refer to especially large rock chambers.

CBD *See* central business district.

cedar-tree laccolith An igneous intrusion composed of a series of laccoliths, one above the other, in which the molten rock was intruded along several bedding planes as it penetrated upwards. The top of the intrusion is domed and the lower intrusions resemble the branches of a tree. *See also* laccolith.

celestial horizon *See* horizon.

cell, atmospheric †A closed circulation of air within the atmosphere, which can either be global in scale or extremely localized. The major cells in the atmosphere are the *Hadley cell*, which produces the trade winds; the *Ferrel cell*, which surrounds the polar front; and the less well defined polar cell, which produces the polar wind. There are many types of smaller circulations producing cells that are of localized importance. For example, thunderstorm cells, which are formed in each individual thunder cloud, and *Bénard cells*, which are shallow circulations between the Earth's surface and the stable layer of air that forms at cloud level under certain conditions. *See also* Hadley cell.

Celsius scale A temperature scale in which the temperature of melting ice is taken as 0°C and the temperature of boiling water as 100°C. In meteorology the scale is more commonly known as the *centigrade scale*. The name was officially changed to the Celsius scale in 1948. [It was named after the Swedish scientist Anders Celsius who in 1742 was the first to divide the interval between the freezing point and boiling point of water into 100 parts]

Cenozoic (Cainozoic, Kainozoic) The last geological era, beginning about 65 million years ago. During this era mammals became dominant (the Cenozoic is often known as the age of mammals) and vegetation similar to present-day forms developed. It includes the Tertiary and Quaternary periods; some geologists regard the Cenozoic as being synonymous with the Tertiary.

census An official count of a population. Population counts vary from the simplest forms, in which the total number of people at one place and one point in time are counted, to the sophisticated listing of a range of demographic and social characteristics. In the UK a national census normally takes place every ten years. Three categories of data are collected in the UK: demographic information; details of household characteristics; and migration and education data.

centigrade scale *See* Celsius scale.

centimetre *See* metre.

central business district (CBD) The commercial, social, and cultural core of a city, where the chief shops and offices are concentrated. The shops are larger and more prestigious than those found elsewhere in the city and include those selling luxury goods and the main departmental stores. The central business district is also the focus of the urban transport network but, although it is physically the most accessible part of the city, it is often the area of maximum traffic congestion. It is also the area of highest land values so buildings are concentrated at high densities and are built to maximum heights. Residential land use has been largely squeezed out as a result so the central business district is characterized by a large daytime population and a significantly smaller nighttime population.

central goods/services/functions †Those goods, services, and functions that tend to be grouped together in urban centres or central places. Each serves an area called the functional area and the size and number of the functions will determine the position of the central place in a hierarchical classification. The concept of central places, services, and functional areas was first developed by Walter Christaller (1933) in Germany. *See* central place theory.

centrality †The degree to which a town serves its surrounding area. In the development of central place theory and locational analysis in geography a number of indices and coefficients have been devised to measure centrality. The first – the *index of centrality* – was produced by Walter Christaller. He based this on the telephone service available at a place relative to its region and population:

$$Zx = (Tn - En)Tg/Eg$$

where Zx is the index of centrality, Tn is the number of telephones in the central place, En is the population of the central place, Tg is the number of telephones in the region, and Eg is the population of the region.

central place An accessible location from which goods and services are provided for the surrounding area. The term was introduced by Walter Christaller when developing central place theory. He termed the services performed at a central place *central functions*. *See also* central place hierar-

chy, central place system, central place theory.

central place hierarchy In central place theory, the ranking of central places in terms of their populations, number of establishments providing goods and services, and the size of their trade areas. Different types of services require different populations to sustain them; the minimum population necessary for the economic provision of a service is termed the *threshold population*. Different goods or services also have different *ranges*, i.e. the distance consumers are prepared to travel to obtain the good or service. *Low order central places* provide low order services, which have low thresholds and ranges; *high order central places* provide high order services and serve larger market areas. A number of different orders of centres (a hierarchy) may be identified according to the services they offer the surrounding areas.

central place system The spatial distribution of a system of central places. It is composed of a complex hierarchy, each order of which is made up of discrete groups of centres with each group being distinctive in the order of services it performs. According to Christaller the number, distribution, and hierarchical order of the settle-

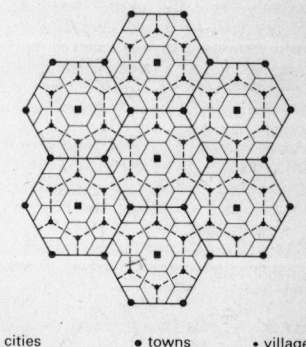

■ cities   ● towns   • villages

Central place theory   Christaller's $k=3$ hierarchy according to the marketing principle

ments in a system is determined by one of three controlling principles: the marketing principle (k = 3), the transport principle (k = 4), and the administrative principle (k = 7). *See also* central place hierarchy, central place theory.

central place theory The theory that there is a pattern in the number and distribution of towns, cities, and villages (central places) and in the ways in which the central places provide goods and services for their surrounding areas (hinterlands). This regularity is hierarchical with central places at different orders because specialist goods found in cities need larger threshold populations than do the day-to-day goods supplied in small towns and villages.

The theory was first outlined by Walter Christaller in 1933 and was later significantly modified by A. Lösch. The original theory is now considered to be over-simple and too inflexible. *See also* central place, central place hierarchy, central place system.

centre–periphery model †*See* core–periphery model.

centripetal drainage A pattern of drainage in which the streams flow towards a central point. This is commonly the case in arid and semiarid regions where the central point may be provided by a lake or inland sea; such regions are also referred to as areas of interior drainage. The central point may also be a main stream, e.g. the Bagmati River in Nepal. *Compare* radial drainage.

centrography †The calculation of the centres of distributions, which are then plotted on maps. By examining the movement and relationships of these centres attempts are made to establish laws of distribution of phenomena.

cereal A cultivated grass used as human or animal food. The most important cereal crop in terms of human consumption is rice, which is the staple diet of over one quarter of

the world's population. Maize (corn) is also important as a food for both humans and animals. The mid-latitude grasslands produce wheat, barley, rye, and oats, which are processed in a variety of ways for human consumption.

chain A mountain complex composed of a series of roughly parallel ranges. The term is also applied to linear groups of lakes, islands, and other physical features.

chain surveying A method of surveying. The instruments used are a *chain*, which consists of a metal chain now usually 20 m (formerly 22 yards) long made up of 100 links, and a set of markers, usually steel pegs, which are used to mark the ends of chains as measurements are made. Chain surveys are made by measuring a series of triangles and offsets from the edges of the triangles.

chalk A soft white very pure limestone, consisting of about 90% calcium carbonate. It is the characteristic rock of the Cretaceous period in W Europe. In the UK it forms the uplands of the North and South Downs, the Chiltern Hills, Salisbury Plain, and the Yorkshire Wolds, and also the well-known landmarks of the white cliffs of Dover and the Needles of the Isle of Wight. It consists largely of the tests of marine micro-organisms, such as coccoliths and foraminifera, and shell fragments. Nodules of flint are often contained within the chalk. Study of the fossil content suggests that the chalk was laid down in a medium depth of the sea, which was surrounded by low-lying, possibly desert coast.

chalybeate Water impregnated with iron salts. It is usually applied to mineral water or spring water.

champagne (champaign, champian, campagna) An area of fairly level open countryside, without woods or hedges. The word is generally limited to describing the characteristic landscape of NE France, where the fields are large and not separated by hedges as they are in Britanny. *See also* bocage. [French]

Chandler wobble †The variation in the Earth's rotation due to a slight wandering of the poles relative to the Earth's surface. Its cause is unknown but it may result from changes in the shape and centre of gravity of the Earth due to fluctuations in ice and snow cover, ocean currents, air masses, and from earthquake activity. It was discovered in 1891 and found to have a periodicity of about 14 months and an amplitude of 0°0′5″.

channel 1. A stretch of water confined between banks, as in a river or stream channel.
2. A narrow stretch of water (larger than a strait) connecting two larger areas of sea as in the English Channel.
3. The navigable part of a waterway.

chaparral A form of vegetation dominated by thick-growing evergreen oaks and other trees, with aromatic shrubs. Chaparral occurs in North America, in California and NW Mexico, where there is a Mediterranean-type climate. Like the equivalent maquis of areas bordering the Mediterranean in S Europe, chaparral has evolved in response to mild wet winters and summer drought. *Compare* maquis. [Spanish]

chart A map that is designed for a specific use. For example, navigation charts for use by sea (e.g. Admiralty charts) or air, weather (synoptic) charts, and star charts.

chase A tract of unenclosed land reserved for breeding and hunting wild animals. This word sometimes occurs in place names and so provides a clue as to how the land was used in the past, e.g. Cranborne Chase.

chelation †In soils, a process of chemical bonding of metallic ions (notably iron, aluminium, and magnesium) to organic compounds derived from the

decomposition of humus. The organic compounds, known as *chelating agents*, are extracted from the litter layer by water passing down through it. Chelating agents are richest in heath plants and conifer needles. As the chelating agents in solution percolate down through the soil they remove many of the minerals of the upper part of the soil profile leaving a grey eluviated sand horizon. The sesquioxides are redeposited in the lower illuvial part of the soil profile, sometimes forming an ironpan. These processes are responsible for the formation of soils with distinct horizons known as podzols. *See* podzol.

chemical weathering The disintegration of rock by chemical processes, most of which involve the action of water. These processes may cause the removal of cementing materials in a rock, weakening its structure so that it tends to crumble, or they may lead to the formation of weaker secondary minerals, which are more easily eroded. Chemical weathering is most effective in wet environments that experience high temperatures. *Compare* mechanical weathering. *See* carbonation, hydration, oxidation, solution, †hydrolysis.

chernozem (black earth) A black or dark-brown zonal soil that is rich in humus and contains lime. It is found in a belt extending from Manchuria in China through the USSR into the Ukraine and into Romania and Hungary; it also occurs between North Dakota and Texas, USA. A fertile soil, it is often used to support cereal crops. †The origin of chernozem soil is strongly influenced by climate; it has developed under natural grasslands in the cool temperate climatic zones with warm summers, cold winters, and a maximum of rainfall in summer.

chestnut soil A zonal soil of mid-latitude grasslands that occurs in drier regions than chernozems. Chestnut soils occur in the USSR in the S Ukraine, in the Great Plains of the

USA, and in the S African veld. The low rainfall limits the growth of vegetation and as a result the supply of organic matter is considerably less than in chernozems. The A horizon is chestnut brown in colour. It merges at about 25 cm into the B horizon, which contains calcium carbonate. The calcium carbonate may form a distinct horizon at depths of about 50 cm. The parent material is frequently loess. *Compare* chernozem.

chili The dry hot sirocco-type wind of Tunisia.

chimney A narrow vertical cleft in a rock face, formed where a joint plane has been expanded by weathering.

china clay *See* kaolin.

chine A deep narrow ravine cut in soft rock by a stream running steeply to the sea, especially in the Isle of Wight and Hampshire, e.g. Blackgang Chine, Isle of Wight.

chinook A warm dry southwesterly wind, similar to the föhn, that blows down the E slopes of the Rockies in Canada and the USA. It is adiabatically warmed and often causes a large rise in temperature; in spring this leads to rapid snowmelt and subsequent avalanches.

C horizon The soil horizon at the base of a soil profile, comprising mainly parent material in a partly weathered state. C horizons supply the minerals to the soil. They are usually clearly distinct from the B horizon above, to which they graduate as disintegration progresses. *See* soil horizon.

chorography The practice of describing or delineating particular regions or districts. The central concern of this approach is areal differentiation, i.e. the examination of the arrangement and association of phenomena that distinguish one area from another. Chorography is sometimes seen as having a more limited scope than regional geography but less limited than topographical description.

chorology 1. The causal study of the distribution and relations of phenomena in an area.
2. The scientific study of the geographical extent or limits of phenomena, e.g. the local distribution of species. This word is not in general use today.

choropleth A map that uses colours or shading to show area density patterns. When drawing choropleth maps shading and colours must be carefully chosen to give the desired visual effect. Examples of choropleth maps are those that show the distribution of population by parishes in a county or the number of cases of malaria in Africa, by countries.

chott *See* shott.

chute cutoff †*See* cutoff.

cinder cone A volcanic cone composed of small fragments of lava. It is formed by volcanic cinders that have fallen back to the ground surface after an eruption accumulating around the volcanic vent. Cinder cones tend to be steeper than ash cones as they are composed of larger fragments, although more common are cones made up of a combination of ash and cinders. An example of a cinder cone is Paricutín in Mexico, which first erupted in 1943.

circumference of Earth Any plane passing through the centre of the Earth would cut through its surface along a circumference or great circle. The equator is the longest circumference (40 075 km). The polar circumference is about 75 km shorter than that around the equator because the centrifugal force resulting from rotation causes the equatorial belt to bulge slightly.

cirque (corrie, cwm) A deep amphitheatre-shaped rock basin with steep sides and an opening downstream found in glaciated upland areas. Cirques originate as small hollows where snow accumulates. The

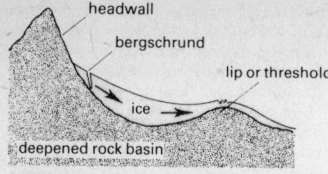

Cirque (cross section)

snow becomes compacted to firn and glacier ice, forming a cirque glacier, and eventually flows downslope under the influence of gravity to feed valley glaciers. The characteristic shape of cirques is a result of freeze-thaw erosion on the headwall and the rotational slip of the ice within the concave floor of the hollow, which is widened and deepened by the plucking of rock fragments by the ice. Many cirques contain small circular lakes, known as tarns. *See also* arête, bergschrund, headwall, †nivation. [French]

cirque glacier A small body of ice, wide in relation to its length, that occupies a rounded rock basin. It may be confined to the basin or may overflow this to form the head of a valley glacier. *See also* cirque.

cirrocumulus (Cc) A type of high cloud (above 6000 m) consisting of ice crystals. It has a globular rippled appearance with blue sky showing between groups or lines of cloud and is sometimes known as mackerel sky.

cirrostratus (Cs) A thin high sheet cloud (above 6000 m). The Sun or Moon may be seen through the cloud surrounded by a halo.

cirrus (Ci) A high wispy cloud (6–12 000 m) consisting of ice crystals. Where the cloud is drawn out, forming 'mare's tails', it indicates strong winds in the upper atmosphere.

citrus fruit Varieties of tropical and subtropical fruit that contain citric acid, of which lemons, limes, oranges, and grapefruit are the most common. These fruit grow on trees that require Mediterranean-type climatic conditions. Cultivation is localized to warm areas suitable for irrigation. The crops are grown under commercial fruit farming conditions with intensive cultivation and specialization. The main producing regions are California, Arizona, Texas, and Florida in the USA, Spain, Italy, Greece, W and S Australia, and South Africa.

city A large urban settlement. There is no strict definition of the term according to size of population. In the UK, the title is conferred on a large town by the Crown and a city usually has a cathedral and the seat of a bishopric.

city region The area that focuses upon and is functionally dependent upon a city. The term emphasizes the close ties between a city and the surrounding area that it serves. The central city and the surrounding area are envisaged as a single functional unit interdependent in terms of the movement of people (commuting), goods (distribution), and information (local radio, newspapers). The city region has two components: the core (the city) and the periphery (the hinterland). *See also* hinterland.

clastic Denoting rocks that are formed from fragments resulting from the breakdown of pre-existing rocks. Examples include sandstones, mudstones, and conglomerates. This term can also be applied to recent sediments, such as the sand on a beach or deposits on the bed of a river. [From Greek *klastos*: broken]

clay 1. (*Geology*) A fine-grained argillaceous clastic sediment consisting mainly of minute flakes of crystals, produced by chemical weathering of the parent rock. It becomes plastic when wet and hardens and cracks when dry. Under the changes occurring during compaction into rock, clay passes into mudstone, shale, and slate. Clay is deposited under a variety of conditions, in lakes, deltas, and ocean depths. Grain size, on the Wentworth scale, is 1/256 mm to 1/1024 mm. 2. (*Soil Science*) A soil in which the particles have a diameter of less than 0.002 mm (0.005 mm in the USA).

3. (*Mineralogy*) A group of minerals consisting of hydrous silicates, mainly aluminium and magnesium, which result chiefly from the chemical breakdown of feldspars. They have a layered structure and the ability to take up or lose water.

clay–humus complex †The close association between the colloids of clay (mineral) and finely-divided particles of humus (organic) in the soil. The clay–humus complex is chemically active in capturing exchangeable nutrients, which are then available for plants. *See* colloid.

claypan A compact layer within the soil consisting of clay particles that have been redeposited after eluviation. Claypans become impervious with waterlogging. *See also* hardpan.

clearing An area of land that has been cleared of trees and undergrowth, usually to provide additional land for cultivation. *See* shifting cultivation.

cleavage 1. (*Geology*) The tendency of some rocks to split into thin sheets or slabs along parallel planes. Cleavage occurs in some sedimentary rocks such as shales where the minerals lie parallel to the layers of deposition. Frequently cleavage results from great metamorphic pressures in the past, which have caused a realignment of the minerals (e.g. in slates) and may make any angle with the bedding planes.
2. (*Crystallography*) The tendency of crystals to split along planes of weakness in the molecular framework.

cliff A very steep or vertical rock face. Sea cliffs are formed by waves undercutting the rock causing its eventual collapse. The form of the cliff depends on such factors as rock type, structure, resistance to erosion, and the presence of bands of weakness. Cliffs are also found in inland areas above lake shores and rivers, in mountainous regions, and marking the site of former coastlines.

climate The average weather conditions and variations in these conditions in both space and time over a large area. Weather conditions over a specific length of time, usually a period of at least 30 years, are taken into consideration. The main elements of climate are temperature, atmospheric pressure, wind, and humidity (including precipitation). On a large scale, the climate of a particular region is determined by:
(1) Latitude and the tilt of the Earth's axis, which determine the amount of solar radiation received by the area.
(2) The distribution of land and sea and proximity of ocean currents.
(3) The altitude and topography of the area.
(4) The location of the area in relation to the main circulation belts of the Earth.
On a smaller scale other factors such as aspect may locally influence climate.

climatic climax vegetation †*See* climax vegetation.

climatic geomorphology (climatomorphology) The branch of geomorphology that concentrates on the influence of climate in the development of landforms. Climate is seen as directly influencing landforms as it determines the processes that operate, e.g. the type, nature, and extent of weathering. On the basis of climate various zones may be recognized where particular processes and forms predominate, e.g. tropical, arid, glacial, and periglacial zones. However, changes in climate result in landscapes exhibiting features derived from a variety of environmental conditions.

climatic region One of the areas into which the Earth can be divided on the basis of climate. There have been many different divisions of the Earth into climatic regions using different climatic criteria but the four basic climatic regions are recognized as tropical, subtropical, temperate, and polar. These may be subdivided in different ways using different criteria.

climatology The study of climate. It includes the study of climatic phenomena, their causes, and their influence on the natural environment.

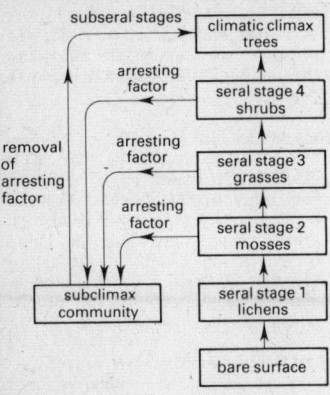

Stages in the development of climatic climax vegetation

climax vegetation †The form of vegetation that has become established in response to natural conditions prevailing in a particular environment. It is the final stage, achieved over many years, of the natural process of succession within the plant community. Climax vegetation is governed most strongly by regional climatic factors (and hence is often called *climatic climax vegetation*) but local variations of soil or topography produce different climaxes. Climax vegetation consists of long-lived and dominant climax species (e.g. oak trees in deciduous forest). It is a relatively stable and permanent community, capable of self-perpetuation and recovery after disturbance, if sufficient time is allowed. It is in equilibrium with its physical and biotic environment, and in particular demonstrates an ecological balance between production and consumption. *See also* plagioclimax vegetation, subclimax vegetation.

climograph (climogram) †A diagram in the form of a graph in which two climatic elements at any one place are plotted against each other. The resultant shape of the graph indicates the character of the climate at that particular place. It is most commonly used to demonstrate humidity, the climatic elements plotted being the mean monthly wet-bulb thermometer values and the mean monthly relative humidity values.

clinometer An instrument used in surveying to measure vertical angles. The *Indian clinometer* is used in plane-table surveys. It consists of a rule fitted with a spirit level and adjustable foot. At one end of the rule is a pinhole sight while at the other end is a slotted sight along which a scale of degrees is marked. A distant object is sighted through the pinhole and its angle of elevation or depression read from its position in the slot sight. The Abney level is another type of clinometer. *See also* Abney level.

clint A ridge of exposed limestone, usually in the form of a rectangular block. It is formed by chemical weathering enlarging joint planes in the limestone to produce the fissures known as grikes. The surface of a clint is usually heavily weathered, resulting in a deeply pitted appearance. Clints are a typical feature of limestone pavements, such as occur in the Ingleborough area of Yorkshire. *See also* grike, limestone pavement.

closed system †*See* system.

cloud A mass of minute water droplets and/or ice crystals formed by the condensation of water vapour and held in suspension in the atmosphere. Condensation, which results from cooling, usually takes place around nuclei such as dust, smoke particles, and salt. The cooling may be caused by convection, uplift over mountains, or ascent in depressions. Clouds may be present at heights ranging from ground level (fog and mist) up to over 13 000 m.
Several classifications of clouds exist but the one internationally agreed on and most used is based on cloud appearance and height. It comprises

ten principal forms. The low clouds (up to a height of about 2000 m) comprise stratus (St), stratocumulus (Sc), and nimbostratus (Ns). Above these are the medium clouds (2000–7000 m), which comprise altocumulus (Ac) and altostratus (As). The high clouds (7000–13 000 m) are cirrus (Ci), cirrostratus (Cs), and cirrocumulus (Cc). Some clouds grow vertically and cannot be classified solely by height; these are cumulus (Cu) and cumulonimbus (Cb).

cloud seeding The modification of weather by the introduction of artificial nuclei into clouds to induce the minute droplets of water or particles of ice to coalesce and form raindrops or ice pellets. The idea was first put forward in the late 19th century but' experiments began only in the 1940s. The most effective materials so far discovered are pellets of solid carbon dioxide (dry ice), fine powdered salt, and a 'smoke' of silver iodide. The main aim of the research is artificially to supply greater quantities of rainfall to semiarid areas, but a basic requirement is that suitable clouds must be present. The technique has also been used to attempt to reduce the destructive power of hurricanes, and results to date are promising.

cluse A narrow steep-sided valley or gorge that cuts transversely across a mountain ridge. The term is generally used in the context of the French pre-Alps in Haute-Savoie and the Jura Mountains. [French]

coal A carbonaceous rock. It is formed from the compaction and heating under pressure of large quantities of partially decomposed vegetable matter such as peat that accumulated in vast coastal swamps, probably similar to those of Florida and New Guinea today. Various types are classified according to the kind of vegetable matter they contain or the proportion of volatile matter. †There are two major groups: *humic coals*, which are derived from wood and include bituminous coal, lignite, and anthracite;

and *sapropelic coals*, which are derived from spores, algae, and fine particles of plant material, and include cannel coal and boghead coal. The place of a particular coal in a classification is called its rank. Coal is mined for use as a fuel and raw material for the plastics and chemical industries throughout Europe and North America.

coalescence In meteorology, the process whereby droplets of water in a cloud are enlarged by combination with other smaller droplets. The size of the resultant droplet will depend on several factors, including the liquid content of the cloud, the amount of variation in the sizes of the droplets, and the length of time the droplet remains in the cloud.

Coal Measures The series of rock layers comprising a number of coal seams, together with associated sedimentary rocks, that were formed during the Upper Carboniferous period. The coal seams were formed in continuously subsiding basins throughout North America, the British Isles, Europe, and Asia. The sequence of layers is: coal, shale, mudstone, grit, sandstone, mudstone, fireclay, and coal. This sequence represents a successive deepening and shallowing of the basins throughout the Upper Carboniferous. *See also* coal.

coast The boundary between land and sea, including the strip of land that borders the sea shore.

coastal plain A flat low-lying expanse of land between the coast and higher ground inland. Coastal plains may be produced by a relative fall in sea level, which exposes a stretch of land previously under water, or by the deposition of alluvium by rivers. Man can also produce coastal plains by dyking and draining areas previously under water.

coastline The boundary between the land and the sea; it may be defined as the line followed by the cliffs or the line reached by the highest storm waves.

cobalt A malleable metal that is used in the production of alloys such as stainless steel, in electroplating, and in the manufacture of blue pigments for the glass, enamel, and pottery industries. It occurs naturally as cobaltite and is also found in copper, nickel, iron, silver, and lead ores. The chief producers are Zaïre, Zimbabwe, and Canada.

cobble A smoothly rounded piece of rock that is larger than a pebble but smaller than a boulder, with a diameter of 60–200 mm. In the USA a cobble is defined on the Wentworth scale as having a diameter of between 64 and 256 mm.

cockpit karst †See polygonal karst.

coefficient of dispersion †A measure of the degree of nucleation or dispersion of rural settlement in an area. The formula for its calculation is:
$$C = p \times n/P$$
where C is the coefficient of dispersion, p is the population of the parish excluding the largest settlement, n is the number of settlements in the parish, and P is the total population of the parish.

coffee The seeds (called beans) of a tropical evergreen tree of the genus *Coffea*, the two main species of which are *Coffea arabica* and *Coffea robusta*. When dried, roasted, ground, and brewed in hot water these produce a beverage. The tree is grown in the tropics in areas with a long hot wet season and a short cool dry season. Location is particularly important as the crop can be affected by frost, drought, and wind; valley bottoms and high altitudes are avoided as the tree is damaged by temperatures below freezing. The most important growing region is the central plateau of E Brazil. Coffee for export is also grown in the Caribbean, Colombia, SE Asia, and E Africa.

coke A fuel consisting chiefly of carbon, obtained from heating coal in an oven from which air is excluded, thereby removing some of the chemical substances as gases. Coke is an essential ingredient in the manufacture of pig iron and is also used in the production of steel. Some coalfields, such as the Saar, do not possess suitable coal for conversion into coke. Good coking coal is found in a number of coalfields in the UK, including the S Wales field. *See also* Bessemer process.

col 1. (*Geomorphology*) A pass between two mountain summits at a high level, formed by glacial or fluvial action. It is often of significance for communications, providing natural routes through mountainous regions, e.g. the Col du Grand St Bernard in the French Alps. [French]
2. (*Meteorology*) An area of relatively uniform atmospheric pressure lying between two opposing anticyclones and two opposing depressions, the pressure being lower than that of the anticyclones and higher than that of the depressions. The weather associated with a col is variable with generally fine weather but frequent thunderstorms in summer and generally dull and foggy conditions in winter.

cold desert 1. The tundra and polar regions where plant and animal life is inhibited by low temperatures.
2. The continental interiors poleward of 50°N.

cold front A front separating a retreating warm air mass and an advancing cold air mass, which forces its way underneath the warm air causing it to rise. A cold front is steeper than a warm front and the weather associated with it tends to pass more rapidly. At a cold front pressure rises, temperature falls, the wind veers, and there are heavy showers often accompanied by thunder. *See also* front.

cold pole The point with the lowest mean annual temperature in each hemisphere. In the N hemisphere this is at Verkojansk in NE Siberia (USSR) where the mean annual temperature is −16.3°C. The mean January temperature is −50°C and −70°C has been recorded. In the S hemisphere the lowest recorded temperatures have been at the Soviet research station of Vostok on the Antarctic ice plateau; −90°C has been recorded. Both these poles are in areas where radiational cooling is extreme under clear skies and low relative humidity.

cold temperate *See* subpolar.

cold wall The plane of contact between a mass of cold water and a mass of warm water in the ocean, similar to a front in meteorology. Differing water masses do not mix freely and this can lead to a considerable temperature difference within a very short distance. For example, in winter on the Grand Banks off the SE coast of Newfoundland water temperatures may vary by as much as 11°C between the warm waters of the Gulf Stream and the cold waters of the Labrador Current.

cold wave A sudden surge of cold air experienced in temperate latitudes, usually caused by polar air being drawn equatorwards behind the cold front of a passing depression. This is most prominent in the N continents where winter temperatures reach much lower levels than in the mainly oceanic S hemisphere. It is most effective and penetrates furthest where there is no E−W relief barrier to block its passage, e.g. in North America. The importance of the cold wave is that it introduces very cold air into areas that are generally much warmer and thus may seriously damage crops. Examples of such winds are the southerly burster, pampero, norther, norte, papagayo, and friagem. In the USA the term is defined more narrowly as a specific fall in temperature below a fixed figure in a specific period of time (usually 24 hours); this depends on the season and the place. *See* polar outbreak.

collective farming A system of farming in which the land is jointly owned or occupied by people who organize and run it according to a plan that the community has accepted or approved. The underlying concept of collective farming is the belief that there is a greater satisfaction and level of efficiency to be obtained from working for the community, rather than for oneself. The system originated on a large scale in the USSR in the late 1920s. In the USSR and China the collective farms are state controlled. In Israel the kibbutzim (collective villages) are voluntary organizations not controlled by the state. This type of farm organization also occurs in Italy, Mexico, India, Pakistan, Japan, and elsewhere. *See also* kibbutz, kolkhoz, sovkhoz.

collectivism The belief that economic planning should be organized and managed by the state and not by private enterprise. The people collectively own the land and the means of production and share the rewards of their corporate enterprise and effort. Collectivism is part of the communist doctrine and practised in countries such as the USSR and Cuba. However, like collective farming, collectivism is not confined to communist countries and may be found as the basis for production in various parts of the world.

colloid †In soil science, a minute division of the soil, either mineral (clay) or organic (made up of finely divided humus). Without a colloidal fraction, soils are physically dry and chemically inert. The vast surface area of colloids has electrical attraction for mineral ions released during weathering or contained in soil water. As a result colloids hold a vital reservoir of exchangeable plant nutrients. Their water retention allows plant roots access to moisture that would otherwise drain away. Colloids swell on wetting, thus giving plasticity to the soil structure. *See also* clay−humus complex.

colluvium A deposit consisting of a wide variety of particle sizes, ranging from rock fragments to fine-grained soil, which accumulates at the base of slopes under gravity.

colony 1. A settlement of people in a country distant from their homeland; also the territory that they settle in.
2. (*Political Geography*) A territory, initially underdeveloped, which is subject to the sovereign rule of a foreign power for strategic or economic reasons, and settled by representatives of the foreign power. Most colonies have achieved independence from their former colonial rulers since World War I.
3. (*Biogeography*) A group of organisms, usually a single species, living together in close proximity. Examples of colonial animals include some birds (e.g. rooks, gulls), some insects (especially bees and ants), and corals, which are structurally linked.

columnar structure The polygonal jointing resulting in columns of rock that occurs in lava flows of basalt and other fine-grained rocks. On cooling, the rock contracts in such a way that the rock cracks into long polygonal columns, most of which are hexagonal. Examples include the Giant's Causeway in Antrim, Northern Ireland, and Fingal's Cave on the Isle of Staffa, Scotland.

combe (coombe) **1.** A short valley with a steep head, partially or wholly dry, found in the chalklands of S England. **2.** A short steep valley that runs down to the sea, e.g. Combe Martin, N Devon.

comfort zone The climatic conditions that are physiologically the most comfortable to human beings. The two key determinants are the range of temperature and the relative humidity. In middle latitudes the comfort zone is usually defined as having temperatures of $20°-25°C$, with relative humidity between 25% and 75%. In the UK it is around 15°C with 60% relative humidity.

commercial centre A town or city that has as its major function the buying and selling of goods and services. Some of the earliest commercial centres were the market towns that traded in the agricultural produce of the locality. Present-day commercial centres, such as Norwich and Inverness, are important regional centres with banking, insurance, and other commercial facilities, as well as a range of retail outlets.

commercial core The area of a town or city that is the focus of business activity. It houses the major department stores and shops as well as banks, offices, hotels, and entertainment centres of the town.
†In 1927 E. W. Burgess introduced the concentric-zone growth theory based on his studies of urban growth in Chicago. The theory suggests that a city expands radially in the form of a series of concentric zones, of which the inner zone is the central business district (CBD) or commercial core.
See also central business district, †concentric model of urban land use.

commercial geography Those aspects of geography that are concerned with production and trade. For example, the major products of a region and the parts of the world to which they are exported. This study is one facet of economic geography.

common (common land) An area of undivided land, often unenclosed, held in joint occupation by a community or manorial waste. The word is now applied to wasteland to which the public has right of access. Common lands are often of quite poor quality and were of little interest to farmers. Today commons are an important component of our system of open spaces.

common market An agreement by a group of countries to form a single market within their group. The term is most commonly applied to the group of countries that form the European Economic Community (EEC), set up by the Treaty of Rome on 1 January, 1958. The primary aims of the treaty are the elimination of all obstacles to the free movement of goods, services, capital, and labour between the member countries, and the setting up of common policies for external trade, agriculture, and transport. The original signatories – Belgium, France, Italy, Luxembourg, the Netherlands, and West Germany – were joined in 1973 by Denmark, the Irish Republic, and the UK, and in 1981 by Greece.

communications The established links between locations along which people, goods, and information travel; these include road, rail, water, and air links, pipelines, and telephone cables.
†A recurrent theme in geography is the distance separating locations and how to traverse the distance. Changes in the pattern of communications result in adjustments to spatial relationships, which can have considerable geographical significance.

community 1. (*Human Geography*) A group of people who have developed a sense of togetherness and who associate with each other more than with outsiders. It is a socially cohesive group often characterized by common social and demographic characteristics. **2.** (*Biogeography*) A group of organisms, both plant and animal, sharing the same environmental conditions and forming a distinct unit. Community is a term of wide applicability, ranging from the biotic components of a large-scale ecosystem, such as rainforest, to the inhabitants of a small habitat, such as a rotting log.
†The members of a community are generally associated in a series of complex interactions, for example they depend on one another for food.
See also association, †ecosystem.

commuter A person who travels a considerable distance regularly (usually daily) to and from work. The most common form of commuting is between a city workplace and a suburban or country home. [The word originated in the USA where it was applied to holders of a commutation ticket (roughly equivalent to a British season ticket)]

comparative advantage †The ability of a country or region to concentrate on the production of a particular item or items that it can sell profitably to other countries or regions, because it is in some ways better endowed. For example, the USA has a comparative advantage over the UK in the production of maize, which grows less well in the UK than in the Midwest of the USA. However, David Ricardo (1817) showed that provided the relative costs of production of different commodities were favourable, trade could be beneficial even though in absolute terms a country or region is poorly endowed.

comparative cost analysis †A method by which costs of production between countries, regions, or specific locations can be compared. Detailed analysis of costs will help to explain the reasons for regional specialization of production. The method can be used to evaluate the efficiency of locations for particular types of production and can also assist in making a choice between alternative locations for the siting of a particular activity.

compass An instrument used for determining the position of magnetic north. It consists of a magnetized needle free to rotate in a horizontal plane. One end of the needle points towards the magnetic north pole.
In the *mariner's compass* the needle is fixed to a circular card, which can rotate. The card is marked with the 32 points of the compass. *See also* gyrocompass, prismatic compass.

composite cone A volcanic cone composed of alternate layers of ash, cinder, and lava. Composite cones result from the accumulations over time of material from a number of eruptions. Some of the best examples of this type of volcano are found in S Italy, including Stromboli, Etna, and Vesuvius. In the USA a composite cone is also known as a *stratovolcano*.

compression The squeezing action that produces folds in the rocks of the Earth's crust; the tightness of the folds is an indication of the strength of the squeezing.

concavity A slope, or section of a slope, in which the angle of slope is reduced (i.e. it becomes less and less steep) progressively downslope. †Under conditions of normal erosion the concave section of a slope occurs at the foot and is referred to as a basal concavity.
Compare convexity.

concealed coalfield A coalfield or part of a coalfield in which the coal seams are not on or close to the surface. The W part of the Yorkshire, Derbyshire, and Notts coalfield is exposed, i.e. the seams are on or close to the surface. This part of the coalfield was worked first and the pits that remain are old or the seams have been worked out. To the E the coalfield is concealed where the seams dip under more recent rocks and mining is carried on at depths of over 1000 m.

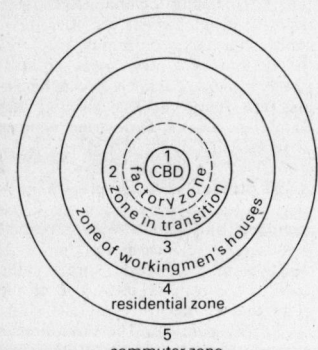

Concentric model of urban land use
(after E. W. Burgess)

concentric model of urban land use (concentric-zone growth theory) †A model describing the arrangement of functional zones within a city. It is based upon the concept that within the city people and services compete for limited space – those who are best able to pay obtain what they desire and those with the least ability to pay have little or no choice. The result of this competition is the creation of land-use zones forming *concentric rings* around the city centre.
The model was developed by E. W. Burgess in the 1920s and was one of the earliest attempts to analyse urban structure and the significant relationships that exist between broad types of land use. It was based upon empirical research in US cities, especially Chicago. Burgess identified the following concentric rings from the city centre outwards: the central business district (CBD), the zone in transition, the zone of workingmen's houses, the residential zone, and the commuter zone.

concordant coast (longitudinal coast, Pacific type of coastline) A coastline that is parallel to the structural grain of the land. Concordant coasts are usually straight and regular. A rise in relative sea level may cause the flooding of valleys to form sounds, leaving ridges exposed as isolated and often elongated islands. Submerged concordant coastlines are commonly known as *Dalmatian coasts* after the Dalmatian coast of Yugoslavia on the Adriatic Sea where this type of coastline occurs. Other examples include Cork Harbour in the Republic of Ireland.

concretion A concentration of a particular mineral within a sedimentary rock, usually as an irregularly shaped nodule. The composition of a concretion is generally very different from that of the surrounding rock, e.g. flint nodules found in chalk. Concretions are formed at or near the surface while the deposit is being transformed from loose sediment into rock, and as a concentration during the deposition of the sediment. Manganese oxide nodules are forming today in deep sea environments.

condensation The physical process of transformation from the vapour to the

49

liquid state. In the atmosphere condensation occurs either when the temperature drops sufficiently for moisture to be cooled to its dew point, or when there is enough water vapour within an air mass for it to reach saturation point. For condensation to take place *condensation nuclei* are necessary. These are present in large quantities in the troposphere and include smoke particles, dust, and salt held in suspension in the atmosphere.

condensation trail (vapour trail) A long white cloud-like trail that forms behind an aircraft flying in cold clear humid air. It forms from the condensation of water vapour produced by the combustion of the aircraft's fuel. In the UK, condensation trails usually only occur above 8500 m in summer and 6000 m in winter. In tropical regions they only form at heights above about 11 000 m.

condominium A territory ruled jointly by two or more countries. For example, between 1896 and 1956 the Sudan was a condominium ruled jointly by Egypt and the UK; the New Hebrides (now the Vanuatu Republic) was ruled by both the British and the French between 1906 and 1980.

conduction The process by which heat is transferred from a region of high temperature to one of lower temperature, through a substance or between two substances in contact, without any movement of matter. Radiation from the Sun is absorbed by the ground surface, which heats the air in contact with it by conduction. As air is a poor conductor of heat this heating effect is confined to a very shallow layer near the ground surface, the heat being spread through the rest of the atmosphere by convection or turbulence.

cone A landform with a circular base tapering upwards to a point. It is the shape most commonly assumed by volcanoes. *See also* alluvial cone.

confluence The point at which two streams meet and unite or a tributary joins a main stream. A stream that does this is described as a *confluent*.

conformal projection *See* orthomorphic projection.

congelifraction †The weathering process by which frost action breaks up rocks. It results from the growth of ice crystals within pore spaces in the rocks; this expansion increases internal pressures to the extent that the rocks will split, shatter, and fragment. The process is particularly effective on schists and some limestones, such as chalk.

congeliturbation †The movement of weathered rock debris or soil by the action of frost in periglacial areas. A wide variety of processes is included in this general term, among them frost heaving, solifluction, and stone sorting.

conglomerate A clastic sedimentary rock consisting of more or less rounded pebbles held together in a cement, usually sandstone or limestone. †Normally comparatively thin deposits, conglomerates are often associated with an unconformity and form the bottom of a new series of sedimentary rocks. Such basal conglomerates are found at the base of the Cambrian and the Ordovician systems of rocks.

conical projection (conic projection) A map projection constructed as if the meridians and parallels are projected onto a paper cone capping the globe, which is then unrolled. Normally the apex of the cone is above one of the poles and it touches the globe along one of the parallels. The projection is accurate along the parallel that the cone touches but away from it distortion is considerable. It is therefore only used for relatively small sections of the Earth's surface. Distances can be made accurate if the cone is made to touch the globe along two standard parallels (*conical projection with two* *standard parallels*), one near the top and one near the bottom of the map. This projection is used for maps showing a considerable E–W extent, for example Eurasia. *See also* Bonne's projection.

coniferous forest A vegetation type dominated by trees whose seeds are carried in cones and which are chiefly evergreen. Coniferous forests usually consist of stands of relatively few tree species, such as pine and spruce, of uniform appearance. Their close-growing nature restricts undergrowth and the development of stratification. Coniferous forests occur under many conditions but are particularly extensive in high N latitudes (e.g. the circumpolar boreal forests of Eurasia and North America) and also occur above deciduous forest in mountainous areas. The trees are adapted to the harsh climates of both environments with thick bark and flexible branches to withstand freezing temperatures and heavy snowfall; thin leaves (needles) to slow down transpiration; and evergreen foliage to permit immediate use of the brief growing season. The trees provide commercially valuable softwood timber; much natural coniferous forest is managed for the construction and wood-pulp industries, or has been replaced by new plantations.
†Conifers generate an acidic litter that decomposes slowly and integrates poorly with mineral matter. This is associated with the development of podzol soils.
See also taiga.

connectivity †In network analysis, the degree to which the nodes of a network are directly connected to each other. A number of different methods known as *connectivity indices* have been developed for measuring the degree of connection between the vertices of a network. The simplest, the beta (β) index is found by dividing the total number of edges (*e*) by the total number of vertices (*v*)

$$\beta = e/v$$

For a given number of vertices, the more edges there are that connect, the greater the connectivity and the higher the beta index.
When complex networks are considered the alpha (α) index is more useful for comparing networks. The alpha index compares the observed number of circuits (the numerator) with the maximum number of circuits for a given number of vertices (denominator):

$$\alpha = e - v + 1/2v - 5$$

The alpha index value is normally expressed as a percentage of the maximum. *See* network.

consequent stream A stream whose course follows the original slope of the land. *See also* subsequent stream, †obsequent stream.

conservation The protection of natural resources and the natural environment for the future. This includes the careful management of natural resources such as minerals, forests, landscapes, and soils to prevent their destruction or overexploitation. Although primarily concerned with natural resources conservation may also involve the protection of certain man-made environments of acknowledged value. †Modern *nature conservation* is concerned with the preservation and protection of whole habitats as well as individual rare species, and practises careful management to achieve an ecological balance, rather than passive preservation. In the UK, the Nature Conservancy Council is the main body responsible for conservation.
See also nature reserve.

conservative plate margin *See* plate tectonics.

constant slope (debris slope) One of the four basic elements of a hillslope, which lies below the free face and above the waning slope. It is thought that constant slopes retain their original angle because they are debris-controlled slopes; the angle of slope is determined by the size of rock fragments lying above bedrock so the slope angle will remain the same provided the size of debris is constant.

constructive plate margin *See* plate tectonics.

constructive wave A wave that contributes to the aggradation (the build-up of material) on a beach. For this to occur the swash of the wave, which pushes material up the beach, must be more efficient than the backwash. Constructive waves break at a rate of between six and eight per minute; this frequency permits the swash of one wave to take place without interference from the backwash of the previous wave. *Compare* destructive wave. *See also* backwash, swash.

consumer (*Biogeography*) An organism (plant or animal) that feeds off organic material, i.e. it assimilates energy already created by producers. Primary consumers (herbivores) depend on green plant matter, while secondary consumers (carnivores and parasites) eat primary consumers. †Primary consumers occupy the second trophic level and secondary consumers occupy the third trophic level. *Compare* producer. †*See also* trophic level.

consumer goods Products that satisfy human needs and will therefore sooner or later be used up. For example, food, clothing, washing machines, cars, and television sets.

container port/terminal A location that is specifically equipped to deal with the loading and off-loading of containers from ships, trains, or lorries. For example, Felixstowe is a container port equipped with the special lifting gear required to move containers on and off ships; Cardiff is a container terminal with special rail facilities for dealing with containers from other parts of the country.

contiguous zone The area of water beyond the immediate territorial seas of a country over which it claims certain rights. International law states that a country may exercise jurisdiction in customs, health, and security matters within this zone, which is usually taken to extend not more that 39 km (24 miles) from the shore.

continent One of the parts into which the Earth's crust is divided that rises above the depressions of the ocean basins. The marginal areas around the exposed landmass of a continent, such as the continental shelf, structurally form part of it. About 29% of Earth's surface is made up of continental areas. Seven are commonly recognized: Africa, Antarctica, Asia, Australia, Europe, North America, and South America.

continental air mass (c) An air mass that has its source region over a continental area and its properties are characteristic of a continental climate, i.e. a tendency to be dry all the year round with a considerable temperature difference between summer and winter.

continental climate The characteristic climate of continental interiors. As these regions are considerable distances from the sea they have low rainfall and a large annual and daily range of temperatures; summers are warm and winters cold. The temperature extremes are most marked in temperate latitudes.

continental divide A divide that separates the rivers of a continent flowing in opposite directions to different oceans. It is usually a major mountain range, such as the Andes of Latin America, which separates the river draining W to the Pacific from those draining E to the Atlantic.

continental drift The hypothesis that continents have moved relative to each other across the surface of the Earth. The idea was originally suggested by Antonio Snider-Pellegrini (1858) and developed by the German geologist Alfred Wegener from 1910. Wegener proposed that there was once just one supercontinent, Pangaea, which began to break up about 200 million years ago, since which time the continents have drifted to their present positions.

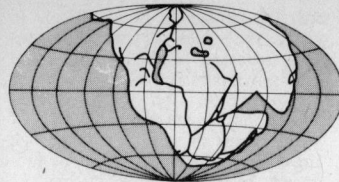

Upper Carboniferous

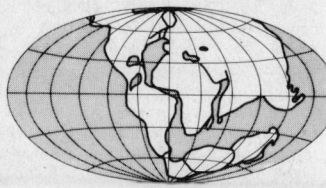

Eocene

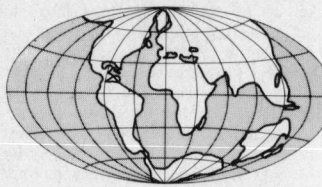

Lower Pleistocene

Continental drift

His evidence for this was the similarity of the shape of the coastlines of the continents on either side of the Atlantic Ocean and of the distribution of geological structures and formations, flora, and fauna. The hypothesis was generally rejected until during the early 1960s when the evidence of sea-floor spreading suggested a mechanism by which continental drift could take place. *See also* plate tectonics, sea-floor spreading.

continental platform The land areas of a continent together with the continental shelf that surrounds it. These areas have a varied relief of plains, plateaus, and mountains, the maximum height being reached at Mount Everest at 8848 m and the greatest depth being under the Dead Sea at approximately 330 m below sea level.

continental shelf The gently sloping part of a continent that lies submerged below the sea. It extends from the shoreline to an average depth of about 200 m. The North Sea and the Baltic Sea are examples of seas that lie on the continental shelf; these are known as epicontinental or shelf seas.

continental slope The comparatively steep slope that descends from the edge of the continental shelf to the deep-sea platform.

contour (contour line) A line drawn on a map joining all places with the same height above or below mean sea level. The relief of an area can be shown on a map by drawing a series of contours representing different heights. The area between contours is sometimes shown in different colours to represent gradations of relief. Contours are also used to show the depths of the oceans.
The *contour interval* is the vertical distance between two consecutive contours on a map. The contour interval depends on the scale of the map, the amount of vertical height involved, and the purpose for which the map is to be used.

contour ploughing A technique of ploughing parallel to the contours of a hillslope rather than up and down the slope. It is widely used in semiarid areas to reduce soil erosion and gullying.

conurbation An extensive urban area formed by the expansion and coalescence of a number of neighbouring towns and cities. In the UK the 1951 census first identified six conurbations – SE Lancashire, W Midlands, W Yorkshire, Merseyside, and Tyneside in England, and Clydeside in Scotland. [Coined by the urban geographer Patrick Geddes in 1915]

convection A process of heat transfer within the atmosphere (or within any gases or fluids), which involves the movement of the medium itself. *Free convection* (or natural convection) in the atmosphere occurs when the air is heated from below by being in contact with a warm surface; this makes it less dense, causing the air to rise allowing cooler denser air to take its place. It usually occurs over heated continental areas and the air currents are often indicated by cumuliform clouds. *Forced convection* or turbulence occurs when a moving current of air flows over a rough surface, thus forming eddies.

convectional rainfall Rainfall caused by the heating of moist air in the lower layers of the atmosphere which rises, expands, and is cooled adiabatically to its dew point. Towering cumulonimbus clouds may form. Convectional rainfall occurs in regions near the equator in the afternoon as the result of the constant high temperatures and high humidity.

convenience goods and services Goods and services that can be obtained or used quickly and with little effort on the part of the consumer. For example, factory-prepared meals, which only need to be heated before being eaten. A number of services can now be obtained from machines reducing the time and effort involved for the consumer, e.g. passport photographs and travel insurance.

convergence †In meteorology, a condition caused when the horizontal inflow of air into an area is greater than the horizontal outflow. If the convergence is at a low level the excess air escapes upwards, while if it is at higher altitudes the excess sinks towards the ground surface. Most areas of convergence are produced when two air masses move towards each other (*streamline convergence*) and at the surface this results in the formation of frontal zones, e.g. the intertropical convergence zone and Atlantic polar front. An *isotach convergence* is a less common form of convergence; it occurs when the velocity of a single air stream progressively slows down. *See also* divergence.

convexity A section of a hillslope in which the angle of slope decreases with increased height, usually found near the summit of a hillslope. †The dominant process on convex slopes is believed to be soil creep, the rate of which increases with increased slope angle. Convex slopes will also be produced by rapid downcutting by streams, which occurs when base levels fall. Coastal cliffs will adopt a convex profile when undercutting by waves affects hard resistant strata. *Compare* concavity.

cooperative (cooperative system) An association in which a group work together to achieve economic advantages that might not be achieved working individually. A number of cooperative organizations have been set up by such groups as farmers to act as buying and selling agencies and to exercise a greater control of the market. Cooperative dairies, such as those in Denmark, the UK, and the Irish Republic, help producers to market their milk and provide high quality dairy products for consumers. By buying seed and fertilizer in bulk direct from the manufacturers, cooperative farming organizations can reduce costs to their individual members.

coordinates A series of fixed lines or points that can be used to define the position of a place. On a global scale latitudes and longitudes can be used as coordinates; for example, the coordinates for Edinburgh are 55°5'N 3°13'W. On Ordnance Survey (OS) maps a numbered grid is used. The coordinates of a place are found by reading the eastings, which are lines numbered from W to E, and the northings, which are lines numbered from S to N. *See also* grid.

copper A reddish-brown metal. It is widely distributed and found commonly in both the pure native state and as an ore in combination with other elements. It is a comparatively soft but very tough metal and has widespread uses in industry. As it is an excellent conductor of electricity and is also malleable the most important use of copper is in the electrical industry. Other uses are in the manufacture of alloys such as different types of bronze, gunmetal, and brass, and in a variety of chemical processes, in disinfectants, and in fungicide. The major producers include the USA, USSR, Zambia, Zaïre, and Chile.

coppice A small group or plantation of trees that is grown so that the wood can be cut from time to time, preventing the trees from growing to their full size. For example, young hazel and sweet chestnut trees are cut close to the ground to encourage the growth of new shoots; after a few years these are cut to be used as posts, poles, and fencing materials.

copra The fleshy kernel of a coconut that has been dried. This is used in confectionery as dessicated coconut, and is also crushed to produce coconut oil, which is used in the preparation of cosmetics and soap. The main exporting regions are the Philippines, Indonesia, Sri Lanka, and the islands of the SW Pacific.

coral A small marine polyp that occurs in colonies, chiefly in warm shallow sea waters. The coral polyps secrete a hard rock-like substance, composed chiefly of calcium carbonate, to form supporting skeletons. As the polyps die their skeletons remain anchored to the surface that they colonized forming coral limestone. Over time succeeding generations of coral polyps result in the accumulation of large masses of coral forming coral reefs. Corals are limited in their geographical distribution as they will only grow in silt-free salt water, the temperature of which is above 21°C. As they also require sunlight their maximum depth of growth is about 45–55 m. Corals are usually only found between 30°N and 30°S; they are absent on the W coasts of continents where cool currents prevent their growth.

coral reef A ridge of coral and other organic material consolidated into limestone lying near sea level. The majority of reefs are relatively narrow platforms with a steep seaward edge. Wave action may throw up broken fragments of coral and other material onto the platform to produce a low mound. The landward edge is usually less steep and covered with a thin layer of sand, shell fragments, and coral, which plants can colonize. *See also* atoll, barrier reef, fringing reef.

cordillera A series of mountain ranges consisting of roughly parallel ridges, plateaus, and intermontane basins. The term is widely used for the W mountains of the Americas, including the Andes (the Cordilleras de los Andes) and the Rocky Mountains. [Spanish]

core The central portion of the Earth's interior. It consists of a sphere bounded on its outer limits by the Gutenberg discontinuity, which lies at about 2900 km below the Earth's surface. It is separated into the inner core and the outer core. The outer core does not transmit the S waves of earthquakes, which suggests that it is liquid. However, there is some evidence that the inner core, below about 5000 km, may be solid. The core is composed of a mass of nickel-iron, at a temperature of about 2700°C in the inner core, and at very high pressure. *See also* Gutenberg discontinuity.

core area 1. (*Political Geography*) The nucleus from which a nation evolved, e.g. the Buenos Aires region of Argentina.
2. (*Urban Geography*) The central business district of a city.

core–periphery model (centre–periphery model) †A model of a pattern of economic development in which major contrasts develop between the level of economic progress achieved within one region in a country, the 'core region', and that in the remainder of the country, which continues to be underdeveloped and almost totally outside the economic activities usually associated with a modern industrialized society. This pattern of spatial inequalities is typical of many Third-World countries.

core sample A sample of rock, soil, or ice obtained by drilling into such material with a hollow bore and withdrawing the sample intact. The technique is used to obtain samples of material from the ocean floor for oceanographic studies. Core samples are also widely used in geological exploration for minerals and fuels.

Coriolis force †A 'fictitious' force that is needed to relate the movement of masses over the Earth's surface to its rotating coordinate system (the latitude and longitude grid). For example, a missile projected from the centre of rotation would appear to an outside observer in space to travel in a straight line while the Earth itself rotated. To an observer on Earth the missile would appear to follow a curved path, being deflected to the right in the N hemisphere and to the left in the S hemisphere. The force is responsible for example for the formation and direction of movement of anticyclones and whirlpools. *See also* Ferrel's law. [It was named after the French mathematician Gaspard de Coriolis, who first described it in 1835]

corn The grain or seeds of any of the cereal grasses used as food, such as wheat, rye, barley, oats, and maize. The word is used in certain regions or countries for the most important cereal crop of that area, e.g. corn is used in England for wheat, in Scotland and Ireland for oats, and in the USA and Canada for maize or Indian corn.

corniche A scenic coastal road along the side of a mountain that has precipitous slopes. The most famous example is the road from Nice (France) to Genoa (Italy), which overlooks the Mediterranean Sea. [French]

corrasion The wearing away of a solid rock surface by material that is being transported (i.e. by water, ice, wind, or waves) or that is moving under gravity. The most common form of corrasion is that produced by a river and its load. Vertical downcutting by a stream eroding its bed is referred to as *vertical corrasion*; horizontal erosion of the banks of a stream is referred to as *lateral corrasion*.

correlation †A statistical technique that determines the extent to which a relationship exists between two variables. This can be calculated by measuring the *correlation coefficient*, i.e. the degree of association between two paired variables. One formula used by geographers is Spearman's rank correlation coefficient. *See* Spearman's rank correlation test.

corridor A strip of territory transferred from one state to another, through negotiation, to give the second state an access to the sea or to an international waterway. One historical example is the Polish Corridor, agreed at the end of World War I, which gave the landlocked state of Poland access to the Baltic ports of Gdynia and Danzig. An *air corridor* is an internationally recognized air route across a country or an air route from a state to an exclave, e.g. the Berlin Corridor, which gives West Germany air access to its exclave of West Berlin across the territory of East Germany.

corrie *See* cirque. [Scottish]

corrosion The wearing away of rocks by chemical action. This covers a wide variety of processes, including solution, hydration, and oxidation. *See* chemical weathering.

cost benefit analysis (benefit cost analysis) †A technique that is used to evaluate the social losses and gains that will be incurred by a particular project. To do this effectively the advantages and disadvantages to society as a whole must be identified and measured. For example, the benefits of building the M25 motorway around London must be considered alongside such costs as the loss of farmland, reduction in areas of scenic beauty, and the increase in noise levels to places near the motorway. A major problem in any such analysis is the evaluation of some of the costs and benefits.

cost space (cost distance) †A measure of relative distance. The concept recognizes that the time or cost involved in overcoming distance is more important than actual mileage. It is possible to draw maps based on cost distance, for example, the location of places in relation to their travel time · from London. Such a map would show cost space in contrast to the area shown by conventional maps.

cotton A plant native to tropical and subtropical regions, several species of which are cultivated for the white fibre (cotton) that surrounds its seeds. The seeds are contained within the boll, or seed pod, and it is shortly after this has split open that the cotton is harvested to avoid rain or wind damage. The fibre is spun, woven, and dyed to produce cotton cloth. It is also blended with artificial fibres to make cloth. The seeds are separated from the cotton fibre on a machine called a gin; these are crushed to produce *cottonseed oil*, used in margarines, cooking oils, soaps, etc., and the residual pulp is used as an animal feed. The main cotton producing countries are the USA, USSR, China, India, Mexico, and Egypt.

counterdrift †A planning strategy that sets out to enable people who live on the fringe of an economic growth area to · share in the benefits of that growth. In so doing it counters the drift of people into the growth area from other parts of the region or country.

country park An area of countryside in the UK that has been designated and is designed to cater for large numbers of people taking part in a variety of recreational activities. Sited close to urban areas, country parks

are intended to ease the pressure on the remoter places, such as national parks. Grants are available from the government to local authorities and private individuals who wish to establish country parks. *See also* national park.

county An administrative division of a state. In the UK, England is divided into 39 non-metropolitan counties and 7 metropolitan counties; Wales is divided into 8 counties. Each county is a unit of local government with political, legal, fiscal, and administrative significance. In the USA and Canada a county is a political and administrative unit that is next in order below a state.

cove 1. A small bay or inlet in the coast, e.g. Lulworth Cove, Dorset.
2. A recess or hollow in the side of a mountain.

cover crop A crop that is grown to cover the soil and protect it from the dangers of soil erosion. Cover crops are particularly important in tropical and subtropical environments where heavy downpours can cause devastating soil erosion, especially on sloping ground. The crop must be quick growing and may eventually be ploughed in to enrich the soil. Ground vines, for example, are planted between the rows of rubber trees in Malaysia to prevent erosion.

crag 1. (*Geomorphology*) A steep rugged outcrop of bare rock. It is often produced by mechanical weathering in high mountainous areas.
2. (*Geology*) The compacted deposits of sand, largely composed of shell fragments, which are widespread in parts of East Anglia.

crag and tail A landform of glacial origin, which consists of a steep rocky slope on one side (the crag) and a relatively gentle slope trailing away in its lee (the tail). The crag is composed of hard resistant rock, such as an old volcanic plug, which formed an obstruction in the path of advancing ice. In the lee of the crag softer rocks were protected from erosion and may also have received a covering of glacial material. A good example is found in Edinburgh, where the castle is built on a crag, composed of a basalt volcanic plug, and the Royal Mile slopes gently away to the E down the tail.

crater 1. A funnel-shaped depression at the top or on the side of a volcanic cone. It may be produced by an explosive eruption or by the collapse of the cone following the withdrawal of underlying lava.
2. A large bowl-shaped depression in the Earth's surface caused by the impact of a meteorite, e.g. Meteor Crater in Arizona, USA, which is 1200 m in diameter.

crater lake A body of water that collects in the crater of an extinct volcano, the best-known example being Crater Lake, Oregon, USA, which occupies a crater about 10 km in diameter.

creek 1. A small inlet on a sea coast.
2. In Australia and the USA, a small stream or tributary.

creep The slow gradual downslope movement of soil and rock fragments, largely under the influence of gravity. The water content of the material and other processes such as alternate wetting and drying, freeze-thaw processes, rainsplash impact, and the activity of animals and plant roots will also contribute to creep. †Solifluction is a special form of creep that occurs in saturated soils in periglacial areas. *See also* mass movement, solifluction.

creole 1. In the West Indies and Latin America, a native-born person of European ancestry, of mixed European and Negro ancestry, or of Negro ancestry.

2. In the S USA (especially Louisiana), a French-speaking descendant of early French or Spanish settlers.
3. A patois based in part upon French and Spanish.

crescentic dune *See* barchan.

Cretaceous The final geological period of the Mesozoic era and the system of rocks laid down during this period. It extended from about 136 million years ago to about 65 million years ago and is divided into the Lower and Upper Cretaceous. The Lower Cretaceous rocks were deposited in shallow-water conditions, after which there was a rapid increase in the size of the ocean as a widespread marine transgression took place. This led to the conditions which allowed the deposition of the chalk. On land dinosaurs and other giant reptiles were the dominant life forms and reached their peak of development but by the end of the period most were extinct. Flowering plants became prominent during the period but mammals, although known, were insignificant. [From Latin *creta*: chalk]

crevasse A deep crack or fissure found in a glacier. †Crevasses are formed by shear stresses set up by differential movements in the ice. When the surface over which a glacier is flowing down increases in steepness crevasses form transversely across the glacier in a series of arcs curving across the surface of the ice, convex side facing upslope. Longitudinal crevasses are formed when the valley widens and the glacier is able to spread out laterally. The friction of the valley walls results in the sides of a glacier moving more slowly than ice in the centre; this causes oblique crevasses to develop along the glacier margins. [French]

critical path analysis †A method of determining the particular sequence of operations that must be followed in order to complete a complicated task in the minimum time. This procedure is used, often with the aid of a computer, when planning a large programme of work.

critical temperature In biogeography, a temperature that brings about a given response in a plant or animal. Freezing point (0°C), for example, is a critical temperature for plants as it prevents photosynthesis. Critical temperatures, especially winter minima, are often crucial in determining distributions of species. *See also* growing season.

croft A small area of farmland worked by a tenant farmer. The term is normally applied to the system of farming in the Highlands and Islands of Scotland. A croft usually consists of a cottage, some cultivated land for such crops as oats and potatoes, and several hectares of rough grazing. Since many crofts are near the coast the crofter is often a fisherman as well as a farmer.

cross profile A transverse section of a river's channel or valley, usually cutting across at right angles to the stream. Cross profiles can show such features as river terraces, floodplains, and the symmetry or asymmetry of the valley sides. They can therefore provide a great deal of information about the type of valley, the nature of the processes that formed the valley, its underlying structure, and its past erosional history, e.g. valleys with a V-shaped cross profile are typically produced by streams in a mountain environment, U-shaped valleys are usually formed by glaciation.

cruciform village A type of village that has grown up around a crossroads. Settlement spreads out in four directions along the routeways radiating from the intersection to form a cross shape. A cruciform village is a form of street village. *See also* street village.

crude birth rate A measure of fertility; it is usually the number of live births per 1000 of population in a particular year. Throughout the world the birth rate varies in different countries from 12 to 50. The lowest rates are found in western industrialised countries with high standards of living. The highest rates are found in the underdeveloped countries of the Third World. The highest birth rate does not necessarily produce the highest rate of population growth for the birth rate has to be related to the death rate. *See* crude death rate.

crude death rate The number of deaths per 1000 of the population over a given period, usually one year. The death rate is significantly influenced by the age and sex structure of the population and by social factors such as living standards and the existence of health programmes. A wealthy advanced country may, however, have a higher crude death rate than a less-developed country because the population of the less-developed country shows a much younger age structure. *See also* crude birth rate.

crumb structure A physical property of soil in which particles join together in loosely-held spheroidal masses, which resemble bread crumbs in appearance. The crumb structure provides the best conditions for plant growth: it allows air and water to move freely between the aggregates, retains nutrients, and is more easily penetrated by plant roots. This structure is therefore aimed at in agricultural practice. It is particularly associated with the surface horizons of humus-rich cultivated soils.

crust The outermost shell of the Earth. It lies above the Mohorovičić discontinuity and varies in thickness from about 6 km below the oceans to about 60 km under the high mountain ranges of continents. It comprises two dominant groups of rocks: the discontinuous sial (continental crust) and sima (oceanic crust). *See also* sial, sima.

crustal movement *See* earth movement, epeirogenesis.

crystal A solid substance with a geometrical form resulting from the regular arrangement of its constituent atoms or molecules. It is bounded by flat surfaces (*crystal faces*) in a pattern that is characteristic of the substance of which the crystal is made. Well-developed crystals show a symmetry in which the faces are either arranged around a line that can be passed through the centre of the crystal or placed so that each face is balanced by a similar face opposite. The study of crystals is *crystallography*. [From Greek *krustallos*: clear ice]

crystalline rock A rock formed by the process of crystallization; i.e. an igneous or a metamorphic rock. In an igneous rock the shape and size of the crystals depends upon the rate of cooling of the magma and the order in which the minerals crystallized. In general, the slower the cooling and the longer the crystallization time, the larger the crystals. In metamorphic rocks the shape and size of the crystals depends on the mineral itself: some minerals take their crystalline form while others never do.

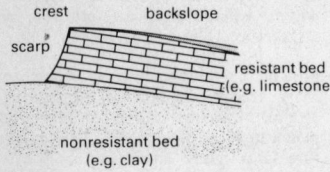

Cuesta

cuesta A ridge with a steep scarp slope (escarpment) and a comparatively gentle dip slope, closely reflecting the structure of the underlying rocks. It is formed by the differential erosion of gently dipping strata. The more resistant rocks of the cuesta (e.g. sandstone or limestone) stand out above the eroded vales (frequently clay) at its foot. Examples include the North and South Downs of S England. *See also* dip slope, scarp. [Spanish]

cultural geography The systematic branch of geography that studies the distribution of cultural groups and the interaction between them and their environments.

cultural landscape †The components of the geographical character of a region that are not attributable to 'natural' causes. The cultural landscape is the product of the choices made and the changes worked by men as members of particular cultural communities as manifested in geographical space.

cumulative causation model (Myrdal's model) †A theory put forward by the Swedish economist Gunnar Myrdal (1957) in which he argued that economic development leads to an increase rather than a decrease in differences between regions. Initially, development takes place in a region as a result of the advantages it can offer. The region moves ahead of other surrounding areas and secondary advantages are established in a process of cumulative causation in which one development leads to another. Surrounding areas suffer a loss of labour and capital to the growth region (or pole) and an inflow of imports, which prevent the development of local industry – these are known as the *backwash effects*. Over time these detrimental effects may be lessened by *spread effects* in which benefits generated by the growth region are transferred to the surrounding areas; for example, the relocation of manufacturing plants, spread of innovation, and the decentralization of population. *See also* growth pole.

cumulonimbus (Cb) A cloud formation that has a low base level (500–2000 m) but great vertical development, often reaching up to heights of 10 000–11 000 m. It is a development of the cumulus formation, being massive and often with an enlarged and flattened top forming a characteristic anvil shape. Its presence indicates strong convection and there is frequently heavy precipitation and thunder.

cumulus (Cu) A cloud formation with a low base level (500–2000 m), the base being flat and the cloud masses isolated, either rounded or towering, and with a clear outline. When these clouds are sunlit they are brilliantly white and are often called 'cotton wool clouds'. They occur mainly in summer and are caused by convection. Cumulus clouds may grow from the shallow 'fair-weather cumulus' to great heights and can develop into cumulonimbus.

current 1. The permanent or seasonal flow of water in a defined direction in the surface water of an ocean, e.g. the North Atlantic Drift, Labrador Current, and Benguela Current.
2. The flow of water in a river channel.

cuspate delta †A sharply pointed delta with curved sides, formed by the even deposition of material on either side of a river mouth. It is usually found on straight coasts, which are exposed to relatively strong wave action. Cuspate deltas have very few distributaries, or none at all. The delta of the River Tiber on the W coast of Italy is an example of this form. *See also* delta.

cuspate foreland †A triangular-shaped accumulation of beach materials, typically shingle, the apex of which extends out to sea. It is most often produced by two spits, roughly at right angles, being built up towards each other by two different sets of constructive sea currents. When one current is stronger than the other the foreland is liable to be moved along the coast by longshore drift, in the direction of the weaker set of waves. Examples of cuspate forelands include Dungeness, on the coast of Kent, and Cape Canaveral, Florida.

customs union An organization of countries to remove all barriers, such as tariffs and quotas, to the free exchange of goods and services with one another, and the setting up of a common external tariff against non-

members. One of the best examples is the European Economic Community (EEC). *See* common market.

cutoff The abandoned loop of river channel formed when a river erodes a new channel through the neck of a meander. The entrance to the old channel of the river soon becomes blocked with sediment to form an oxbow lake. †A *chute cutoff* is formed when a river breaks through its bank to follow the line of a swale in point-bar deposits.
See also oxbow.

cwm *See* cirque. [Welsh]

cycle A completed series of events that follows or is followed by another series of similar events occurring in the same order. The concept has wide application throughout the natural sciences, e.g. the circulation of elements within the Earth and the atmosphere, such as in the carbon cycle and hydrological cycle. In geography, the concept has been used particularly in the context of the cycle of erosion, developed by W. M. Davis. *See* carbon cycle, cycle of erosion, hydrological cycle.

cycle of erosion A model of the development of landforms, based on the idea that they pass through a series of changes from youth to maturity to old age. The concept was first proposed by W. M. Davis in 1899, and is sometimes referred to as the Davisian cycle. The cycle begins with a recently uplifted land mass of little relief. This is eroded by fast-flowing streams with many waterfalls and rapids to form a tableland dissected by V-shaped valleys. This is referred to as the youthful stage. In the mature stage valleys become wider, rivers flow less rapidly, and slopes are less steep. By old age the land surface has been eroded to a virtually flat featureless plain – a peneplain – with very slow moving rivers.
At the present time the main criticisms of this idea are that land masses do not remain stable for sufficient

time for the cycle to be completed and that climatic changes make the situation far more complicated.

cyclone (tropical cyclone) An area of low atmospheric pressure in the tropics 80–400 km in diameter. It is a storm system of great intensity with winds circulating at speeds of 120–280 km per hour around the 'eye' (centre) of the cyclone and accompanied by torrential rain. Winds rotate in an anticlockwise direction in the N hemisphere and in a clockwise direction in the S hemisphere. Cyclones can cause considerable damage to buildings and are a hazard to shipping.

In mid- and high latitudes low-pressure systems are now commonly referred to as depressions or lows.

cyclonic rainfall (frontal rainfall) Precipitation along the frontal surfaces of a depression in mid- and high latitudes. It is caused by the warm moist air of one air mass moving over the cold heavier air of another or by air rising through horizontal convergence in an area of low pressure. Drizzling rain falls in a broad belt followed by more concentrated squally rain as the cold front passes.

cylindrical equal-area projection A cylindrical map projection in which meridians are shown as parallel straight lines spaced at their true-to-scale distance from the equator. Lines of latitude are also parallel but are closer near to the equator and wider apart towards high latitudes. As a result the map has equal-area properties but is highly distorted towards the poles. It is used for showing distributions in the tropics; for example, of rubber and sugar cane growing areas.

cylindrical projection A map projection constructed as if a rectangle of paper formed into a cylinder surrounds the globe, onto which the details of the Earth's surface are projected. The paper is then rolled out flat. Meridians and parallels appear as straight parallel lines. On this type of projection the polar regions cannot be shown accurately and territories near the poles are exaggerated in size. *See also* Mercator projection.

D

dairy farming The rearing of cattle for the production of dairy products, such as milk, cream, butter, and cheese. Commercial dairy farming is, in many ways, the most advanced type of farming. It requires high inputs of capital, skilled labour, and scientific methods, and is often combined with arable farming to produce fodder crops. The main regions are located in temperate lands, such as E and central North America, W Europe, New Zealand, and E Australia.

dale An open river valley, mainly in N England, e.g. the Yorkshire Dales.

Dalmatian coast *See* concordant coast.

dam A structure built across a river. A dam may be constructed to provide water for irrigation, industrial use, and human consumption; to provide a head of water for hydroelectric power (HEP) generation; to divert the water flow for navigation; or to control the water flow to prevent flooding. Frequently a dam fulfils more than one purpose and may also be used for recreational purposes, e.g. Lake Meade behind the Hoover Dam on the Colorado River, as well as providing power and water supplies to a large area. Other large dams include the Itaipu Dam on the Rio Paraná on the Brazil –Paraguay border, the Kariba Dam on the Zambezi River on the Zambia–Zimbabwe border, and the Aswan High Dam on the Nile in Egypt.

datum level (datum plane, datum line) A horizontal line or point (zero reference) from which heights and depths are calculated. It is usually mean sea level. *See also* Ordnance Datum.

deadweight tonnage The difference between the weight of a ship when it is loaded and when it is empty, i.e. the full weight of the load, including packaging and containers.

death rate, crude *See* crude death rate.

debris The loose assorted fragments of rock material, sand, and clay produced by weathering and rock disintegration, consisting of a wide range of particle shapes and sizes.

decentralization The movement outwards from the central areas of cities and metropolises of urban population and/or employment in the form of suburbanization, urban sprawl, or widespread population dispersal. For example, office decentralization is the relocation of offices away from the city-centre locations in which they were formerly concentrated.

deciduous Denoting vegetation that sheds its leaves at the same season each year. For example, temperate deciduous woodland loses its foliage in autumn (called the fall in North America). Many broad-leaved trees are deciduous. The loss of leaves is an adaptation that has evolved as a response to climate and counteracts the effects of seasonal drought or cooler temperatures. *Compare* evergreen.

declination 1. In astronomy, the angle between the plane of the equator and a line joining a star or other heavenly body to the centre of the Earth. **2.** *See* magnetic declination.

decomposer An organism that feeds on dead organic material breaking it down, either chemically or physically, into simpler substances. Decomposers include animals (e.g. earthworms), bacteria, and plants (fungi). †Decomposers share the same multiple trophic levels as consumers. In certain ecosystems, such as forests and tundra, the energy flow and recycling of vital nutrients is governed to a large extent by decomposer activity. *Compare* consumer, producer. †*See also* trophic level.

decomposition The weakening and disintegration of rock through chemical weathering. This may occur through the washing away of cementing materials or through the alteration of constituent minerals, e.g. in the decomposition of granites, feldspars are decomposed to produce kaolin (china clay).

deep, ocean *See* trench.

deep-sea plain *See* abyssal plain.

deep weathering The process by which rocks that lie at some depth in the ground are broken down through chemical weathering to produce a thick layer of regolith (weathered material). It is most effective in humid tropical regions in which only limited erosion takes place; the high temperatures and high rainfall combine to give high rates of chemical weathering. Vegetation contributes to the process with the production of humic acids from decaying plant matter. Deep-weathered material is also found in temperate and desert areas where it may be a relict of former warmer and wetter conditions. Average depths of deep weathering are about 15 m, although in some areas such as in the crystalline granites around Rio de Janeiro, Brazil, depths of 80–100 m have been recorded. *See also* tor.

defensive site A site for a settlement that was chosen as a result of its physical features that could provide natural protection for the inhabitants. These sites included hill tops, river meanders, and islands (e.g. in a river, lake, or marsh). For example, Shrewsbury was sited within a meander of the River Severn; Lincoln was sited on top of the scarp of Lincoln Edge overlooking the Witham gap.

defile A narrow gorge, pass, or ravine between mountains, usually of quite a small scale.

deflation (*Geomorphology*) The transportation of fine particles of sand and dust by wind action. The finest material may be carried considerable distances by duststorms to be finally deposited as loess. Where deflation is concentrated in one area a *deflation hollow* may form. One of the largest, the Qattara Depression in Egypt, has been eroded to below sea level and an estimated 3000 cubic kilometres of material has been removed.

deforestation The removal of the tree cover of an area by felling or burning. The process is usually deliberate in order to make the land available for other uses (e.g. agriculture). *See also* reafforestation.

deformation A change in a layer of rock after it has been deposited, including folding, faulting, and tilting, caused by earth movements and igneous activity.

degradation The wearing down of a land surface, most commonly applied to the action of rivers. *Compare* aggradation.

degree Symbol: ° **1.** A unit of angle; 1/360 of a complete turn. **2.** A unit of measurement of latitude and longitude, which is most frequently used to indicate the position of a point on the Earth's surface. A degree of latitude is equal to 1/360 of the Earth's circumference (110.569 km at the equator). A degree of longitude is equal to 1/360 of the Earth's circumference at the equator but becomes progressively smaller polewards until at the North and South Poles it is zero. Each degree is divided into 60 minutes (') and each minute is divided into 60 seconds ("). **3.** An interval on a scale of measurement, such as a temperature scale.

dejection cone *See* alluvial cone.

dell A small hollow or valley, usually wooded.

delta A tract of alluvium, usually fan-shaped, at the mouth of a river where it deposits more material than can be carried away. The river becomes divided into two or more channels (distributaries), which may further

divide and rejoin to form a network of channels.

†A delta is formed by a combination of two proceses: (1) sediment is deposited when the load-bearing capacity of a river is reduced as a result of the check to its speed as it enters a sea or lake; and (2) at the same time fine clay particles carried in suspension in the river coagulate in the presence of salt water and are deposited. The finest particles are carried furthest to accumulate as bottom-set beds; coarser material is deposited as a series of steeply sloping wedges forming the foreset beds; and the coarsest material is deposited on the braided surface of the delta as topset beds. *See also* arcuate delta, bird's foot delta, †cuspate delta. [From the Nile Delta, so named by the Greeks because of its resemblance to the Greek letter delta (Δ)]

demersal fish Those species of fish that live on or near the sea bed, e.g. cod, haddock, and plaice. These fish are normally caught in a trawl net, a bag-shaped net towed behind a trawler. Over 90% of the UK catch consists of demersal fish.

demographic equation (basic demographic equation) †An equation that defines the content of population studies. For any area it is:

$$p^{t+n} = p^t + B^{t,t+n} - D^{t,t+n} \pm NM^{t,t+n}$$

where, given a population at a particular time (p^t), then that population after a period of time, t to $t+n(p^{t+n})$, will be the result of increase due to births during the period ($B^{t,t+n}$), decrease due to deaths ($D^{t,t+n}$), and either increase or decrease due to net migration ($NM^{t,t+n}$).

The equation has practical value because if four of the components are known the fifth can be estimated. It is also used to check the quality of data by comparing expected values with enumerated values.

demographic transition theory †A theory that attempts to explain the change from a low total population experiencing high birth rates and high

death rates to a high total population experiencing low birth and low death rates. The theory was based on the pattern of population changes that took place in Europe from about 1700 up to the present day. There are four main stages recognized in the process:
(1) *High fluctuating (pre-industrial) stage* A period of high birth and high death rates (over 30 per thousand) resulting in little natural increase in population size.
(2) *Early expanding stage* A period during which the death rate declines while the birth rate remains high resulting in a high rate of population growth.
(3) *Late expanding stage* The birth rate is now declining more rapidly that the death rate so the population grows slowly.
(4) *Low fluctuating (industrial) stage* Both birth and death rates are low or moderate so natural population growth is slow or there may be a decline in numbers.

demography The empirical, statistical, and mathematical study of population. It has two main branches – formal (or pure) demography and social demography. Pure demography is essentially concerned with the collection, evaluation, analysis, and projection of population data. Social demography explains demographic patterns and processes.

dendritic drainage A drainage pattern consisting of a single main stream with tributaries resembling the branches of a tree. Dendritic drainage patterns are developed most fully where the underlying rock is of a uniform type and structures are relatively simple. From laboratory experiments it seems that the dendritic pattern is the most efficient form that a stream can adopt to drain a given area of uniform rock type.

dendrochronology A method of inferring past environmental conditions from the width of annual tree rings. It is particularly useful to climatology as each ring reflects the temperature and

precipitation of the year in which it was formed. In the SW USA long-lived conifers, such as the Douglas fir and bristlecone pine, have enabled dating to as far back as 4000 years. The method is also of use to archaeologists dating prehistoric Indian cultures.

density The mass of a substance per unit volume. It is expressed in $kg\ cm^{-3}$.

density, population *See* population density.

denudation The wearing away of the land surface; it includes the processes of weathering, mass movement, erosion, and transportation. (The term is sometimes used synonymously with erosion but this technically excludes weathering.)

dependency 1. A territory subject to the rule of another country with usually only minimal autonomy, for example the Cook Islands are a dependency of New Zealand.
2. †(dependence) A theory used by some political geographers to explain the underdeveloped nature of many former colonies and dependencies. It suggests that the rule of a foreign power, under the guise of being beneficial, in fact leads to exploitation of the economy and to economic, social, and political instability of the country after independence.

depopulation The reduction or decline of population in a geographical area. Depopulation may be the result of outward migration, for example, as occurred in Ireland in the 19th century; a reduction of the gross reproduction rate, as a result perhaps of the ageing of the population; or of political decisions such as the mass removal of people from a particular location.

deposition The laying down of material that has been weathered, eroded, and transported by natural processes, such as water, wind, and ice. In terms of volume of material being deposited, the most important depositionary process is that carried out by rivers, which deposit solid material on the land, in estuaries, and on continental shelves. The action of the sea, ice, and the wind produce such depositional features as spits and bars, glacial moraines, and dunes respectively. Deposition also includes materials deposited from chemical solutions, e.g. by evaporation or precipitation, and organic matter, such as the remains of vegetation (e.g. peat) and growth of coral.

depressed region A region in a state of economic depression. The unemployment rate is often high, usually because of the closure of basic heavy industries. *See also* development area.

depression 1. (*Meteorology*; disturbance, low) An area of relatively low atmospheric pressure found mainly in temperate regions. On a weather chart it shows up as a series of enclosed isobars, more or less circular in shape, with pressure decreasing towards the centre. It is formed when a cold air mass meets a warm air mass along a front and the weather associated with it and its attendant fronts forms a large proportion of the weather experienced in temperate regions. It can vary greatly in size (150–3000 km across), speed (stationary to 1000 km per day), and intensity. When the pressure difference between the centre and the surroundings is great it is said to be 'deep'; when the pressure difference is small it is said to be 'shallow'. Winds tend to be stronger than those associated with an anticyclone as the pressure gradient is generally steeper. *Compare* anticyclone. *See* cold front, warm front, occlusion. †*See also* Bjerknes polar front model.

2. (*Geology*) A hollow sunken area of the Earth's surface; it may be produced by erosional processes or by the structure of the underlying rocks.

deranged drainage †A system of drainage that shows no clear pattern but consists of an irregular network of streams, lakes, and marshes. These forms tend to occur on recently glaciated surfaces, e.g. the Canadian Shield and Siberian tundra, where erosion

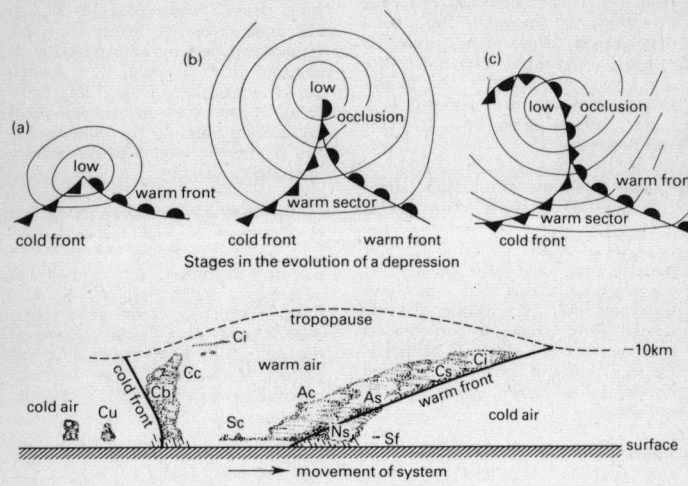

Stages in the evolution of a depression

Cross section through a depression

Depression

and deposition have disrupted the preglacial drainage network.

desalinization (desalination) The removal of the salt content of soil or brine. Desalinization of the polder lands in the Netherlands after sea water has been drained off consists of leaving the soil to weather under the action of rain. In some countries where fresh water is scarce special plant is installed to desalinize sea water, e.g. in the United Arab Emirates.

desert An area of the Earth's surface where vegetation and animal life is considerably limited or nonexistent as a result of low precipitation. One criterion for desert conditions to exist is sometimes taken as being a precipitation level of less than 250 mm per annum. Deserts occur as a result of a variety of conditions in the following locations:
(1) In areas of persistent high pressure, such as the Sahara.
(2) In areas on the W coast of continents where natural atmospheric instability is emphasized by surface cooling due to cold offshore currents, e.g. the Atacama Desert, Chile.
(3) In continental interiors of the mid-latitudes with high summer and low winter temperatures, and where mountain barriers prevent the passage of moisture-bearing winds. For example, the Gobi Desert in Asia.
Distinctive landforms exist in deserts; mechanical weathering takes place to a greater extent than chemical weathering as a result of the lack of moisture. The wind is an important agent of erosion and transportation. Rainfall, when it does take place, is usually in the form of violent downpours, which can move weathered debris. The characteristic landscape is thus stony scrubland with occasional resistant rock uplands and areas of shifting sand dunes.
The polar lands are also sometimes regarded as being ice and snow deserts.
See also erg, hamada, reg.

desert pavement A comparatively flat smooth area that occurs in desert regions, consisting of closely-packed angular pebbles and gravels, whose upper surfaces have been ground and polished by abrasion from wind-borne sand. The pebbles and gravels may be cemented together by salts drawn to the surface in solution by capillarity. Such pavements are referred to as reg in Algeria and serir in Libya and Egypt.

desert varnish (patina) †A thin coating, black, brown, or matt red in colour, that occurs on some boulders and outcrops in arid areas. It is made up of oxides of iron and manganese. It was thought that this varnish was produced by solutions being brought to the surface of rocks by capillarity and then evaporating, leaving a residue of oxides. It is now thought to be more likely that desert varnish is a product of weathering, although the exact process remains unclear.

desiccation A process of drying up, by which the moisture is removed from a material. The term is most often applied to bodies of water, such as lakes and streams, or to deposits that have been waterlogged, such as muds and clays.

desiccation cracks (mud cracks, sun cracks) †Cracks that develop in fine-grained material, such as mud and clay, as it dries out through evaporation. These form polygonal patterns, particularly when the material is homogeneous.

desilication The process by which silica (together with many bases) is removed from a soil profile by intense weathering and leaching. It is characteristic of the humid tropical areas and leads to the development of ferralsol soils.

desire line The straight line between the point of origin of a trip and the destination; i.e. the shortest distance between the two points. A *desire line map* shows the route a person would like to take if such a way were available and provides an indication of the pattern of routes people prefer. For example, desire lines from Hull to the S and vice versa across the River Humber indicated the need for the new Humber Bridge.

destructive plate margin *See* plate tectonics.

destructive wave A wave that causes removal of material from a beach. Destructive waves tend to have a relatively high ratio of height (i.e. the difference between the crest and trough) to length (i.e. the length between two successive crests) and thus break onto the beach near vertically. The backwash is more powerful than the swash and this moves material down the beach. *Compare* constructive wave. *See also* backwash, swash.

determinism †The doctrine that human action is not free but is essentially determined by a chain of causation. In geography, therefore, environmental control is held to be dominant in determining the ways in which man uses space. Determinists believe that the environment can only be modified to a very limited degree and that in formulating theories about human phenomena the fundamental causal factors are environmental ones.

detour index A measurement of the directness of a route, which is obtained by comparing the straight line distance between two locations and the actual distance by road or rail. The formula is: 100 × shortest distance by road or rail/direct distance. The more direct the link, the lower the index. A maximum efficiency, i.e. when the route and the shortest distance were identical, would result in a detour index of 100.

development In British town planning, the carrying out of building, engineering, mining, or other operations in, on, over, or under land, or the making of any material change in the use of any buildings or other land.

development area In the UK, an area where economic growth is encouraged by grants and other benefits available to industrialists in these areas. It forms part of an attempt to reduce differences in employment/unemployment levels and per capita incomes between regions in the UK with restrictions on development in the more prosperous regions. Changes in government policy have recently reduced the area covered by assisted-area status.

development control The process, in the UK, by which local authorities control the use and development of land and buildings. All development, with some exceptions, requires approval from the local authority; this involves the submission of a planning application covering the proposed development, its determination, and a decision. The decision can be either to grant permission unconditionally, to grant permission with conditions, or to refuse permission. *See also* development.

Devonian The fourth geological period of the Palaeozoic era and the system of rocks laid down during this period. It followed the Silurian and preceded the Carboniferous, extending from about 395 million years ago to about 345 million years ago. In the British Isles rocks laid down in the sea in this period are found in S Devon and Cornwall. Other Devonian rocks are continental and known collectively as Old Red Sandstone; they occur in Scotland, Wales, and SW Ireland. Fishes developed considerably during this period and by the end of the Old Red Sandstone had developed lungs. Plants had developed into tree ferns.

dew The deposition of water droplets on the ground and objects, such as plants, near the ground which occurs when the temperature of the ground surface falls and the air in contact with it is cooled below its dew point. Water vapour in the air or diffused from the soil then condenses and is deposited as droplets. The conditions favouring dew formation are moist air, light winds, and clear night skies to ensure maximum cooling by radiation.

dew point The temperature of air at which it becomes saturated with water vapour; below this point water vapour starts to condense to form water droplets. It can be determined by using a hygrometer with tables based on dry- and wet-bulb temperatures or by using a special dew-point hygrometer with a polished metal surface that can be cooled until a film of moisture appears at a temperature that is recorded.

dew pond A man-made pond, usually lined with puddled clay, found chiefly on chalk downlands in S England. The pond can hold water for long periods, even through periods of drought, therefore providing an important source of water for livestock. Most of the water is in fact derived from rainfall with only small amounts from dew.

D horizon In soils, the solid unweathered rock below the soil profile. †Some D horizons are unrelated to the soil development above. For example, a calcareous bedrock over which acidic drift deposits generate podzols.

diamond The transparent crystalline form of carbon. It is the hardest known mineral and as such is used as an abrasive and in gem cutting. The highest quality diamonds are prized as gemstones. They form under conditions of intense heat and pressure during the cooling of magma of basic or ultrabasic composition. Some diamonds are mined from the igneous rock itself, as at Kimberley in South Africa, where the diamond-bearing rock forms pipes of kimberlite in the surrounding rocks; other diamonds are recovered from stream gravels formed by the erosion of the diamond-bearing rock. Diamonds can be white or colourless, sometimes red, green, or yellow, and very occasionally blue or black. Those most free from colour

are considered to be the most valuable. The main producers of diamonds include South Africa, Namibia, and Brazil.

diastrophism The movements causing the large-scale deformation of the Earth's crust that produces the continents, oceans, mountains, etc. *See also* epeirogenesis.

diatom ooze A siliceous ooze formed from the minute particles of silica that remain when microscopic diatom algae, which are found in the colder surface waters of the ocean, die and decompose. It is found mainly on the ocean floor beneath these colder waters; for example, in a broad belt beneath the Southern Ocean. *See* ooze.

differential denudation A process that occurs in rocks that have varying resistance to the action of weathering and erosion. More resistant rocks, such as granite, will be weathered and eroded less quickly than comparatively soft rocks, such as clay and mudstone, producing landforms and landscapes strongly determined by rock type. Differential denudation at a coastal location can be seen at Lulworth Cove, Dorset, where the sea has breached hard resistant Portland and Purbeck Limestone to erode weaker beds inland, forming a wide bay with a narrow opening to the sea.

diffluence The branching of a glacier, so that part flows away from the main body of ice. A common occurrence is for ice to be pushed into distributary channels from a main valley, sometimes producing overflow channels into an adjoining valley that breach preglacial watersheds. This may occur where the main glacier is blocked further down the valley.

diffraction In meteorology, the bending of rays of light from the Sun as they pass very close to water droplets in a cloud. The red part of the light is diffracted most and the result may be a coloured ring centred on the Sun, bluish on the inside and reddish on the outside. Such a ring is called a corona.

diffusion In meteorology, the mixing of the air between two different air masses. This tends to be a slow process of molecular diffusion but it may be speeded up if there is turbulence between the air masses, when eddy diffusion occurs.

dike *See* dyke.

dip The angle made between a bed of sedimentary rock and the horizontal plane, given in degrees. This is always at right angles to the strike. *See also* strike.

dip slope A surface or slope that dips in the same direction as the underlying rock strata, usually roughly parallel to them. The term is most commonly applied to the back slope of a cuesta.

disappearing stream A stream that flows over a land surface for a short distance and then disappears underground; such streams are commonly found in areas of limestone or other pervious rock. The stream usually sinks underground down a swallow hole where it passes from an impermeable rock to the permeable rock. In limestone the water flows down through joints in the rock, enlarging these through carbonation and erosion, to re-emerge on reaching impermeable rock strata as a resurgence. *See also* karst, swallow hole.

discharge of a river †The quantity of water passing through any cross section of a stream in a given unit of time. Discharge is calculated by multiplying the velocity by the cross-sectional area and is usually expressed in cubic metres per second (cumecs). *See also* hydrograph.

discontinuity A boundary between two layers within the Earth that possess very different physical properties. The Earth has two major boundaries though minor discontinuities have been discovered in both the crust and

the mantle; these are the Mohorovičić discontinuity between the crust and the mantle and the Gutenberg discontinuity between the mantle and the core. These boundaries were discovered by study of the behaviour of seismic waves as they travelled through the Earth.

discordant coast *See* Atlantic type of coastline.

discordant drainage A pattern of drainage that bears no relation to the structure and dip of the underlying rock. Two sets of explanations have been suggested:
(1) The drainage pattern developed on rock strata that have since been removed by erosion; a process known as superimposition.
(2) The drainage pattern was already present before a period of uplift and folding that formed the present structure. As the uplift took place the rivers were able to cut down at approximately the same rate and so maintain their courses. This process is known as antecedence.
Compare accordant drainage. *See* antecedent drainage, superimposed drainage.

dismembered drainage †A pattern of drainage in which streams that were once tributaries of a single river system now form separate streams flowing into the sea individually. This results from a rise in sea level, or submergence of the land, which has drowned the lower part of the river system, e.g. on the S coast of SW England, where it is thought that the Rivers Dart, Teign, and Exe were all tributaries to a river now lying under the English Channel.

dispersal The moving away of industry and/or people from concentrations or centralized localities. For example, the dispersal of vital industries from major industrial concentrations for strategic reasons or the dispersal of people from cities in time of war (evacuation). One of the original intentions behind the building of the

British new towns was to disperse population from London and other major cities.

dispersed settlement (dispersion) A pattern of rural settlement in which most of the population live in scattered farms, houses, and cottages and there is a general absence of villages and other nucleations. It is normally associated with regions of high land, poor soils, and an abundance of available sources of water. The pattern is also associated with the break-up of large estates, low population densities, agricultural specialization such as market gardening, and with livestock farming. *Compare* nucleated settlement.

dispersion, coefficient of †*See* coefficient of dispersion.

dispersion diagram †A diagram that shows the spread of a series of values, sometimes as a series of dots with each dot representing a specific value, for example each day in a year with measurable rainfall. The dots are plotted against a vertical scale graduated from zero to the highest monthly rainfall in the series. From these diagrams the median and percentile values can be derived and also shown as a dispersion diagram. These diagrams are particularly valuable for estimating rainfall variability and classifying rainfall regimes. They are also useful for determining critical values of density ranges when drawing choropleth and isopleth maps.

dissection In geomorphology, the process by which a land surface is cut into by rivers. The term is generally used descriptively, e.g. dissected plateau, where an uplifted flat surface (peneplain) has become transformed into a surface of steep-sided river valleys and intervening mountain ridges.

distance decay †The principle that there is a decrease in interactions as distance increases. This occurs as a result of (1) a decrease in movement as costs increase; and (2) the effects

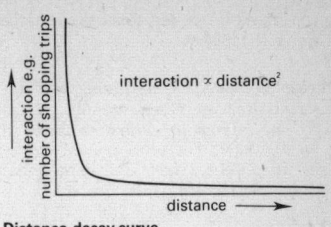

interaction $\propto$ distance^2

Distance-decay curve

of distance on levels of information. The general pattern is for contacts to fall off sharply over short distances and to decline more slowly over longer distances.

distributary A river channel that branches off from the main stream without rejoining it. Distributaries are commonly found on deltas, but are also important in the formation of alluvial fans and cones.

diurnal range The difference between the maximum and minimum of weather elements, especially temperature, recorded at a place during any 24-hour period.

divergence 1. (*Meteorology*) †A condition caused when the horizontal outflow of air from an area is greater than the horizontal inflow. If the divergence is at the surface the outflowing air is replaced by air sinking from higher altitudes, while if it is at high altitudes air rises to replace it. The horse latitudes are an area of surface divergence situated between two areas of surface convergence (the intertropical convergence zone and Atlantic polar front). At high altitude there are complementary areas of convergence above the surface divergence, and divergence above the surface convergence. *See also* convergence.
2. (*Oceanography*) The movement apart in different directions of two or more ocean currents. This often coincides with areas of atmospheric divergence as winds are a major factor in the generation of ocean currents. The main effect is to cause upwelling of water from depth, which is usually

considerably colder than the surface water it replaces.

divide A boundary separating two different drainage basins, usually formed by an intervening ridge or area of high ground. The term is used chiefly in the USA. *See also* continental divide, watershed.

divided circle diagram *See* pie graph.

djebel *See* jebel.

doab A low-lying area of land composed of alluvium that lies between two converging rivers, chiefly in the Indo-Gangetic plain of N India, e.g. Sind Saga Doab lying between the Indus and Jehlam rivers. The term is also used for such areas elsewhere in the world.

doctor A local name given in various parts of the world (e.g. W Australia and Cape Province, South Africa) to a dry and relatively cool breeze that blows off the sea during daylight hours in summer bringing relief from the normal hot humid conditions. *See also* harmattan.

doldrums The zone of light and variable winds in the oceanic equatorial regions where the northeast and southeast trade winds converge; it forms part of the intertropical convergence zone (ITCZ). It is an area of low pressure and the weather is characterized by thick clouds, frequent thunderstorms, and heavy rain showers with sudden squalls. Its position and extent is variable, moving N and S between latitudes 5°N and 5°S lagging behind the apparent seasonal movement of the Sun.

dolomite 1. A mineral consisting of calcium magnesium carbonate ($CaMg(CO_3)_2$). **2.** A sedimentary rock that resembles limestone and is made up chiefly of the mineral dolomite. It is formed through the replacement of limestone.

dome 1. An anticlinal fold in which the rocks dip away from a central point, e.g. the Lake District. **2.** A dome-shaped rock mass. **3.** A landform with a rounded top and steep sides.

dominant In biogeography, a species that is the most abundant within its community or that exerts a major influence on its environment. Oak trees, for example, are frequently the dominant vegetation of deciduous woodland in the British Isles. Dominants attain their position in the community by out-competing neighbours and simultaneously changing environmental conditions in their own favour, for example by producing shade. They are usually the largest and strongest plants in the community.

dominion 1. Formerly, a self-governing country in the British Empire and, later, in the Commonwealth of Nations that had, before its independence, been a colony. Canada was given dominion status in 1867. Others included Australia, New Zealand, and South Africa. **2.** The territory ruled by a ruler or government.

dormant volcano A volcano that has not erupted in recent times but is not regarded as being extinct. *See also* active volcano, extinct volcano.

dormitory town (residential town) A town in which a significant proportion of the working population commute to a city to work. Suburbs and country districts where city people live are dormitory areas, for example, the towns and villages of the home counties (e.g. Dorking and Reigate in Surrey) that provide residential areas for those who work in London.

dot map A map showing distribution by means of dots, each dot of a uniform size representing a specific value. The dot is the simplest form of symbol used on maps and the result is very effective if the value represented by each dot is appropriate. Examples include maps showing the distribution of farm animals such as cattle and sheep in England and Wales.

downs (down, downland) A stretch of open gently undulating land, especially in the chalk areas of S England, e.g. the North and South Downs.

downthrow See throw.

downtown The popular name for the central business district (CBD) in US cities. See central business district.

downwearing †The erosion of a land surface that results in a reduction in the overall height of land and a progressive reduction in the angles of slopes. The concept was introduced by W. M. Davis as a major element of the cycle of erosion, and is thought to be most appropriate in temperate humid environments. Compare backwearing. See also cycle of erosion.

drain A channel or pipe to carry away excess water or sewage from the land. In the Fens of E England the term is used for the wide canals used to carry water to the sea or a river, e.g. the Hundred Foot Drain, which joins the River Ouse near Downham Market.

drainage basin An area of the Earth's surface drained by a single river system and bounded by a watershed, which separates it from adjoining drainage basins.

drainage pattern The arrangement of streams within a drainage basin. This reflects the relief and the geological structure of the area, e.g. the arrangement of rocks, folds, and faults, and is also influenced by past and present climatic conditions. See also accordant drainage, annular drainage, dendritic drainage, discordant drainage, parallel drainage, radial drainage, rectangular drainage, trellis drainage, †deranged drainage, dismembered drainage, inconsequent drainage.

Dreikanter A pebble or stone worn and polished by wind-blown sand into a roughly triangular shape with smooth flat facets and sharp edges, commonly found in desert areas. See also ventifact. [German: three faces]

drift 1. The deposits of material laid down by glacial and fluvioglacial activity. Thick layers of drift accumulated during the Pleistocene period but these have since been largely eroded. **2.** In geology, the superficial deposits that lie above the solid bedrock. These are shown on drift editions of maps issued by the British Geological Survey. (The solid editions of these maps show only the underlying solid rock.) **3.** In mining, a passage underground that follows a vein of the mineral.

drift ice A floating body of ice, formed by the break up of an iceberg or ice floe, which is carried by ocean currents far from its source area, usually being found as an isolated fragment in the open sea.

drilling The method by which valuable materials can be obtained from beneath the earth's surface by boring a hole. A number of important minerals and chemicals are obtained by this method, e.g. oil, natural gas, salt, and sulphur.

drizzle A form of precipitation in which the water droplets are very fine (less than 0.5 mm) and are close together. Normally drizzle is produced by stratus and stratocumulus clouds.

drought A period of dry weather, especially one lasting long enough for vegetation to begin to wither. In the UK an *absolute drought* is defined as a period of at least 15 consecutive days, none of which receives more than 0.25 mm of rainfall; a *partial drought* is a period of 29 or more consecutive days with an average daily rainfall of less than 0.25 mm. In the USA a drought is defined as a period of 21 days or more when the rainfall is 30% or less of the average for the time and place.

drowned valley A valley that has been submerged, either as a result of a rise in sea level or by the subsidence of the land surface. Drowned valleys result in coastlines that are indented by inlets representing the former valleys. *See also* concordant coast, fiord, ria.

drumlin A smooth streamlined mound, usually oval or egg-shaped in plan, that is composed of glacial till, sometimes with a core of bedrock. Drumlins commonly occur in groups or clusters referred to as swarms or fields, and produce a land surface frequently described as 'basket-of-eggs' relief. They are usually aligned with the direction of ice movement, the long axis lying roughly parallel to the direction of flow and the blunter end facing the direction of ice advance. †The process by which drumlins are formed is still not clear and a number of theories have been advanced although it is agreed that drumlins result from glacial streamlining. Two theories as to their origins are:
(1) They are formed by deposition by the ice around a nucleus of frozen till or rock.
(2) Those consisting chiefly of rock or glacial drift that was already deposited were formed by erosion by the ice.
Drumlins occur widely in Ireland, Scotland, N England, Wales, and the N states of the USA. [Irish Gaelic]

dry adiabatic lapse rate (DALR) †The constant rate at which a rising 'parcel' of unsaturated air will cool as a result of adiabatic expansion. This rate is constant at about 1°C per 100 m no matter what the original temperature may be, providing no water vapour condenses within the air. It tends to be greater than the environmental lapse rate, i.e. the rising air cools faster than its surroundings are cooling, therefore conditions within the atmosphere are said to be stable. *See* lapse rate. *See also* saturated adiabatic lapse rate.

dry-bulb temperature The shade temperature as measured on a simple mercury thermometer (*dry-bulb thermometer*). Simultaneous readings of this and a wet-bulb thermometer are used to calculate the relative humidity. *See* wet-bulb temperature.

dry farming The farming methods that are practised in semiarid regions to catch and store the limited moisture available so that crops can be grown. The main method used is to grow a crop every other year, breaking the surface to prevent it cracking at intervals during the fallow year to conserve the moisture for the crop of the year that follows. This type of farming is found in Mediterranean lands and the Columbia Plateau of the USA.

dry-point site An elevated site on which a settlement was located in a region where flooding was likely to take place or where the surrounding areas were waterlogged. For example, settlements on alluvial fans on glaciated valley floors and 'island' sites in fenland. *See also* defensive site, wet-point site.

dry spell In the UK a period of 15 or more consecutive days, in which no day receives more than 1 mm of rainfall. The term however is often used generally to describe periods that have a withering effect on plant life.

dry valley A river valley, occurring chiefly in chalk and limestone areas, which has no permanent stream flowing in it. The formation of dry valleys has been a source of controversy for many years; suggested explanations include the fall of the water table, river capture, and periglacial erosion by mud glaciers. At the present time it is generally thought that they were produced by surface streams flowing over partially frozen ground, both before and during the last ice age, with falling sea levels contributing to steep downcutting by these streams.

dumpy level An instrument used in surveying to determine height. It consists of a spirit level and a telescope that are fixed so they can only move horizontally. †*See* levelling

dune A mound or ridge of sand found in deserts and some coastal areas, resulting from the deposition of particles of sand that have been transported by the wind. The major control over the growth of a sand dune is the availability of a continuous supply of sand.

The majority of dunes in deserts are either crescentic (barchans) or elongated and narrow (seifs), their precise form being dependent on the direction and strength of the dominant winds. Dunes may vary greatly in size; some seif dunes can reach heights of several hundred metres and stretch for many kilometres. The source of sand supply is areas of erosion.

Coastal dunes, which characteristically back extensive sandy beaches, are built up from wind-blown sand which has initially been transported by longshore drift. They tend to migrate inland but are often partially fixed by the growth of vegetation.
See also barchan, seif dune.

duricrust A hard surface layer in the soil that occurs chiefly in hot arid areas such as W Australia. It consists of various minerals, including iron and aluminium, usually strongly cemented by silica. It is believed that duricrusts are formed in areas of extensive flat terrain through capillarity, which draws strong solutions to the ground surface where they evaporate forming a hard deposit. *See also* hardpan.

dust Fine particles of material that are small and light enough to be carried over long distances by the wind. Dust may be produced by volcanic activity (*volcanic dust*), by weathering and erosion, and by human activity (e.g. the burning of fossil fuels). *Hygroscopic dust* acts as nuclei in the condensation of water vapour within clouds. *Cosmic dust* is of meteoric origin.

dust bowl A semiarid area in which the removal of vegetation (e.g. by drought, deforestation, or overgrazing) has resulted in extensive wind erosion of the topsoil. The name Dust Bowl was originally applied to parts of the SW USA, especially Kansas. During the 1930s a period of severe droughts

in combination with overfarming of the land led to severe erosion of the fertile topsoil in these areas. *See also* soil erosion.

dust devil A short-lived whirlwind or small tornado in a semiarid area that whips up sand and dust into the atmosphere, sometimes to heights of up to 3000 m. It is most likely to form in dry areas during the hot season when the ground is intensely heated during the day, forming localized convection currents and gusty surface winds. Dust is carried upwards within the convection currents forming whirling columns, which normally collapse fairly quickly. If a dust devil grows very large it is called a simoom.

duststorm A storm in which dense masses of dust are carried by the wind. It is a common occurrence in desert or semidesert areas where the surface deposits are loose and dry and the wind blows with sufficient force to raise the dust, e.g. at the passage of a depression. In certain circumstances the dust can reach a great height affecting areas well away from the desert with such phenomena as fiery sunsets and blood rain or snow, which is coloured red by the dust.

dyke (dike) A wall-like intrusion of igneous rock formed as magma rose up through a near-vertical crack, forcing the rock apart. As the magma cooled it formed a vertical sheet of rock with parallel sides cutting across bedding planes. A dyke can vary in thickness from a few centimetres to hundreds of metres, but 2–6 m is the average. Dykes tend to occur in large numbers, known as swarms, e.g. on the coast of the Isle of Arran and on the Isle of Mull, Scotland, from where some of the dykes can be traced across S Scotland into N England. *See also* sill.

dyke spring A type of spring formed by the presence of an impermeable igneous dyke acting as a barrier to the normal flow of underground water, so forcing it to emerge at the surface as a spring. *See* spring.

dynamic equilibrium †In geomorphology, the state of balance between erosion and deposition towards which rivers tend to adjust. Rivers never in reality achieve a balance of no erosion or deposition and are in a state of constant adjustment due to changes in load and discharge. In systems theory, where natural systems are viewed in terms of inputs, throughputs, and outputs, a state of dynamic equilibrium is said to exist when the output of a river system is in balance with the input due to internal changes to the throughputs.

dynamic rejuvenation †*See* rejuvenation.

dyne Symbol: dyn †The former unit of force in the c.g.s. system. It is equal to 10^{-5}N.

E

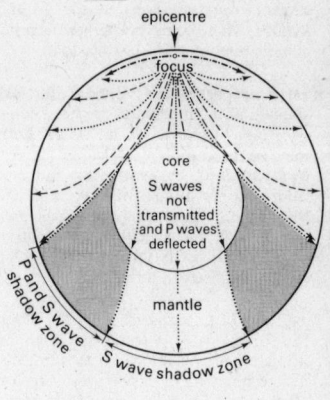

crust

←— Mohorovičič discontinuity

←— Gutenberg discontinuity

upper mantle

lower mantle

outer core

inner core

←— inner core

— 5150 km

— 2900 km

— 700 km

Internal structure of the Earth

Earth The third planet of the solar system and the only one on which life is known to exist. It is a flattened sphere in shape with an equatorial diameter of 12 756 km, and a polar diameter of 12 714 km. It makes one rotation on its axis every 23 hours 56 minutes,

and orbits the Sun in 365 days, 6 hours, and 8 minutes.

The Earth is believed to be about 4600 million years old. It is made up of three internal layers: the crust (the outermost layer), mantle, and core. About two thirds of the surface of the Earth is covered with water (the hydrosphere). The Earth is surrounded by a gaseous envelope known as the atmosphere.

earthflow A mass movement of waterlogged material down a slope, which usually leaves a bowl-shaped scar on the slope from which it has become detached. The material is deposited as a tongue at the foot of the slope. *See also* mass movement.

earth movement The movement of a section of the crust of the Earth both horizontally and vertically, including the relative movement of land and sea, movement along faults, and the folding of rocks.

earth pillar A tall column of soft materials, capped by a boulder, which can reach heights of about 10 m. It is formed by the erosion of the surrounding material by rain water or

epicentre

focus

core
S waves
not
transmitted
and P waves
deflected

mantle

P and S wave shadow zone

S wave shadow zone

---→ L wave   ----→ S wave   • P wave

Earthquake waves

stream action, while the soil immediately beneath the boulder remains protected. Earth pillars are most frequently found in mountainous areas, often produced in boulder clay. Good examples are found in the Austrian Tyrol.

earthquake A series of shocks that result from sudden earth movement, e.g. along a fault, which can be recognized by the passage of earthquake (seismic) waves. The waves spread out from the seismic focus of the earthquake and can cause widespread destruction. In areas of the world that lie along the edges of lithospheric plates, e.g. down the W edge of the Pacific Ocean, earthquakes are very common. The severity of earthquakes is measured on the Richter scale. One of the most severe earthquakes of recent times was the Good Friday earthquake of 27 March, 1964, near Anchorage, Alaska.

Earthquake waves are classified into three main types:

P (*primary* or *push*) *waves* are compressional, i.e. the particles vibrate in the direction of movement of the wave, similar to a sound wave. They can pass through solids, liquids, and gases and are the first earthquake waves to be recorded on a seismogram of an earthquake.

S (*secondary* or *shake*) *waves* are recorded on a seismogram after the P waves. They are distortional waves in which the particles vibrate at right angles to the direction of movement of the wave. S waves travel through the Earth's interior but cannot be transmitted by liquids.

L (*surface* or *long*) *waves* travel along the surface of the Earth and are recorded after the P and S waves. Two types of L waves are identified: Love waves and Rayleigh waves.

earth science Any of the specialized sciences that study the Earth. These include geography, geology, geophysics, seismology, volcanology, geochemistry, meteorology, climatology, hydrology, oceanography, glaciology, biogeography, and pedology (soil science).

easting Any of the N–S grid lines on a map showing the distance eastwards from the point of origin of the grid. In map grid references the easting coordinates are given first, before the northing. On Ordnance Survey maps each grid line is identified by two digits to which a third can be added, using a tenths scale, to give precise coordinates for a point between two grid lines. *See* grid. *See also* northing.

ebb tide The retreat of the tidal stream, following high tide and before low tide. *Compare* flood tide. *See* tide.

echo sounder An instrument used to determine the depth of the sea by recording the time interval required for a sound wave to travel from a transmitter on the surface to the sea bed, and be reflected back to a receiver at the surface. Knowing the velocity of sound waves in water, the depth can be calculated. Readings can be distorted by variations in salinity and temperature, schools of fish, etc. Echo sounders can also be used for measuring the thickness of ice.

Eckert projections A series of six broadly similar map projections in which each pole is represented by a line half the length of the equator, instead of by a point; the parallels are shown as straight lines. The shapes of the continents are good and the projections are used for world maps.

eclipse The passage of all or part of a celestial body into the shadow of another. A *solar eclipse* occurs when the new moon passes exactly between the Sun and the Earth, casting a shadow on the Earth. The shadow is some 45 km wide and crosses a belt of the Earth several hundred kilometres long. A *lunar eclipse* occurs when the Earth passes exactly between the Sun and the full moon. The Earth's shadow is far larger than the Moon and covers the Moon's surface for about 3 hours. Eclipses do not take place every month because the Moon's orbit is inclined at an angle of about 5° to the Earth's orbit.

ecliptic The great circle in which the plane of the Earth's orbit meets the celestial sphere. During one year the Sun appears to move round the ecliptic, passing through each one of the 12 constellations called the signs of the zodiac. Eclipses occur when the Sun, Moon, and Earth lie in a straight line on the plane of the ecliptic.

ecoclimate The climate as it affects the animal and plant life of a particular place.

ecological niche †The role played by a plant or animal in the functioning of an ecosystem. For example, the ecological niche of a pike is as a freshwater predator of other fish. An organism fulfilling its niche exerts an influence on both its neighbours in the community and its abiotic environment. The ecological niche of an organism is a relatively stable position attained through evolutionary adaptation and competitive exclusion, i.e. if two species occupy the same niche competition occurs until one has replaced the other. Similar niches may be occupied by different species in different areas – they are said to occupy *parallel ecological niches*. For example, a bison in North America as a herbivore occupies the same niche as a kangaroo in Australia.

ecological regulation (natural regulation, balance of nature) †A mechanism whereby populations within a community are controlled about an optimum level. Limiting factors within the ecosystem, such as weather, food supplies, or predation, will slow down or stop population growth by suppressing a species' birth rate or increasing its death rate. Such regulation is frequently 'density-dependent' (i.e. it operates progressively more strongly as the population density grows). Ecological regulation is the principle underlying biological (as opposed to chemical) control of pests.

ecology The scientific study of the relationship of a plant or animal to its natural environment. This includes the organism's responses to its physical surroundings (e.g. to the weather or to the changing day length) as well as its interactions with other organisms (such as who eats whom). *Applied ecology* seeks to improve man's use of environmental resources by translating lessons learnt from ecological research to such problems as the population control of pests. †*See also* ecological regulation, ecosystem. [From Greek *oikos*: home]

economic geography Those aspects of geography that deal with the spatial variations in economic activity and the different ways in which wealth is produced, distributed, exchanged, and consumed. It is essentially concerned with the economic activities of people, and the relationships between these activities and the physical environment.

economic planning The planning of a country or region's economy with the objective of increasing the productive capacity and thereby raising the living standard and general wellbeing of the people. A number of countries, such as the USSR and India, have economic plans as blueprints for growth over a number of years. Economic planning is particularly important in Third World countries where limited resources must be allocated efficiently to stimulate development.

ecosphere The parts of the Earth in which life forms can exist, i.e. the land surface, soil, lakes, rivers, oceans, and the lower atmosphere.

ecosystem †A community of plants and animals sharing a given environment, together with the nonliving (abiotic) habitat that they occupy. Ecosystems form the basic units for ecological study. The concept can be applied on a wide range of scales, for example, to the whole world (the *global ecosystem*), tundra, tropical rainforest, or a small unit such as a pond or lake. Ecosystems characteristically derive their energy from sunlight and receive inputs of nutrients, water, and gases. Heat, oxygen, carbon dioxide, and organic compounds form the outputs. The organisms typically comprise producers (green plants), which convert inorganic compounds through photosynthesis; consumers, which feed on plants and each other; and decomposers (e.g. bacteria, fungi), which promote decay. Ecosystems are thus self-sustaining and mutually independent.

ecotone †The transition zone between neighbouring ecosystems, such as between tundra and boreal forest, where representatives of both communities coalesce and environmental conditions merge. The zone contains a greater number of species than the zones on either side and so is of ecological interest. Ecotones may be considered to be ecosystems in their own right. *See also* ecosystem.

ecumene Those parts of the Earth's surface that are settled by people. It has been roughly estimated that about 60% of the world's land surface can be classified as ecumene and about 40% as *nonecumene* – those parts of the Earth that are uninhabited or only sparsely inhabited. [From Greek: the inhabited part of the world known to the Greeks]

edaphic †Of or relating to the soil. For example, edaphic factors such as soil moisture, texture, pH values, and organic content influence plant growth.

eddy A roughly circular movement within the flow of a stream of air, water, or other fluid. Within rivers eddies may be set up by obstructions to the flow of water, e.g. by a bridge. In deserts eddies in the wind play an important role, for example in the formation of barchans, being largely responsible for the production of the steeper downwind slope.

effectiveness of precipitation (effective precipitation) That proportion of the total precipitation that is of use to plants. It is the total precipitation less that lost through runoff, evaporation, percolation to ground water, etc.

efficiency of a stream †A measure of the amount of load a stream can transport in relation to its potential energy. Its potential energy is derived from the discharge of the stream in conjunction with the slope of the stream channel. A stream working to its maximum efficiency is therefore carrying the maximum load possible in relation to the volume and velocity of water in its channel, so that deposition of material is not taking place.

effluent 1. A stream flowing out of a lake, reservoir, or larger stream.
2. (industrial) The waste products from a factory or industrial complex. In many countries effluent is discharged into nearby rivers or the sea causing extensive pollution. In West Germany the lower Rhine is heavily polluted, while in the UK the River Trent and its tributaries receive large quantities of toxic chemicals and other pollutants from the factories of the Midlands.

elbow of capture *See* capture, river.

elevation The height of a point above a given level (usually mean sea level).

eluviation †The downwashing of clay and other fine particles in free-draining soils. The upper horizons are thus depleted of fine clay particles, becoming an eluvial zone. The particles may be redeposited at a lower horizon or may be washed out of the soil altogether. *Compare* illuviation. *See also* leaching.

emergence The exposure of previously submerged land brought about by a fall in sea level or by the uplift of a continental land mass. Emergent coastlines are usually characterized by raised beaches, wave-cut platforms, and coastal plains, and occasionally by spits and bars. *See also* submergence.

emigrant A person who has left one country and settled in another. *Compare* immigrant.

enclave 1. A small piece of territory surrounded completely by another country, viewed from the position of the surrounding territory. For example, Baarle-Hertog, an outlying portion of Belgium, is a small enclave within the Netherlands. *Compare* exclave.
2. A concentrated group of foreigners within a country.

endemic Denoting a feature that is regularly found among people or a group of people in a particular country or region. For example, endemic diseases such as malaria in Sri Lanka.

end moraine *See* terminal moraine.

endogenetic (endogenic) Denoting the processes and materials that originate within the Earth, e.g. earthquakes and igneous rocks. *Compare* exogenetic.

englacial river A stream derived from meltwater that flows through a tunnel within the ice of a glacier.

enterprise zone In the UK, a zone within an urban area that has been given special status in order to encourage industries to establish themselves there. Industries are given incentives in the form of simplified planning procedures and other benefits such as exemptions from general rates on industrial and commercial property. A plan is produced for each zone by the local authority showing which developments will be permitted in each part of the zone; planning permission is not required for developments that conform to this zoning. Each zone operates for 10 years; the first was designated in June, 1981, in the Lower Swansea Valley. The idea was first put forward by Peter Hall in 1977 as a means of helping solve the problems of the inner city. *See also* inner city.

entrenched meander †*See* intrenched meander.

entrepôt A place to which goods are brought from one part of the world for distribution to other parts, or a place where goods are temporarily stored before distribution. Goods entering or leaving an entrepôt are usually free of any duty. Ports frequently perform an entrepôt function, e.g. Rotterdam for goods to and from central Europe and Hong Kong for goods passing to and from SE Asia, especially China. [French]

environment In biogeography, the sum of physical, chemical, and biological conditions experienced by an organism, including climate, soil, water, light, neighbouring vegetation, individuals of its own sort, and members of other species. Environmental conditions vary according to the time of day, season, weather, and other factors.

environmentalism †A philosophical concept that stresses the influence of the environment on the life and activities of man. Extreme environmentalism is known as environmental determinism. *See also* determinism.

environmental lapse rate (ELR) The rate of decrease of temperature with increased altitude for a particular place at a particular time. This varies according to air mass conditions but is generally about 6°–8°C per 1000 m. *See* lapse rate.

environmental perception †Part of the behaviourist approach to geography that recognizes that the individual's perception of the opportunities offered by the environment differs according to the cultural background, educational standards, personal values, and tastes of that individual. It is held that the environment therefore has social as well as physical attributes and that there is a consequent need to interpret human spatial behaviour and the land-use patterns such behaviour produces in the light of what people think is there (their perceptions) and not necessarily in terms of what is actually there (physical phenomena).

Eocene The second geological epoch of the Tertiary period. It extended for about 17.5 million years from the end of the Palaeocene, about 54 million years ago, to the beginning of the Oligocene. It consists of continental rocks in the W grading into marine rocks in the E. Eocene rocks outcrop in the British Isles in the London Basin and the Hampshire Basin. Earth movements in the S of England were slight but in the N and in Scotland there was widespread volcanic activity. The basalt flows of the Giant's Causeway, Northern Ireland, and Fingal's Cave on the Scottish Isle of Staffa, date from this time. There was a marked absence of dinosaurs, which were so prolific during the Cretaceous, but mammals became increasingly important during this epoch. Plants and trees were very like present-day varieties, with grasses making their appearance and contributing to the rapid evolution of hoofed mammals.

Eozoic 1. The earliest division of geological time, synonymous with the Precambrian.
2. The earliest geological era of the Precambrian, followed by the Archaeozoic and the Proterozoic.
The term is now rarely used.

epeirogenesis The large-scale movements of the Earth's crust by which the continents are raised or lowered, accompanied by faulting, but with very little, if any, folding. Emergent land, block mountains produced by the movement of fault blocks, and fluctuating sea level can be attributed to this type of movement. The Black Sea and the Mediterranean Sea are modern examples of depressed areas of crust formed through epeirogenic movements. *Compare* orogenesis. [From Greek *epeiros*: continent]

epicentre The point on the Earth's surface immediately above the seismic focus or origin of an earthquake.

epidemic A disease that simultaneously affects a large number of the people of a country or region at a particular time. For example, the cholera epidemics of the 19th-century industrial towns of the UK.

epigenetic drainage *See* superimposed drainage.

epiphyte A plant that is attached to other vegetation or objects for support, but which is not parasitic. Epiphytes obtain the essential nutrients they require from the atmosphere or from decaying plant matter. Orchids and ferns are examples of equatorial rainforest epiphytes. Examples in temperate regions include many mosses and lichens.

epoch A subdivision of geological time. A number of epochs together make up a period and each epoch comprises a number of ages. During one epoch one series of sedimentary rocks is deposited. *See also* geological time.

equal-area projection A map projection in which the regions shown have the same area in proportion to the corresponding regions on the globe. Correct representation of area can be achieved easily on maps but in many cases the shape must suffer, particularly in the outer areas of the projection. *See* Bonne's projection, Mollweide projection, zenithal equal-area projection.

equator The great circle around the Earth with the parallel of latitude 0°, lying equidistant from the two poles and dividing the N hemisphere from the S hemisphere. It has a length of 40 076 km and is the longest great circle.

equatorial climate A climatic type found in low-lying areas between latitudes 10°N and 10°S. In these areas seasonal variations are slight: temperature, humidity, and rainfall are high all year round. Mean monthly temperatures are 25°–27°C and the rainfall is heavy and well distributed, falling mainly as heavy convectional showers and often accompanied by thunder. Dense equatorial rainforests are a response to this climate. The main areas in the world that have a true equatorial climate are the Amazon Basin (South America), the Zaïre Basin and Guinea Coast (Africa), Malaysia, Indonesia, and New Guinea.

equatorial current The westerly movement of surface oceanic water in tropical areas. There are in fact two currents, one to the N of the equator (the north equatorial current) produced by the northeast trades, and one to the S of the equator (the south equatorial current) produced by the southeast trades. Between these there is frequently an easterly drift of surface water, known as the equatorial countercurrent.

equatorial rainforest (tropical rainforest) A type of vegetation dominated by very tall trees (up to 60 m) that grows in equatorial regions. The plant life is rich, varied, and quick growing, in response to the climatic conditions of heavy all-year-round rainfall (above 1500 mm) and constantly high temperatures (25–30°C). The trees form a dense canopy and restrict sunlight reaching the lower levels. They have straight slender trunks, usually supported by buttresses at the base. The leaves are leathery and have drip tips (extensions at the points of the leaves) enabling water to be rapidly shed. The vegetation shows a marked stratification with three layers: the canopy layer, a layer of smaller trees and saplings, and the poorly developed undergrowth. Epiphytes and lianes are common. There is a wide diversity of plant species and associated fauna and a lack of seasonality in flowering and fruiting. Such vegetation occupies equatorial lowlands in South America, central Africa, and SE Asia. Equatorial rainforests produce valuable hardwoods, such as mahogany, but the resource is endangered by overexploitation. *See also* monsoon forest, selva.

equatorial trough (intertropical trough) A trough of low pressure found in equatorial latitudes where the trade winds converge. It corresponds to the intertropical convergence zone (ITCZ).

equatorial westerlies A zone of generally westerly winds that occur between the two trade wind belts in summer, especially over the continental areas. They occur when the intertropical convergence zone (ITCZ) extends more than 5° from the equator.

equidistant projection A map projection that attempts to show distances accurately between different points. No projection can avoid errors in showing distances on a world scale but in some, such as the conical with one standard parallel, scale along the meridians and along the standard par-

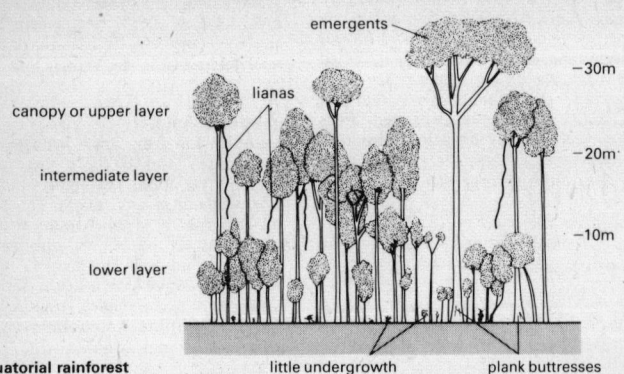

emergents

canopy or upper layer

lianas

intermediate layer

lower layer

—30m

—20m

—10m

Equatorial rainforest   little undergrowth   plank buttresses

allel is correct, but it is exaggerated along all other parallels. *See* zenithal equidistant projection.

equilibrium The condition of a natural system or ecosystem when it is in a state of balance. For example, a river channel is said to be in equilibrium (or a state of grade) when it is neither eroding nor depositing. †*See also* dynamic equilibrium.

equilibrium line †The line on a glacier where the accumulation of ice is equal to the loss of ice by melting and evaporation, i.e. it separates the accumulation zone from the ablation zone. As snow and ice being accumulated in the upper reaches of the glacier exceeds ablation, and ablation in the lower reaches exceeds accumulation, there will tend to be a continuous transfer of ice over the equilibrium line under the influence of gravity. The position of the equilibrium line is not fixed and will vary according to weather conditions.

equinox One of the two dates in the year when day and night are of equal length throughout the world. It occurs on about 21 March (the vernal or spring equinox in the N hemisphere) and 23 September (the autumnal equinox in the N hemisphere). On these dates the Earth's axis lies at right angles to the line joining the centres of the Earth and Sun. Both poles then lie exactly on the line where the sunlit half of the Earth meets the half in shadow. Therefore, as the Earth makes its daily rotation round the poles, every place will experience 12 hours daylight and 12 hours darkness. The Sun rises due E and sets due W, and is seen directly overhead on the equator.

equipluve †A line drawn on a map joining all places with the same pluviometric coefficient. *See* pluviometric coefficient.

era A division of geological time. Following the Precambrian three eras – Palaeozoic, Mesozoic, and Cenozoic –

are commonly recognized by geologists. (The Tertiary and Quaternary periods were formerly regarded as eras in the UK but this usage is now obsolescent.) Each era is subdivided into a series of periods. *See also* geological time.

E-region (Heaviside layer, Heaviside-Kennelly layer) A band of high ion density within the ionosphere, occurring at about 95–120 km. Its main importance is that it reflects long radio waves back to Earth so that these may be transmitted over considerable distances, but short radio waves pass through it and are subsequently reflected by the Appleton layer at about 240 km.

erg A type of arid desert with an extensive cover of sand in the form of dunes and sand sheets, especially in the Sahara. *See also* hamada, reg. [Arabic]

erosion The wearing away of the land surface by agents that involve the transportation of material. The agents of erosion – water, ice, and wind – are only capable of having a marked effect on the land surface when they are transporting fragments of weathered material. In a strict sense, erosion is distinct from denudation in that erosion does not include weathering, although the terms are often used interchangeably. *See also* cycle of erosion, denudation, weathering.

erratic (erratic block) A rock fragment, ranging in size from a pebble to a boulder, that has been transported a long distance from its source. The term is most commonly applied to material that has been transported and deposited by glaciers and ice sheets. It is often possible to trace erratics back to their source outcrop of rock; they have thus proved useful in indicating the source and direction of movement of glaciers and ice sheets. Erratics found in East Anglia, for example, have been traced back to S Norway. It is, however, possible that an erratic may have been moved by several ice masses in different directions.

eruption The process by which material from within the Earth's crust is forced upwards in the form of solids, liquids, or gases, and extruded onto the surface as a result of volcanic activity. *See also* fissure eruption, volcano.

escarpment *See* scarp.

esker A long narrow winding ridge composed of coarse sand and gravel, deposited by glacial activity. The sands and gravels are well sorted and stratified, being very similar to river deposits. Some eskers are only tens of metres long but others may extend for hundreds of kilometres and reach heights of over 30 metres. Examples can be found in Ireland, Sweden, and Maine (USA). †Eskers form in contact with ice from materials deposited by meltwater streams flowing in, on, or below a glacier or ice sheet. Although the deposits may be formed by the meltwater streams of active glaciers it is assumed that they are only laid down as eskers by slow-moving or stagnant glaciers because otherwise the glacier movement would disrupt their form. The courses of the eskers are generally aligned at right angles to the ice front, thus providing an indication of the direction of ice flow. *See also* beaded esker.

estancia A cattle ranch or station in South America, e.g. on the pampas of Argentina. [Spanish]

estuary The tidal mouth of a river where fresh and saline water are mixed. It may be flanked by mud flats, marshes, or swamps, especially when it is sheltered from erosion by the sea. Most estuaries are drowned valleys resulting from a relative rise of sea level. Estuaries tend to make ideal sites for the development of fishing, port, and industrial activities, providing access to deep water and 'greenfield' locations; many such examples exist in the UK, among them Thamesside, Merseyside, and Clydeside.

étang A shallow body of water that is found among coastal dunes, especially on the SW coast of France. It is gradually filled in with silt and colonized by vegetation. For example, the Étang de Berre near Marseilles. [French]

etesian winds (etesians) The northerly or northwesterly winds that blow over the E Mediterranean area between about mid-May and mid-September. They are caused by a steady southerly drift of surface air over the area towards the thermal low pressure over the Sahara. They tend to be strongest in the early afternoon and die out at nightfall.

ethnography The scientific description of nations, races, or groups of men focusing upon their customs, habits, and characteristics that make them distinctive and different from others.

ethnology A science that is concerned with the study of races and people and of their relationships to one another. It forms a branch of anthropology.

eustasy (eustatism) A change in sea level on a global scale. It is the result of either a change in the size and capacity of the ocean basins through tectonic movements of the sea floors or landmasses or a variation in the volume of sea water.
†The volume of water in the sea is influenced by changes in the size of the ice sheets. The extraction of water, during the growth of ice sheets, will lower sea level; conversely, the melting of the ice will raise sea level. [The mechanism was termed glacial-eustasy by the Austrian geologist Edward Suess in 1888]

eutrophic †Denoting an environment, especially a lake, that is rich in nutrients, notably phosphorus and nitrogen, and as a result supports a rapid growth of plants, algae, and phytoplankton. Oxygen is depleted by the lake's organisms, particularly in summer, and this may ultimately lead to the death of fish and other aquatic life. This process of *eutrophication* may be caused by leaching of fertilizers into rivers and lakes or by sewage effluent. *Compare* oligotrophic.

evaporation The process whereby a liquid changes into a gas. Evaporation from the surface of the Earth together with transpiration (the release of water vapour from plants) provides nearly all the water vapour in the atmosphere; this is very important as it is water vapour that forms the basis for most types of weather. The rate of evaporation is dependent on the air temperature, vapour pressure (the amount of moisture already present in the atmosphere), the nature of the surface, and the wind. Evaporation is most effective when temperatures are high, when there is a wind blowing, and where there is a large water surface; the highest evaporation rates occur in desert areas and the lowest in regions with equatorial climates or cool climates.

evaporimeter (atmometer) An instrument used to measure the amount and rate of evaporation of a liquid. There are two main types: one measures the loss in weight of a known quantity of liquid and another measures the change in level of the surface of the liquid.

evaporite A sedimentary rock that has been formed by the precipitation of minerals from a concentrated solution, which have then been dried out by evaporation, e.g. rock salt (sodium chloride) and gypsum. This occurs chiefly under desert conditions where there is prolonged evaporation from a shallow stretch of water. These conditions are found today around the Great Salt Lake, Utah (USA), which is currently depositing large quantities of salt, and in the Dead Sea, where crystals of salt and gypsum are also being deposited.

evapotranspiration The total moisture returned to the air by the combined processes of evaporation from land surfaces and transpiration by plants. The rate of evapotranspiration depends mainly on temperature. Its relationship to precipitation is a very important factor in climatology, since more water is recycled by this method than by runoff. †*Potential evapotranspiration* is the maximum amount that is theoretically possible assuming a constant water supply. *Actual evapotranspiration* is the observed amount.

everglades The extensive swampy tracts of Florida, USA, that have permanent standing water around hummocky islands of tall tufted grasses, canes, and swampforest. Part of the area forms the Everglades National Park.

evergreen Denoting plants that keep their leaves throughout the year or through a number of years instead of losing their leaves seasonally. Evergreens are thus never totally bare of foliage. Many conifers (e.g. pines) and tropical broad-leaved trees are evergreen. *Compare* deciduous.

exclave An outlying portion of a country surrounded entirely by a foreign country, viewed from the position of the home country. For example, Baarle-Hertog is an exclave of Belgium lying within the Netherlands (to which state it is an enclave). *Compare* enclave.

exfoliation (onion weathering) A weathering process in which the outer layers of a rock split off into thin concentric sheets or scales. It has been suggested that this results from insolation weathering where solar heating during the day is followed by rapid night-time cooling. The expansion and contraction resulting from this wide diurnal range of temperature would cause stresses in the rock leading to the exfoliation. †Experiments have however proved that rocks can withstand extreme temperature changes for long periods. It is now thought that the presence of water may be important in the process.

exile A person compelled to live away from his or her native land. The compulsion to live as an exile, or in exile, is usually the result of religous or political persecution.

exogenetic (exogenic, exogenous) Denoting the external processes that act upon the Earth's surface. These include weathering, mass movement, erosion, transport, and deposition. Exogenetic processes in combination with the internal, or endogenetic processes, are important in the formation and the development of landforms. *Compare* endogenetic.

exosphere The boundary between the Earth's atmosphere and interplanetary space. It extends from about 400 km above the Earth's surface. †However, recent researches by Chapman, Nicolet, and others suggest that there is no boundary and that above a height of about 400 km the Earth's atmosphere may be continuous with that of the Sun.

exotic A plant or animal that is of foreign origin and not native to the area in which it is found. The term is used of species occurring outside their normal geographical range, and suggests a natural mode of arrival rather than deliberate introduction. *Compare* indigenous.

expanded town A town where planned growth is undertaken under the Town Development Act (1952). This act encouraged the development of towns in country districts to relieve overpopulation and congestion elsewhere, principally in the metropolitan areas. The development can be undertaken by both the local authority from which the pressure of overcrowding is to be relieved and by the receiving local authority (or by the county council in which it is situated). The Greater London Council has entered into a number of such schemes, which include the expanded towns of Andover, Basingstoke, and Kings Lynn. *See also* new town.

expatriate A person who, unlike a refugee or exile, has chosen to live away from his native country. For example, 'tax exiles' in Switzerland and British tea planters in the Kenyan highlands.

exponential growth The situation in which one variable increases by addition (or decreases by subtraction) and the other variable increases by multiplication (or decreases by division). The line graph that expresses this relationship is known as the *exponential curve* or J-curve. When data with an exponential relationship is plotted on semilogarithmic paper the curve is shown as a straight line.

exports Goods produced by one country that are sold to another. In return the first country may receive imported goods and services, gold, or foreign exchange. The main exports of the UK are machinery, chemicals, motor vehicles, food, and textiles.

exposure In geology, an outcrop of rock that is not covered with soil, scree, vegetation, or by artificial constructions. Such an exposed rock surface may vary from a small near-level limestone pavement to a near-vertical sea cliff. The exposed outcrop may be denuded at a faster rate than rock with, for example, a soil cover.

extensive agriculture Farming in which the amount of capital and labour applied to a given unit of land is relatively small. Most forms of livestock farming common in Australia and New Zealand are examples of this type of agriculture. In the Great Plains of the USA some 5 to 10 hectares of grazing are needed to support one cow or five sheep; the farms therefore need to be large and the capital investment per hectare is low. *See also* intensive cultivation, intensive livestock rearing.

extinct volcano A volcano that was once active in the geological past, but which is not active today with no prospect of any future activity.

†The form of extinct volcanoes is often greatly changed by denudation subsequent to their formation. The original cone may be denuded to its resistant lava core and then forms a volcanic plug (e.g. Arthur's Seat, near Edinburgh).
Compare active volcano, dormant volcano.

extractive industry An industry that removes materials from the Earth's crust, such as mining, drilling, and quarrying for ore, fossil fuels, or precious metals. Extractive industries are sometimes considered part of the 'robber economy' since the materials are irreplaceable once they are consumed.

extrapolation †The estimating of the value of a quantity or function outside the known range of values. For example, a linear graph showing population growth can be extended beyond the most recent values on the graph. The further from the known range the line is taken, the greater the risk of error in the extrapolation.

extraterritoriality The condition of being exempted from the jurisdiction of a state, granted to some aliens. For example, diplomats and troops stationed abroad may enjoy immunity to the local laws of the country in which they are living, under agreements reached by their own and their host governments.

extreme climate A climate in which the seasonal range of temperature is very great. Continental climates are the most extreme as they are found far from the moderating influence of the sea, and in areas such as central Asia the mean temperatures for January and July may differ by up to 50°C.

extrusive rock (volcanic rock) Igneous rock formed by the cooling and solidification of magma that has poured out onto the surface of the Earth, e.g. volcanic lava. It usually has a fine-grained or glassy texture resulting from rapid cooling once in contact

with air; it may also be filled with gas holes. Extrusive rock can be acid, basic, or ultrabasic in composition. *Compare* intrusive rock.

eye The central low-pressure area around which the high winds of a tropical storm (hurricane or typhoon) circulate. Within this area, which varies from 20–65 km in diameter, atmospheric pressure is usually very low and it is calm or has light and variable winds.

eyot *See* ait.

F

Fahrenheit scale A temperature scale on which 32° represents the freezing point of water and 212° represents the boiling point of water. The scale is not now used in the UK. To convert between degrees Fahrenheit (F) and degrees Celsius (C) the following formula is used:

$$C/5 = (F-32)/9$$

[Named after the German physicist Gabriel Fahrenheit who introduced the scale in about 1724]

fall *See* autumn.

fall line The line that marks the abrupt change from an upland to a lowland, e.g. between a plateau and a plain, along which there are waterfalls and rapids where rivers descend to the plain. The fall line often marks the limit of navigation upstream, and the waterfalls may be used to produce hydroelectric power. In the E USA the Fall Line is the junction between the resistant rocks of the Appalachian Mountains and the weaker Atlantic coastal plain.

fallow Land that is normally cropped but which is allowed to remain unused over a period of time, in either a tilled or untilled condition, in order to restore the fertility. The time period for which it is rested may be as short as one season, e.g. winter fallow in high latitudes, or for several years, as

fan

in areas of shifting cultivation in the tropics.

fan *See* alluvial fan.

farm An area of land and its buildings used for agriculture. Farms may be owner occupied, rented, or managed on behalf of a company or individual.

fathom A measure of depth of water, equal to 1.83 m. It was based on the distance from fingertip to fingertip of a man's outspread arms.

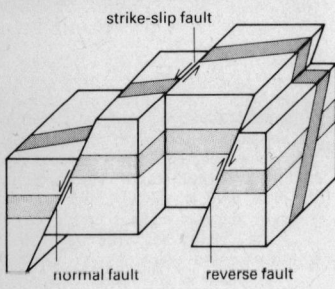

strike-slip fault

normal fault   reverse fault

Faults

fault A linear break in rocks of the Earth's crust along which there has been displacement in a horizontal, vertical, or oblique direction. Faulting results from forces acting on the rocks that cause the rock strata to fracture along lines of weakness known as fault planes and each bed of rock to be displaced along this plane. The extent of vertical displacement is known as the throw; the horizontal displacement is the heave. Many types of fault have been recognized. *See* normal fault, oblique fault, reverse fault, step fault, strike fault, tear fault, thrust fault, transform fault.

fault block A portion of the Earth's crust that is bounded on at least two sides by fractures or faults.

fault-line scarp The rock face formed when faulting brings rocks of different resistances next to each other. The less resistant rock is eroded to leave a scarp, which runs along the line of the

fault. The scarp marks this line, rather than the actual fault plane. †As denudation progresses the fault-line scarp may reach an obsequent stage in which it faces the opposite direction, or a resequent stage when it has reverted to its original direction.

fault plane The surface along which faulting takes place. The rock surfaces of the fault plane may be polished or scored with lines, known as slickensides; alternatively the fault plane may be a zone of crushed rock. It may be vertical or inclined at an angle. This angle is called the dip of the fault and is used in the identification of the type of fault. The horizontal direction of the fault plane at right angles to the dip is called the strike.

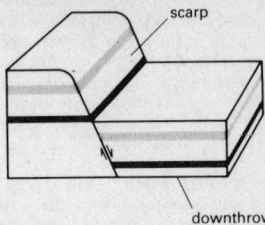

scarp

downthrow

Fault scarp

fault scarp The steep face of rock that is formed by the movement of rocks in a fault. The uplift of one side exposes the scarp, the surface along which movement in the fault has occurred. The presence of a fault scarp indicates recent fault activity.

fault spring A spring that emerges along the line of a fault. The springs are formed where a permeable stratum of rock is lifted above an impermeable stratum on the downthrow of the fault. Water then flows out of the permeable rock to form a spring. The water is channelled away as a stream on top of the impermeable rock.

fauna The animal life of a given area or time. †Fauna may be qualified geographically (e.g. Indian fauna), environmentally (e.g. soil fauna), or by classes of organisms (e.g. invertebrate fauna).

federalism A two-tiered political administration system in which central government divides some of its authority among its regions. Federalism is used by all the larger countries of the world (except China) because it allows rationalization of planning and administration over large areas. For example, the 50 states of the USA, under the federal constitution, administer locally such matters as health and education. Federalism is also often used to maintain the outward unity of a country while allowing a degree of autonomy to the constituent states, a useful device in countries with strong regionalism.

feldspar (felspar) A group of minerals that consist of silicates of aluminium, combined with potassium, sodium, calcium, and barium. It forms one of the most common minerals in igneous rocks and is the most abundant mineral in the Earth's crust. It is divided into two groups: *alkali feldspars*, in which potassium is dominant, and *plagioclase feldspars*, containing either sodium or calcium or both in varying proportions. The crystals are white, pink, or grey. The type of feldspar present is important in the classification of igneous rock.

fell In N England, a mountain or hill that is usually used for rough grazing. Fells are found particularly in the Lake District, where examples include Scafell and Bowfell. [Norwegian]

felsenmeer An area of large angular boulders, formed by the action of frost on well-jointed rock, on either a high plateau or mountain top. [German: sea of rocks]

fen An area of low-lying marshy land that is underlain by a peaty soil. Examples include the Fens of Norfolk and the Somerset Levels. †Fens are characterized by soils that are rich in humus, and usually alkaline. When drained the dark fenland soils are very fertile, e.g. the Fens of E England are now among the richest arable areas of the UK. Drainage often results in shrinkage and, without proper management, can lead to the formation of a dry dust, which can be removed by the wind.

fermentation layer In soils, a layer of partly decomposed litter lying below the freshly fallen organic material and above plant remains that have been fully decomposed to form humus. This stratified structure is typical of mor humus.

ferrallitic soils †A group of red-coloured soils that occur on the old planation surfaces of the humid tropics, formed under forests. Heavy rainfall causes strong leaching to great depths. The red colour derives from iron compounds. The loamy texture and fertility can be easily impaired by agricultural use.

Ferrel's law †A law that states that, as a result of the Earth's rotation, a body moving in any direction over the Earth's surface will tend to be deflected, to the right in the N hemisphere and to the left in the S hemisphere. The effects of this are to be seen most markedly in the movements of air and water. *See also* Coriolis force. [Named after the US meteorologist William Ferrel who first postulated the law in 1859]

ferrisols †A group of red-coloured soils occurring in the humid tropics that resemble ferrallitic soils in structure and development but are less intensely leached and weathered. They are formed where surface erosion is constantly removing weathered material from the surface (e.g. on slopes). They are more fertile than ferrallitic soils.

fertility In population studies, the number of live births within a given period and area. With mortality and migration, it is one of the three components of population change in an area. A number of measures are used to compare spatial variations in fertility, the most commonly used measure being crude birth rate. *See* crude birth rate, †general fertility rate, gross reproduction rate, net reproduction rate. *See also* demographic equation.

fertilizer A substance that is added to soil to make it more fertile by supplying the nutrients required by plants. Natural (organic) fertilizers include manure, bone meal, leaves, and decaying vegetation. Modern artificial (inorganic) fertilizers include ammonium nitrate, potassium (from potash deposits), and phosphate. The latter are chemical fertilizers and are most used in countries practising intensive cultivation.

fetch The distance over which a wave is generated by the wind, or over which it has travelled; it determines the height and energy of a wave. †Waves are small at the upwind end of the fetch and they develop in size as the fetch increases. However, with a constant wind there is a threshold beyond which an increased fetch will not produce larger waves.

ffridd A fenced field of rough pasture on a hillside near a farm; an important part of hill farming. Sheep are collected together in the ffridd during the spring lambing season so that the farmer can crop the lower fields. In early summer, when the mountain grass has grown, the animals leave the ffridd for the higher pastures. [Welsh]

fiard (fjard) **1.** An area of water that is surrounded by small islands known as skerries. The coast of Sweden possesses many inlets of this sort.
2. A coastal inlet that possesses less steep slopes and shallower water than a fiord, but which is deeper than a ria. Small islands are found at a fiard's seaward end. It is formed from a sea-level rise along a glaciated rocky lowland. [Swedish]

field 1. An area of land, usually surrounded by hedges, fences, or a ditch, which is used for recreation, pasture, or cultivation.

2. An area that is rich in a particular natural product, such as natural gas, oil, coal, or gold, e.g. Brent oilfield in the North Sea.
3. An outdoor locality in which exploration and research takes place, away from the laboratory, classroom, office, etc. The work done at such locations is referred to as *field work*.

field capacity †The state of the soil once infiltration has ceased and all gravitational water has drained away. Water remaining in the soil surrounds particles as a film or occupies tiny pores as a result of capillarity. Soil moisture that is available to plant roots lies between field capacity and wilting point. Field capacity is expressed as a percentage of the soil's oven-dried weight. *See* capillarity, wilting point.

field system The system of farming practised in Britain and parts of Europe during the Middle Ages. Arable land was divided into strips and members of the community cultivated strips in different parts of the village. Animals were grazed on an area of common land. When the land was cultivated one year and left fallow the next it is called the two-field system; cultivation for two years and fallow for one is called the three-field system.

finger lake An elongated lake that is narrow and relatively straight, usually formed by the damming of water by moraine deposits in a glaciated U-shaped valley. Examples include those of the Lake District in England, the uplands of N Sweden, and the Finger Lakes region of New York State, USA.

fiord (fjord) A glaciated valley that has been inundated by the sea and forms a deep steep-sided coastal inlet, with a U-shaped cross profile. A submerged ridge or sill often marks the change from the deep fiord to the shallower sea lying beyond it.
†Fiords show many of the characteristics of glaciated U-shaped valleys and moraine deposits have often been laid

down on the sill, which may mark the limit of effective erosion by the ice. A fiord may be extremely deep, e.g. Sogne Fiord in Norway is 1234 m deep. Fiords are found in Norway, on the Pacific coast of British Columbia, Canada, and in S Chile. [Norwegian]

fire clay A soft fossil clay that often occurs below coal seams. It may have been the soil in which the swamp plants that eventually formed the coal seam grew. Fire clay can withstand great heat (up to 1600°C) and is used in the manufacture of refractory bricks for the linings of furnaces.

firn (nevé) Snow that has been compressed, but which is not yet glacier ice. A deposit of firn may show a series of thin layers, which represent successive seasonal snowfalls. †Compaction and recrystallization of snow results in an increase in its density and a reduction in the pore spaces between grains; firn exists when the density is greater than 0.4 mg m^{-3}. With further compaction the density is increased and glacier ice formed when it is greater than 0.82 mg m^{-3}. [German: old snow]

firth A sea inlet in Scotland. The term is widely applied to inlets of various sizes and shapes, and it includes the Firth of Forth and the Dornoch Firth. [Scottish]

fissile Denoting the tendency of some rocks and minerals, such as mica, slate, and shale, to split (cleave) along defined planes. *See also* cleavage.

fissure eruption A volcanic eruption in which the vent is linear. There is usually little or no explosive activity. Fluid basaltic lava wells up from the fissure and may flow for many kilometres to form a lava plateau on solidifying. Fissure eruptions have been found measuring about 30 km in length. The eruption that formed the island of Surtsey, near Iceland, in 1963, began in a sea-bed fissure.

fixed capital *See* capital.

fjard *See* fiard.

fjord *See* fiord.

flash 1. A pool or small lake formed where subsidence has occurred due to mineral extraction below the surface. Flashes are common in Cheshire (e.g. Top Flash near Winsford) where rock-salt mining has caused subsidence and the resultant hollows on the surface have been flooded.
2. A sudden rise of water in a stream.

flash flood A quick and sudden flood that occurs in a usually dry valley. In semiarid areas it often occurs after brief but heavy rainfall and may be channelled along wadis. †Flash floods are characterized by concentrated and rapid runoff giving very high discharges over short periods of time.

flat 1. A stretch of land that is judged to be nearly level when compared to the relief around it.
2. An area of low-lying marsh or swamp in a river valley.
3. *See* mud flat.

flatiron A triangular-shaped mass of rock that is left upstanding when a steep ridge is dissected by streams. †Flatirons are often found on the slopes of hog's back ridges and mountain anticlines, for example, the limestone anticlines of the Zagros Mountains of Iran display many flatirons. [The features are so called because they resemble in shape the flatirons once used for pressing clothes]

flax A plant with a pale blue flower that is grown for the fibre obtained from its stem and for its seed, which when crushed produces linseed oil. Flax stalks are processed to provide the fibre that is used to make linen, fine writing paper, and cigarette paper. The main growing regions are Northern Ireland, Belgium and the neighbouring parts of France, the Netherlands, and the USSR.

flint A hard dark-grey form of silica that occurs in chalk as scattered nodules of irregular shape, tabular sheets, or vertical pipes. Although it is very hard it is brittle and this enabled it to be worked by Stone Age man into tools. It is used as a building material in areas in which it occurs.

flocculation In soils, the joining together of very fine colloidal soil particles to form larger masses (i.e. crumbs). †Flocculation results from electrochemical activity on the surface of clay and humus colloids. In agriculture heavy clay soils are improved by the addition of lime, which causes the clay to flocculate. This improves the soil as spaces are left between the larger particles through which air and water can move more freely.

floe *See* ice floe.

flood The submergence of land not usually covered with water, or an increase in the depth of water on land already partially submerged, such as in areas of wet rice cultivation, through a temporary rise in river, lake, or sea levels. It may be caused by increased rainfall, snowmelt, a high tide coinciding with a storm surge, the collapse of a dam, or by movement of the land.
†A river floods when it can no longer contain the discharge from its catchment and the bankfull stage is exceeded. This may be a seasonal event, such as snowmelt, and part of the river's regime: A certain magnitude of flooding may be expected over a particular time period; a flood can thus be labelled according to the frequency with which it recurs, e.g. a one-hundred-year interval flood.

flood control The prevention and regulation of flooding. This may be achieved by the building of dams, storage basins, and embankments combined with the constant monitoring of water levels. The Thames Flood Barrier at Woolwich, for example, is part of a flood-control system for London. It is designed to hold back storm surges in the tide coming in from the North Sea.

floodplain The low-lying land that borders a river and is subjected to periodic flooding. It is composed of deposits of sediment (alluvium) of variable thickness laid down by the flood waters above the rock floor and is bounded by low bluffs.
†The river channel may meander across the floodplain, and it is often raised above the level of the plain by banks of sediment called levées. Rejuvenation of the river may cause the stream to cut down and leave the old floodplain upstanding as a river terrace. Floodplains are usually very fertile.

flood tide The rising tide, which raises the water level from the low-water mark to the high-water mark along coastlines and in estuaries. The flood tide does not necessarily cause flooding, although exceptionally high tides may do so. It may rise very rapidly over gently sloping coasts; for example, at Morecambe Bay in Lancashire the flood tide flows in faster than a man can run. *Compare* ebb tide.

flora The total vegetation or plant life of a given period of time (e.g. postglacial flora), a region (e.g. British flora), or a habitat (e.g. dune flora).

flow 1. The movement of water in a stream under the influence of gravity.
2. The movement of ice in a glacier.
3. The movement of air, which is generally from an area of high pressure to an area of low pressure.
4. The deformation of rock under intense pressure that causes the structure and arrangement of the minerals to be altered but without the rock fracturing.
5. The movement of molten magma.
6. *See* earthflow.

flow diagram (flow chart) A diagram that shows a sequence of interlinked operations, events, etc.

flume 1. An artificial stream channel constructed for an industrial use, e.g. to drive a turbine or wheel for power,

float logs, irrigation, or to wash away soft rocks in a placer mine.
2. A narrow ravine or gorge through which a stream flows.

fluvial Of or relating to a river. The term fluvial is usually applied to aspects of stream flow and erosion by the stream, hence fluvial erosion. *Fluviatile* is usually applied to the results of river action, e.g. fluviatile deposits, or to the plant and animal life living in the rivers, for example duckweed is fluviatile flora.

fluvioglacial (glaciofluvial) Denoting the action and resulting landforms of streams that are derived from melting ice in glaciers or in ice sheets. Examples of features resulting from fluvioglacial activity include the valleys cut by meltwater streams and outwash plains.

focus *See* seismic focus.

foehn *See* föhn.

fog Droplets of water suspended in the lower layers of the atmosphere resulting from the condensation of water vapour around nuclei of floating dust or smoke particles. A visibility of less than 1 km is the internationally recognized definition of fog. *Compare* mist. *See also* advection fog, frontal fog, radiation fog, steam fog, smog.

fogbow *See* rainbow.

fog drip Precipitation from dense fog formed when the water droplets coalesce and become heavy enough to fall to the ground. It is found where the air is very humid and the fall in temperature considerable, especially along desert coasts with cold offshore currents, e.g. SW Africa, California, Chile, and Peru, where, in the driest areas, it may be sufficient to support some vegetation.

föhn (foehn) A warm dry wind that blows down the valleys of the N-facing slopes of the Alps and is most common in spring and autumn.

†It occurs when a depression to the N of the Alps draws air from the S over the mountains. If, when rising over the S slopes, this air reaches its dew point, then it cools at the saturated adiabatic lapse rate, but on its descent down the N slopes it heats up at the higher dry adiabatic lapse rate; thus at a given altitude the air is warmer and drier on the N slopes than on the S. When the föhn starts to blow rapid temperature rises are experienced, often as much as 10°C in a few minutes. In the spring this can cause rapid snowmelt with resulting floods and avalanches, fires as wooden houses are rapidly dried out, and premature development of plants, but it is useful in clearing spring pastures of snow. In the autumn the warm wind is useful for the ripening of crops, particularly grapes. All mountain areas experience a föhn-type wind. *See also* berg wind, chinook, nor'wester, samun, Santa Ana, zonda. [German]

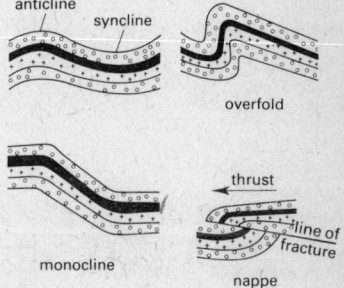

Folds

fold A bend in rock strata resulting from compression of an area of the Earth's crust. The severity of the compression is indicated by the tightness of the folds. Folding occurs along the edge of a continental lithospheric plate where it pushes up against another plate. *See also* anticline, monocline, nappe, overfold, syncline.

folded mountains (fold mountains) Mountains that are formed under compression in which the sedimentary rock strata are squeezed into a succession of synclines and anticlines. Under extremely severe compression the rock folds pile up on top of one another and can break into slices, so that the upper part of the fold moves over the lower part along a fault plane.

food chain A series of links by which food energy is passed from organism to organism within a natural community. Green plants (producers) commonly initiate food chains; they are eaten by herbivores, which may in turn be eaten by carnivores. Other consumers, such as plant and animal parasites, may also form the intermediate links in the chain. An example of a food chain is:
cabbage leaf → caterpillar → blue tit → cat.
†In reality the links are frequently more complex than this. Different food chains are often interlinked to form a *food web. See also* trophic level.

foothills A range of relatively low hills that lie along the margins of a mountain range in an approximately parallel direction. They separate the mountains from a plain or from an area of relatively low relief and may consist of outliers of the rock of the mountains. Examples include the foothills that lie between the mountains of the Andes and the coastal plain of Chile.

Foraminifera An order of microscopic single-celled marine animals (protozoans). The shells secreted by Foraminifera may be calcareous or chitinous, but never siliceous. *See* globigerina ooze.

ford A shallow stretch of river that may be crossed on foot or in a vehicle. Settlements have often developed around the fording points of rivers and the location of a present, or frequently a past ford, is often indicated in place names, e.g. Oxford and Ashford.

forecast In meteorology, a prediction of the weather for a set period, usu-

ally a day, based on observations of present weather patterns and knowledge of the normal development of the systems producing the present weather.

foreland 1. A promontory of land that projects into the sea, e.g. North Foreland and South Foreland in Kent. †*See also* cuspate foreland.
2. †The land that lies adjacent to a folded mountain range, e.g. the Alpine Foreland. The folds of the foreland characteristically lie approximately parallel to those of the mountains.

foreshore The area that lies between the low- and high-water marks along a coastline.

forest A large area of land covered with trees and undergrowth. It is generally more extensive than woodland; consists of either natural or artificially planted vegetation with tall mature trees, which are often of commercial value; and is dense growing forming a continuous canopy. *See also* coniferous forest, equatorial rainforest, gallery forest.

forked lightning An electrical discharge from a thunder cloud observed as downward-pointing and branching flashes.

formal region †An area of the Earth's surface with homogeneous features. Systematic mapping of spatial variations and the study of spatial associations by map comparisons provide the basis for delineating formal regions. A traditional basic central concern of the regional geographer has been the recognition and regionalization of places with similar sets of attributes and areal differentiation from other regions with basically dissimilar sets of attributes. *See also* functional region.

formation An individual layer, or series of layers, of sedimentary rock laid down during one geological age, e.g. a coal seam or a sandstone band. Large formations such as the Chalk can be subdivided into zones on the basis of changing fossil content.

fossil The remains of a plant or animal preserved in sedimentary rock. These may be the whole or part of the organism itself, such as a bone, a piece of shell, a seed, or a leaf, which has usually been chemically altered and replaced by mineral matter, or a trace or impression of the organism, e.g. a wormcast or a footprint of an animal. Rocks of a similar age and depositional environment contain similar fossils and therefore the fossil content can be used to date and correlate layers of rock. The term can also be applied to a feature that has been buried by geological processes, such as an erosion surface.

fractus Ragged or shredded clouds that may be formed either by high winds in the upper atmosphere or at the stage when the clouds are just forming or dispersing (e.g. fractostratus).

free face A vertical or near vertical rock face that is too steep for weathered material to accumulate on it. The debris falls to form cones of scree at the base of the free face.
†The term was specifically applied by W. Penck and Alan Wood to that part of a slope profile that lies between the waxing or convex slope and the debris slope. *See also* standard hillslope.

free port (free zone) A port, or zone within a port, where no custom dues are charged on the bulk of the goods entering or leaving. Duty may be payable on certain goods, e.g. Hong Kong is a free port but duty must be paid on such things as liquor, perfume, cosmetics, and tobacco. Other free ports include Singapore and part of Copenhagen.

free trade area An association of states, the members of which have agreed to trade with one another without import tariffs, quotas, export subsidies, or other government measures to regulate trade. Each member country can impose its own regulations when trading with countries outside the association. The European Free Trade Association (EFTA), founded in 1959, consists of Austria, Iceland, Norway, Portugal, Sweden, and Switzerland, with Finland as an associate member.

freeway In the USA, a major road with few entrances and exits on which there is no toll. The term is used extensively in California, particularly for the city routeways, such as the Santa Ana Freeway in Los Angeles and the Bayshore Freeway in San Francisco.

freeze-thaw A weathering process that involves the freezing and thawing of water in joints, crevices, and spaces in rock and soil. The water expands on freezing and may break up the surrounding rock, which can be removed by the meltwater when the ice thaws.
†Freeze-thaw processes are also responsible for the sorting of coarse from fine materials in periglacial areas leading to the formation of patterned ground and nivation hollows, and the movement of material by solifluction.

freezing point The temperature at which a liquid or gas becomes solid. The temperature at which water freezes is 0°C.

free zone *See* free port.

frequency curve †A graph that shows the number of occurrences of values in each of a series of classes. One way of drawing a frequency curve is first to make a histogram showing the frequency distribution of a set of data. A smooth curve can then be interpolated from this diagram to make a frequency curve. For example, a frequency curve could be drawn to show the distribution of the annual rainfall of a place over a number of years, grouped into millimetre classes on the horizontal axis and the number of occurrences in each class on the vertical axis. *See also* histogram.

friable 1. Denoting a rock or mineral that is easily crumbled and disintegrates when pressed between the fingers, e.g. kaolin.

2. Denoting a soil with a well-defined crumb structure.

friagem A cold period experienced during the dry winter season in the campos of Brazil and in E Bolivia; it may last several days during which the temperature may fall below 10°C. It is caused by the development of an anticyclone over the Amazon area, which draws in cold air from the S. It is of particular importance as the inhabitants, unused to such cold conditions, suffer badly.

friction of distance †The effects that increasing distance from a given point have on human activity and spatial patterns. Since higher costs are normally incurred as distance increases many decisions are taken to minimize the frictional effects of distance, e.g. the location of a factory. August Lösch described this concept as 'the law of minimum effort'.

frigid climate A general term for Arctic-Antarctic type climates or for areas where the surface is snow-covered for a large part of the year and where the subsoil is permanently frozen (permafrost). To early geographers it was one of the three global climatic zones, the others being temperate and torrid.

fringing reef An uneven platform of coral that is separated from the mainland by a narrow shallow lagoon. Its seaward edge slopes steeply into deep water. *See also* coral, coral reef.

front The boundary zone between air masses that have originated in different source areas and therefore have differing temperature and humidity characteristics. Where the boundary zone intersects the Earth's surface it forms a line of separation, which is also called a front and represented by a line on weather charts. A front is usually associated with a trough of low pressure and is very important as a considerable amount of 'weather' is generated along it. *See* cold front,

warm front, occlusion. †*See also* Bjerknes polar front model.

frontal fog A type of fog or fine drizzle that sometimes occurs with the passage of a warm front. Rain falling through the cold sector from the warm air above saturates the cold air and under certain circumstances condensation occurs and the fog forms.

frontal rainfall *See* cyclonic rainfall.

frontogenesis †The atmospheric processes that lead to the formation or development of a front. *See also* Bjerknes polar front model.

frost A weather condition that occurs when the air temperature is at or below 0°C. Moisture on the surface of the ground and objects freezes to form an icy deposit. *See also* glazed frost, ground frost, hoar frost, rime.

frost heaving The disturbance of soil and weathered debris in periglacial areas caused by the freezing of water within them. This results from the water expanding as it freezes. Frost heaving is frequently caused by the formation of large bodies of ice known as ice lenses, which force the overlying material upwards into mounds. †*See also* congeliturbation.

full *See* beach ridge.

fumarole A small vent or hole in the Earth's surface from which hot gases escape. The gases include large amounts of steam, sulphur dioxide, and hydrochloric acid, which are usually forced out under pressure as a jet or plume. Fumaroles are commonly found in areas of volcanic activity, such as at Mount Etna in Sicily and in the Valley of Ten Thousand Smokes in Alaska, USA. [Italian]

functional analysis †A statistical technique used in geography in which phenomena are analysed in terms of their function or role within a particular organization. For example, a group of towns can be ranked according to the various services such as shops and banks they offer to the region in which they are located.

functional region †A geographical unit that is delimited as a result of human organization. One example of such organization is the *city region* – the area around a city that is served and influenced by the city (e.g. shopping facilities).

functional zone In a town or city, an area that is dominated by a particular function (e.g. commercial and business activity, residential buildings, or industries). As a town evolves there is a tendency for different functions to occupy different spatial areas within it. Examples of functional zones include the central business district and the zone in transition (e.g. in the concentric model of urban land use). *See also* urban morphology.

funnel cloud A whirling mass of cloud that forms at the heart of a tornado or water spout. It extends downwards from the base of the low-lying cloud and when fully developed reaches the Earth's surface.

G

gale A term commonly applied to any strong wind. In specific meteorological terms it is a wind of force 8 or over on the Beaufort wind scale, the minimum wind speed at the surface being about 18 m per second.

galeria forest *See* gallery forest.

gallery forest (galeria forest) A long ribbon of dense woodland that fringes river banks in otherwise grassy plains such as savanna. The arching tree canopies meet overhead, forming a leafy tunnel. Such woodland develops in response to the plentiful moisture provided by the river in a generally dry climate. [From Italian and Spanish *galeria*: tunnel]

Gall's stereographic projection A cylindrical map projection in which

the cylinder cuts the surface of the Earth along parallels 45°N and S. These parallels are made their true-to-scale length. The meridians are equidistant parallel lines. It is not an equal-area projection and the least distortion in shape occurs in the mid-latitudes. There is less distortion of polar areas than in the cylindrical equal-area projection. *See* cylindrical projection.

gangue (gang) The valueless waste material (e.g. quartz) that occurs in an ore with the valuable mineral. It is often removed at the point of extraction before the mineral is transported in order to reduce bulk. [French]

gap A break or indentation in the line of a ridge that is commonly caused by river capture across a watershed. *See also* water gap, wind gap.

gap town A town that is situated at the mouth of, or in, a gap in a ridge, for example, Lewes in Sussex, which is sited on the River Ouse in the South Downs. Gaps often provided natural routeways for communications so a gap site was advantageous.

garden city A self-contained urban settlement that is planned and built to give an attractive and spacious environment for its inhabitants. The concept was devised by Ebenezer Howard (1850–1928). The first garden city was Letchworth in Hertfordshire (1903); this was followed by Welwyn Garden City, also in Hertfordshire, in 1920.

garigue (garrigue) A scrub vegetation that occurs on limestone (calcareous soils) in the Mediterranean region. Garigue comprises short-lived sweet-smelling undershrubs such as lavender, thyme, and broom, that can exist on poor soil and also survive the Mediterranean summer drought. Areas of bare stony waste may occur within the vegetation. Little of the original evergreen oak forest remains. Garigue has very limited agricultural use. *Compare* maquis. [French]

garua The local name for fog drip along the coast of Peru. It forms a blanket of thick mist or very fine drizzle and is virtually the only source of precipitation in this very dry region.

gas, natural The gaseous form of hydrocarbons, consisting chiefly of methane, that is found in underground reservoirs in the Earth's crust, often above deposits of petroleum. Natural gas is formed under the same conditions as oil and originates from the decomposition of organic matter. It is used for heating, as it gives off great heat when it burns, and as a raw material in the chemicals industry. Large reserves have been found under the North Sea and these are of great economic importance to the UK.

GDP *See* gross domestic product.

geanticline (geoanticline) An upfold in the Earth's crust that has occurred on a global scale as a result of lateral compression. *Compare* geosyncline.

geest An area of heathland that is underlain by deposits of coarse sands and gravels of fluvioglacial origin. Such areas possess soils of low fertility and are found in N West Germany, Denmark, and the Netherlands. [German]

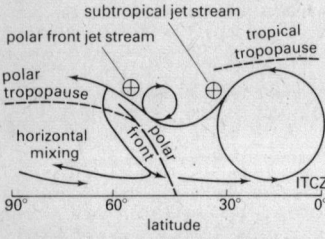

Cross-sectional model of the general circulation of the atmosphere

general circulation of the atmosphere The large-scale movements of air in the atmosphere in the form of pressure cells and wind systems. It is the mechanism by which energy is transferred around the Earth. The driving force of the general circulation of the atmosphere is the differential pattern of radiation received at the Earth's surface. Without the Earth's rotation there would be a thermal gradient between the warm air towards the equator and the cold air towards the poles as more solar radiation is received in low latitudes than at high latitudes. This would result theoretically in air rising along the equator, flowing polewards in the upper levels of the atmosphere, and descending at the poles to return as surface flow in the opposite direction. However, the effects of the Earth's rotation, and the distribution of mountain barriers, land and sea masses, and ocean currents make the circulation pattern far more complex.

†In low latitudes air rises along the intertropical convergence zone (ITCZ), where the trade winds converge, and flows towards the subtropics where it descends to maintain the trade winds. This circulation is known as the Hadley cell. Further polewards the circulation pattern is more complex. It is characterized by the horizontal mixing of contrasting air masses. As the polar front between these air masses varies in its location and strength it is averaged out when represented on models of the general circulation.

See also planetary winds.

general fertility rate (GFR) †In population studies, a commonly used measure of fertility that is based on the number of births per 1000 women in the age range 15–49 (15–44 is sometimes used) during the period under examination.

generative city †A city that stimulates the economic growth of the wider region in which it is located (the opposite of a 'parasitic' city). The city may stimulate growth in a number of ways such as by providing a source of employment, creating a new demand for industrial raw materials, increasing the demand for food from the countryside, and dispersing assembly plants and industries to the surrounding area. As a result, economic development extends over an increasingly extensive area and affects a growing proportion of people outside the city itself. *See also* parasitic city.

geo A long narrow coastal inlet that is formed through marine erosion along a line of weakness (e.g. a joint or fault) in the rock. It may form through the collapse of a cave roof.

geoanticline *See* geanticline.

geochronology †The science of dating geological events. The methods used can be either absolute or relative. Absolute dating gives the actual date BP (before present) that an event took place. The techniques used involve radioactive decay and radiocarbon dating. Relative dating establishes the order of geological events in relation to each other; methods used include varve counting, tree-ring counting (dendrochronology), pollen analysis, and fossil correlation. Other less successful methods have been attempted, based on rates of sedimentation and the increase in the salinity of sea water.

geodesy †The science concerned with the measurement and determination of the shape and size of the Earth, or large portions of the Earth's surface, and of the position of points on the Earth's surface.

geographical horizon *See* horizon.

geographical mile *See* nautical mile.

geographical momentum The tendency for places with well-established installations and services to maintain or increase their importance, even after the conditions that resulted in their original devlopment have disappeared or altered. Once a centre has developed with a large labour force then the infrastructure of roads, housing, and services gives a momentum for continuous development, e.g. Manchester grew when the cotton industry flourished but the city continues to retain its importance today, despite the decline in the importance of the cotton industry.

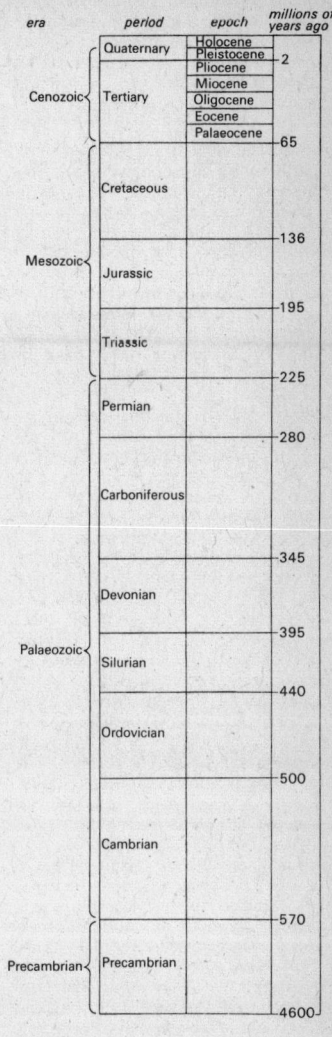

era	period	epoch	millions of years ago
Cenozoic	Quaternary	Holocene	
		Pleistocene	2
		Pliocene	
	Tertiary	Miocene	
		Oligocene	
		Eocene	
		Palaeocene	65
Mesozoic	Cretaceous		
			136
	Jurassic		
			195
	Triassic		
			225
Palaeozoic	Permian		
			280
	Carboniferous		
			345
	Devonian		
			395
	Silurian		
			440
	Ordovician		
			500
	Cambrian		
			570
Precambrian	Precambrian		
			4600

Geological time scale

geography The study of the features of the Earth's surface, including their spatial distribution and interrelationships, and the interaction of man with them. It has conventionally been subdivided into human geography and physical geography. *See also* human geography, physical geography.

geoid The theoretical shape of the Earth based on estimates of its mass, elasticity, and speed of rotation, and ignoring surface irregularities.

geological time The division of time since the formation of the Earth and the relationship of this division to the formation of rocks. Geological time is divided into a hierarchy of time intervals. In descending order these are eras, periods, epochs, and ages.

geology The study of the structure and composition of the Earth. It is divided into a number of branches: cosmology – the study of the evolution of the Earth in space; geophysics – the study of Earth processes; geochemistry – the study of the distribution of minerals and how the rocks formed; mineralogy and petrology – the study of the rocks, mineral deposits, and ores; palaeontology – the study of fossils; stratigraphy – the study of the succession of rocks and fossils in geological time; and palaeogeography – the study of the changes over geological time in surface relief and climate.

geomagnetic field The Earth's magnetic field, which causes a compass needle to align N–S. It is believed that the Earth's crust and mantle are rotating slightly faster than the very dense core, which is composed largely of iron. Following the same principle as a dynamo, this movement generates a magnetic field in and around the Earth. During the history of the Earth there have been periodic complete *magnetic reversals* in which the north magnetic pole became south and vice versa. The alignment of the magnetic field becomes 'frozen' into some rocks at their formation and this relic magnetism has provided strong evidence for the theory of seafloor spreading and hence plate tectonics.

geomorphology The study and interpretation of the origins and development of landforms on the Earth's surface. †It includes the study of the landforms themselves and of the processes creating the landforms. The chief areas covered are weathering, hillslopes and mass movement, fluvial features, aeolian features, glacial and periglacial features, coastlines, and karst landforms. It is a scientific study that involves the increasing use of statistics. Geomorphology involves the study of specific landforms, and an interpretation of how each landform is related to others in space and time. It excludes the study of the major forms of the Earth's surface, such as mountain chains and ocean basins.

geophysics The study of the physics of the Earth, both the crust and the interior, and of the mechanisms and causes of movement (e.g. faulting and folding) in the crust. It also includes the application of physical methods to the study of the Earth, for example the surveying techniques used to search for petroleum, natural gas, and other minerals of economic importance.

geopolitics †The study of the influence of geographical factors on political systems, especially international politics. It was adopted and developed by academics and politicians as *Geopolitik* in pre-World War II Germany. The key concepts were race superiority and the need of a state for *Lebensraum* (living space), and it was used to justify the expansionist aims of the Nazi government. As a result geopolitics became discredited and although it gave rise to a modified form of the subject in the West the term is no longer used in Germany. [Coined by the Swedish political scientist Rudolf Kjellen]

geostrophic flow (geostrophic wind) †Theoretically, a wind produced by the interaction of the forces set up by

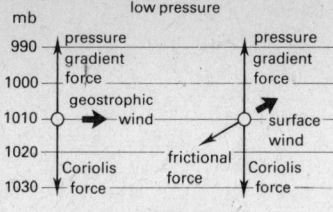

low pressure

mb

990 — pressure gradient force · pressure gradient force

1000

geostrophic wind → surface wind

1010

1020 — frictional force

Coriolis force · Coriolis force

1030

high pressure

(a) geostrophic balance   (b) including friction

Geostrophic flow

the Earth's rotation (the Coriolis force) and the pressure gradient. Such a wind would blow parallel to the isobars obeying Buys Ballot's Law, but this cannot in fact occur because of the effect of friction; near the Earth's surface this causes the wind to blow at an angle to the isobars towards the low pressure area. A close approximation can be found in the upper atmosphere where friction has little effect. The calculation of geostrophic winds can be useful in areas where direct readings are difficult to obtain. [Coined by Sir Napier Shaw in 1915]

geosyncline A downfold or depression in the Earth's crust on a global scale. A geosyncline may be formed by the warping of the Earth's crust over a long time period. The large structural basin formed by a geosyncline may be filled in with sediment derived from the land masses on either side; this infilling process may be accompanied by volcanic activity.

geothermal energy The energy that is derived from the internal heat of the Earth. The temperature increases with depth at approximately 1°C for every 30 m; some of this heat is transported to the surface by geysers, hot springs, and volcanoes. Countries to make use of geothermal energy provided by hot springs and geysers include Iceland, Italy, New Zealand (in Wairakei), and the USA.

geyser A periodic jet of hot water and steam that is ejected under pressure

from a vent in the Earth's crust. Geysers are associated with past or present volcanic activity. Ground water percolates into fissures beneath the surface and in volcanic regions is heated by the surrounding hot rock. Because of the pressure of the water above, the water is heated to above its surface boiling point. Eventually the water changes to superheated steam, which may release the pressure that is built up in the underground fissures or pipes and force hot water and steam out from the geyser with great force. This emission is frequently at regular intervals and the spouting of a geyser may be predicted with great accuracy. Examples include Old Faithful in the Yellowstone National Park, USA, which erupts on average every 65 minutes. [Icelandic]

ghetto Originally, the quarter of a town or city to which Jews were confined by law. The term is now used for an area of a city in which members of a particular ethnic or cultural group are concentrated. *Temporary ghettoes* are segregated areas through which populations become adjusted to new ways of life, for example, immigrants find refuge with people of their own kind before they become adjusted to the country and the city and are diffused through the population. *Permanent ghettoes* are created when cultural groups resist being absorbed into the larger country or community.

gibli (ghibli) A very intense form of the sirocco experienced in Tunisia and Libya. It is a strong, very hot, and very dry southerly wind. [Arabic]

gill (ghyll) A fast-flowing mountain stream. The name is applied locally to such streams in the Lake District of Cumbria and in Yorkshire, e.g. Ashdale Gill and Brow Gill, which flow off Middleton Fell into the River Lune in Cumbria.

glacial drift *See* drift.

glacial lake A stretch of water that is confined between a valley wall and the ice along the margin of a glacier or ice sheet. Examples include the Märjelen See along the margin of the Aletsch Glacier in Switzerland.

†Sediment may be brought into the lake by meltwater and accumulate on the lake bed as distinct seasonal layers, which are known as varves.

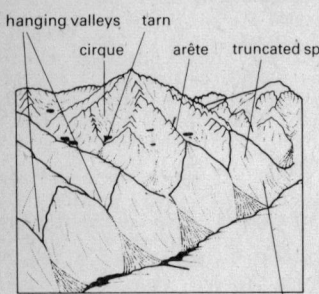

Features resulting from glacial erosion

glaciation 1. The covering of an area by an ice sheet or by a glacier due to a lowering of temperatures, or the processes and landforms resulting from these conditions.
2. A period in the past during which extensive areas were covered by ice, e.g. the Quaternary glaciation. *See* ice age.

glacier A mass of ice that moves under the influence of gravity along a confined course away from its source area. It is formed by the accumulation and compaction of snow, which is transformed to firn (with a density over 0.5) and ultimately to glacier ice (with a density of 0.89–0.90).

†Some glaciers are frozen to their bedrock where their basal layers are below pressure melting point – these are known as *cold glaciers* and are typical of those in Antarctica. They move by internal shearing and deformation. *Warm glaciers* have basal temperatures close to the pressure melting point and are not frozen to their beds. They may contain large volumes of water above, within, and below the ice. This enables the glacier to slide over bedrock and hence have greater potential for erosion and deposition. Warm glaciers include Alpine glaciers, such as the Mer de Glace.

See also cirque glacier, piedmont glacier, valley glacier. [French]

glacier breeze A cold down-valley breeze experienced in mountain valleys occupied by glaciers. The air movement is caused by the chilling and sinking of the air in contact with the ice. It is an example of a katabatic wind.

glacier budget †The relationship between the amount of snow and ice

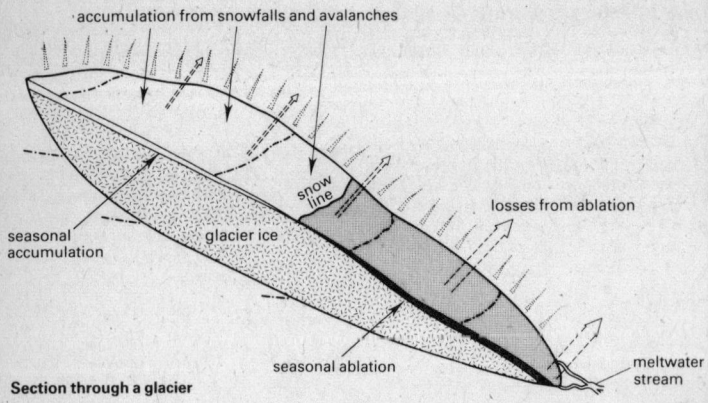

Section through a glacier

that is accumulated by a glacier and that lost through melting or ablation. If the annual accumulation of snow exceeds that lost through ablation the glacier will grow but if the reverse occurs the glacier will stagnate or retreat. The study of the budget of a glacier is important in predicting its future behaviour.

glacier mill *See* moulin.

glacier table A block of rock that is supported on a pedestal of ice on the surface of a glacier. The block protects the underlying ice from the melting effect of the Sun's rays, so that although the glacier surface around it is melted, the ice under the block remains standing as a small pillar. Glacier tables are common features on glaciers but seldom achieve more than one metre in height.

glaciology The study of ice, including the chemical and physical properties, formation, and distribution of ice, and its action on the land surface. It is closely linked to *glacial geomorphology*, which studies the features produced as a result of ice action.

glacis †A gently sloping bank or bench that is found particularly on the sides of mountains. Such banks are also to be found on the walls of fiords in Norway, where they may mark the maximum height that the ice of the glacier that formed the valley attained. [French]

glazed frost (glaze) A covering of smooth clear ice that coats objects and surfaces. The most common way in which this is formed is when rain falls onto a surface having a temperature below freezing point. It also occurs when supercooled water droplets freeze on contact with a surface (e.g. the icing of aircraft wings) and when a fall in temperature causes wet surfaces to freeze. Two effects caused by glazed frost are 'black ice' on roads and the bringing down of electricity and telephone wires by the weight of accumulated ice.

glei soil *See* gley soil.

glen A mountain valley, especially in Scotland, that possesses a relatively flat bottom and steep sides. A glen tends to be narrower and steeper-sided than a strath, although the Great Glen in Scotland is a large-scale feature. The locations of glens are often indicated by place names, such as Glencoe and Glenfinnan.

gleying A soil process that occurs in wet or waterlogged soils. The anaerobic (oxygen-deficient) conditions lead to the process of reduction in which ferric oxide is reduced to ferrous oxide. This gives the soil a blue-grey colour. Mottling in the soil with both bluish (reduced) and reddish (oxidized) patches indicates periodic (e.g. seasonal) saturation with drier periods during which some iron may oxidize back to its ferric state. [From Russian *glei*: clay]

gley soil (glei soil) An intrazonal hydromorphic soil that develops in waterlogged conditions. This may be, for example, because an impervious layer prevents infiltration of rainwater, or in low-lying land subject to flooding. Gleying may affect the whole soil profile or result in a *gley horizon* – a compact layer of bluish-grey structureless clay. *See* gleying.

globe 1. An alternative name for the Earth.
2. A small-scale spherical model of the Earth, usually showing the main surface features, chief towns, and political boundaries.

globigerina ooze A type of ooze deposited on the ocean floor that is composed largely of the shells of globigerina, the most abundant type of Foraminifera. When they die their shells sink to the ocean floor where they mingle with the other deposits. The globigerina give the ooze a calcareous character. It is the most widespread of the oozes, occurring mainly at a depth of between 2750 m and

3500 m, in the Atlantic, Indian, and S Pacific oceans. *See* Foraminifera.

globular projection A map projection that shows each hemisphere as a globe. It has small distortion in shape and is not equal area. Two such projections, one of each hemisphere, are sometimes used side by side in atlases.

gloup *See* blowhole.

GMT *See* Greenwich Mean Time.

gneiss A coarse-grained metamorphic rock with a banded structure. It is formed by the large-scale application of heat and pressure associated with mountain building and volcanic activity. The bands, which are often irregular or poorly defined, may vary in thickness from a millimetre up to several centimetres. Some of the bands may develop concentrations (or 'eyes') of very coarse crystals.

gnomonic projection A type of zenithal map projection in which the projection is made to a plane from a point at the centre of the globe. In the polar case the distance between the parallels increases rapidly from the pole and the equator cannot be shown at all. Any part of a great circle will appear on the map as a straight line, enabling great circle routes to be easily plotted. Once this has been done the route can be transferred to another projection, such as the Mercator. The projection is little used except for charts of polar areas.

GNP *See* gross national product.

gold A rare native metal that is yellow in colour, does not dissolve in acid, and is soft enough to be cut with a knife. It is found in two main types of deposit: (1) in placer deposits, particularly in river beds, derived from the weathering of gold-bearing rock, and (2) in hydrothermal veins, where it is associated with quartz and pyrite (fool's gold). About 60% of the gold output is used for investment purposes and coinage. Other uses include jewellery, in which it is alloyed with other

elements to increase its hardness, and electrical contacts, as it is a good conductor of electricity. Nearly half of the world's gold output comes from the Witwatersrand region of South Africa. Australia, Canada, Ghana, the USA, and the USSR are also major producers.

Gondwanaland The supercontinent of the S hemisphere that is thought to have existed about 200 million years ago following the break-up of Pangaea. It fragmented to form the present-day landmasses of Africa, Australia, India, South America, and Antarctica. *See also* continental drift, Laurasia, Pangaea.

gorge A deep valley with steep and rocky side walls. †A gorge may form through the rejuvenation of a stream; for example, the Rhine Gorge in West Germany was formed by the downcutting of the river during a period of uplift. It may also develop as a result of the underlying rock structure, for example along a fault or as a result of the downcutting of streams through horizontal rock strata, as at the Grand Canyon. Newton Gorge Spillway in Yorkshire originated as a glacial overflow channel. Gorges may also result from cavern collapse in limestone areas (e.g. Cheddar Gorge in the Mendip Hills) or may be dry valleys such as the Winnats Pass in Derbyshire.

Graben

graben A block of rock that has sunk between two roughly parallel faults. †It is frequently regarded as being synonymous with a rift valley but is strictly a structural feature; it thus continues to exist even if the surrounding land is denuded to the level of the graben and no valley remains.

See also block faulting, horst. [German]

grade †The state during which a balance between erosion and deposition exists in a river, and when a balance between the production and removal of debris occurs on a slope. A river may achieve this state by developing a smooth long profile from its source to its mouth. The processes that are active in a river or on a slope tend to work towards establishing this equilibrium state. If the grade of a river is interrupted by base-level change, as in rejuvenation, then the gradient of the river will be adjusted until a new state of grade is reached. Grade in rivers and slopes is closely linked to the base level of erosion and a state of grade exists at a particular base level.

gradient 1. The steepness of a sloping surface, which may be expressed either by the angle in degrees from the horizontal or by representing the vertical distance and the horizontal distance as a ratio as follows:

average gradient between two points = horizontal distance/difference in height
2. *See* pressure gradient.

grain 1. A small particle of rock that is smaller than a granule. A grain (e.g a sand grain) is less than 2 mm in diameter.
2. The general trend or direction of strata, structural features, and relief. For example, the grain of an area such as the Dalmatian coast of Yugoslavia is clear to see in the orientation of its longitudinal valleys.

gram (gramme) Symbol: g A unit of mass in the metric system, defined as 10^{-3} kilogram.

granite A light coarse-grained igneous rock consisting of quartz, feldspar (mainly alkali), and usually mica, with a variety of other minerals in small quantities. It is acid in composition, which results from its high silica content (over 70%). Granites are intrusive rocks and occur in many forms such as sills, dykes, and batholiths. They are hard and resistant and occur as large masses in uplands, which have been exposed by denudation; for example, the tors of Dartmoor are weathered out of the granite batholith that forms the moor.

granular disintegration A weathering process that involves the breaking down of rock into individual grains. Both mechanical and chemical weathering processes may be responsible for the disintegration. Granular disintegration occurs chiefly in coarse-grained rocks. For example, in granite the different minerals that make up the rock expand and contract at different rates when subjected to large extremes of temperature such as occur in deserts. This causes stresses and pressures in the rock leading to its break-up. The process may also occur, for example, in porous rocks such as sandstone containing water that is subjected to freezing. The expansion of the water on freezing causes stresses in the rock leading to its disintegration.

granule A particle of rock that is larger than a grain and possesses a diameter of 2–4 mm. Granules may be either rounded or angular in shape.

graphite (plumbago, black lead) A soft black crystalline form of pure carbon, found in many metamorphic rocks, especially metamorphosed coals and other carbonaceous sediments. It occurs as veins and as lens-shaped masses. In the UK veins of graphite, now exhausted, were worked in Borrowdale in the Lake District. Commercial deposits are found in Sri Lanka, North Korea, South Korea, Mexico, Canada, and the USA. Graphite is used in paint, rubber, and lead pencils, as a lubricant, for making metallurgical crucibles, in electroplating, as electrodes for electrical furnaces, and as a moderator in nuclear reactors.

grassland (natural grassland) A major world vegetation type that is dominated by extensive grassy plains. Natural grasslands occur in both tropical

latitudes (e.g. savanna) and temperate regions (e.g. steppe in Eurasia, prairie in North America, pampa in South America, and veld in South Africa). The widespread growth of trees in such areas is prevented to a certain extent by low rainfall, notably during the dry season. The grasses are adapted to these conditions by temporarily dying back and by long roots that effectively trap available soil moisture. Grasslands merge into forested areas where precipitation is higher and into deserts where it is lower. The temperate grasslands have been largely converted to agriculture, especially grain production. †The grassland ecosystem is structurally simpler than that of forests, having only a field (herb) layer, in which the tallest grasses are dominant. Though soil drought and precipitation levels are important factors in the maintenance of grassland and restriction of forest growth, the grazing by large herbivores and recurrent fire are also contributory factors. Natural grassland is therefore not a truly climatic climax vegetation.
See also pampas, prairie, savanna, steppe, veld.

graticule The network of lines representing the lines of latitude and longitude on a map projection. Only in the simple cylindrical projection does the network form a series of rectangles. The term should not be confused with a grid. See also grid.

gravel A deposit of unconsolidated material, ranging in size from 2 to 60 mm (some authorities give 2–10 mm). In size it lies immediately above sand and below cobbles. The term is, however, frequently used for any loose material in the size range 2–200 mm. The particles are usually water worn and hence rounded, and are derived from more than one type of rock.

gravity model †A model used in human geography that is analogous to Newton's law of universal gravitation, which states that 'any two bodies attract each other with a force that is proportional to the product of their masses and inversely proportional to the square of the distance between them' (i.e. the closer things are to each other and the larger their size, the greater will be the force of attraction between them). The gravity model is used in the investigation of population movements and interaction between places. It is expressed by the formula:

$$Mij = PiPj/(dij)^2$$

where Mij is the potential attraction or population movement between the places i and j, Pi and Pj are the populations of the two places, and dij is the distance between the two places.

great circle A circle on the Earth's surface, the plane of which passes through the Earth's centre dividing it into two hemispheres. The shortest distance between any two points on the Earth's surface is the arc of the great circle that passes through them. Each meridian is half of a great circle and the equator is also a great circle. Compare small circle.

great circle route A route that follows a great circle between any two points; it is therefore the shortest distance on the Earth's surface between the two points. Great circle routes are used whenever possible by aircraft and ships since they cut down travelling time and reduce costs. An example is the great circle route between the UK and Japan, which takes aircraft close to the North Pole.

greco See gregale.

green belt An area of open land and farmland that surrounds an urban area and within which development such as housing and industry is strictly controlled. Green belts are designed to prevent the joining of adjacent urban areas, to preserve the character of historic towns, and to provide truly rural areas close to towns. London is surrounded by an extensive green belt.

greenhouse effect An insolation effect produced by the atmosphere that helps control the temperature of the Earth. Short-wave radiation from the Sun passes relatively easily through the Earth's atmosphere but much of the outgoing long-wave reradiation from the Earth is absorbed by the water vapour and carbon dioxide in the atmosphere. Therefore the effect is very like that of the glass in a greenhouse, maintaining the temperature at the Earth's surface at a higher level than would otherwise be the case. This explains why clear nights are much cooler than cloudy ones and why increasing atmospheric pollution is giving rise to fears that the heat balance of the Earth may be disturbed in the future.

Green Revolution The increase in agricultural productivity of cereals that has taken place since the late 1960s, mainly as a result of the introduction of high-yielding varieties of wheat and rice. Through research and plant breeding scientists produced plants that have significantly higher yields, in some cases doubling the yield and enabling double cropping. The use of fertilizers was also made more efficient. This increase in productivity was aimed to be beneficial primarily to Third World agriculture. It has been mainly limited to wheat and rice crops grown in areas where water supply can be controlled. Increases in yields have been remarkable with the result that some countries, like Mexico and the Philippines, which once had a grain deficit, now have a surplus.

green village A common type of village form in which houses and other buildings are clustered around an open area – the village green. In England, this village form dates from Anglo-Saxon times when woodland was cleared to provide timber for building and fuel as well as grazing room for animals. The grazing of the cleared land destroyed fresh seedlings and prevented regeneration. The village houses were grouped around the clearing and surrounded by fields. *Compare* street village.

Greenwich Mean Time (GMT) The mean local time at Greenwich, located on the 0° meridian. This is used as the standard time for the UK and parts of W Europe. The standard times of different areas of the world are calculated from this; most countries have standard times that are an exact number of hours or half hours ahead of or behind Greenwich Mean Time (15° longitude represents one hour in time).

Greenwich meridian (prime meridian, standard meridian) The 0° meridian of longitude that passes through Greenwich Observatory in London. All other meridians of longitude are based on it. *See also* meridian.

gregale (grégal, greco) A strong north-easterly wind that blows mainly in the winter in the S and central Mediterranean areas. It is associated with relatively high pressure over central Europe and the Balkans and the passage of depressions to the S over Libya.

grey-brown podzolic soil †A zonal soil of cool moist climates that is intermediate in character between true podzols and brown earths. The soil is less leached and hence is slightly less acid than podzols and as a result contains more soil fauna, which incorporate more humus into the A horizon. The B horizon generally has a brown blocky structure and an accumulation of illuvial clay. Grey-brown podzolic soils are found in the NE USA, the British Isles, and W Europe. *Compare* brown earth, podzol.

grey desert soil *See* sierozem.

greywether *See* sarsen.

grid A network of parallel horizontal and vertical lines forming squares superimposed on a map in order that places may be precisely located. On Ordnance Survey (OS) maps the lines are numbered eastwards and northwards from an origin and form the national grid, which enables the position of any place in the UK to be identified by a series of coordinates known as the *grid reference*. For example, the grid reference of the Royal Pavilion at Brighton is TQ 313042. *See also* coordinates, national grid.

grike (gryke) A crevice of variable depth that dissects a limestone pavement. These mark the boundaries of the flat pavement surfaces known as clints. Grikes are solution features that have been produced by the carbonation of limestone. The joints in limestone concentrate runoff and so are the sites of maximum solution along which grikes develop, often up to 3 m deep. Grikes and clints are characteristic of karst scenery and are found, for example, in the limestone pavements near Malham Cove in West Yorkshire and in the Burren region of W Ireland. *See also* clint, karst, limestone pavement.

grit A coarse-grained sandstone in which the particles are either angular or of unequal size. The term is also used as a proper name for Millstone Grit.

gross domestic product (GDP) The total value of goods and services produced by a country over a period of time, normally a year. Goods and services are valued at market prices and goods used for final consumption are included; goods used in the production of other goods, such as steel, are excluded. The word gross is used because there is no deduction for replacement capital goods such as new machinery.

gross national product (GNP) The gross domestic product of a country plus the income obtained from investment abroad by domestic residents, less income earned in that country from investments made by foreigners abroad.

gross reproduction rate (GRR) †A measure of fertility based on the average number of female babies born per 1000 women during their reproductive lifetimes. The reproductive period is usually taken as 15–49 inclusive. The GRR is expressed as a ratio. A GRR of over 1.0 should mean that the female population is more than capable of reproducing itself with an equivalent number of daughters. However, this generalization is not applicable to regions with high infant mortality rates since many daughters will not reach the age their mothers were when the daughters were born. *See also* net reproduction rate.

grotto A cave or cavern, which is found in either rock or ice. The term is often applied to caves created artificially for ornament.

ground fog A type of radiation fog usually found in hollows and valleys, the top level of which does not reach the cloud base level.

ground frost A temperature of 0°C or less on the surface of the ground, although the air temperature may be above freezing. It is particularly dangerous to new plants.

ground moraine A sheet of till deposited when the ice of a glacier or ice sheet melted. It has a gently undulating upper surface and may be tens of metres thick. It may be formed partly from the moraine that was carried at the base of the glacier or ice sheet and partly from moraine that was carried within or on the ice surface and let down when the ice melted. *See* moraine, till.

groundnut (peanut) A plant that is cultivated in subtropical countries for its seeds, which can be eaten or crushed to produce an oil (used in margarine). The seeds are contained in pods, which develop just below the surface of the ground. The green leaves of the plant serve as animal fodder. The main producing countries are the USA, India, SE Asia, Nigeria, The

Gambia, and the Central African Republic.

ground water Water that is contained in the soil and underlying rock. †Ground water may be derived from rain water that has percolated down or from water that was trapped within the rock during its formation. The water percolates down to collect above impermeable layers of rock; eventually all the pore spaces above this layer become saturated with water forming the *ground-water zone*, the upper surface of which is the water table. This may emerge at the surface as a spring.

growing season That part of the year in which plants maintain growth. Temperature is a major determining factor and the growing season is often defined as the period between lethal frosts. Maize, for example, requires 150 frost-free days for its growing season, and cotton 200 frost-free days. At the equator plants are able to maintain growth all year and no growing season can be recognized. The growing season is about two–three months in polar latitudes. *See also* critical temperature.

growth pole †An area that develops and expands economically more rapidly than the surrounding areas. The concept recognizes that economic development will not occur uniformly over a particular region but concentrates in particular locations. The initial growth attracts other linked industries and commercial activities because of the external economies created. This leads to expansion of the growth pole and intensification of its locational advantages. The concept has been used in economic development plans on the assumption that economic concentration will stimulate greater economic development than would spreading resources more widely.

groyne A construction that is built at right angles to a coast to prevent the movement of sand and shingle across a beach by longshore drift. †Groynes

are important elements in coastal protection schemes and are common on beaches in S England where they cause an accumulation of sand and shingle on their W sides. This is because the prevailing winds are from the SW and cause the sea to move material eastwards.

gryke *See* grike.

guano The accumulated excrement of certain animals, especially seabirds and bats, which can be used as a fertilizer. Large deposits of seabird excrement are found on the Peruvian and Chilean offshore islands where there are desert conditions with little or no rain to wash the material away. Guano is rich in phosphorus and extraction is carefully controlled to conserve the diminishing resources. [Spanish]

gulch In the W USA, a deep narrow rocky gully. The name is usually applied specifically to those gullies where gold prospecting has taken place among the alluvial (placer) deposits.

gulf A sea inlet that is larger and more enclosed than a bay and penetrates further inland. Many rivers may empty into a gulf and cause complex depositional features to form within it. For example, the Gulf of Mexico.

gully An incised water-worn channel, which is particularly common in semi-arid areas. It is formed when overland flow down a slope, especially following heavy rainfall, is concentrated into rills, which merge and enlarge into the gully. *Gully erosion* is the erosion of soil and rock by the concentration of runoff into gullies. Large volumes of soil and debris may be removed, especially where the soils and underlying rock are soft. Removal of protective vegetation and ploughing downslope can also contribute to this form of erosion. Its effects can be prevented by contour ploughing. Gully erosion occurred on a large scale in the Midwest of the USA during the 1920s when large areas were ploughed up and the vegetation removed for agriculture.

Gutenberg discontinuity (Weichert-Gutenberg discontinuity) A boundary surface within the interior of the Earth that separates the mantle from the core. It lies at about 2900 km below the surface of the Earth and marks the depth at which the S waves of earthquakes cease to be transmitted. [Named after the US seismologist Beno Gutenberg who discovered it]

gyrocompass A type of compass that consists of a gyroscope rotated electrically so that the axis of rotation is parallel to the Earth's axis of rotation. Since the Earth's axis of rotation is the imaginary line between the poles, the axis of the gyroscope will point to true north and south. The gyroscope is more accurate than the magnetic compass and is used for navigation by ships and aircraft.

H

haar A sea mist or fog experienced along the E coast of Great Britain, usually in summer, when an easterly wind is blowing.

habitat That part of an environment in which a plant or animal lives and which offers the conditions favourable for its existence. It includes the climate, vegetation, topography, and other conditions of the area. Very small-scale divisions of the environment, such as under the bark of trees, are called *microhabitats*. *See also* environment, †ecological niche.

haboob A duststorm experienced in the Sudan, mainly during the summer months. The swirling dust and strong winds are often accompanied by a rapid fall in temperature and heavy rainfall with thunder. [Arabic]

hachures Lines sketched on maps to give an impression of the relief. The lines run down the slopes, and are

closer together or thicker where the slopes are steeper.

hacienda A large privately-owned agricultural estate, with a house for the owner or manager. They are found in Spain and existing or former Spanish colonies and employ landless peasants who live on the estates. In Peru some of the estates have been taken over by the government and operate as workers' cooperatives, e.g. the Paramonga haciendas producing sugar N of Barranca. [Spanish]

Hadley cell One of the main circulation cells in the Earth's atmosphere, extending from the equator to about latitudes 30°N and S. Warm air rises at the equator, moves poleward at high altitude, descends at about 30°N and S, and returns towards the equator at the surface. [Named after G. Hadley who attempted an explanation of the movements of the trade winds in the 18th century]

haematite (ferric iron oxide, Fe_2O_3) The principal ore of iron, containing over 70% iron. It occurs chiefly in sedimentary rocks, in which it is derived from altered iron carbonates and silicates, and produces the red coloration in sandstones. Haematite occurs in a number of different forms, such as crystalline (specular iron ore), powdery (red ochre), and in the form of lumps (kidney ore). The world's greatest production comes from the deposits around Lake Superior in North America; other major deposits include those of Minas Gerais (Brazil), Cerro Bolívar (Venezuela), and Labrador and Quebec (Canada). In the UK it is found in Lancashire, W Cumbria, and the Forest of Dean. [From Greek *haema*: blood]

haff A shallow brackish coastal lagoon, especially along the Baltic coast of East Germany, Poland, and the USSR. The feature is formed by the extension of a sandspit, known in Germany as a *nehrung*, across the mouth of a river and by the flooding of nearby low-lying land by the sea. [German]

hail Precipitation in the form of pellets of ice (*hailstones*) that develop in and fall from a cumulonimbus cloud, either at a cold front or where intense heating of the surface causes rapidly ascending convection currents. The hailstone develops in the updraught of air as water vapour freezes onto the surface of a nucleus or 'embryo' of ice in the cloud. When it has grown sufficiently its weight overcomes the force of the updraught and it falls. There is a great variation in size and shape from the normal of about 5 mm in diameter to giants weighing over 1 kg, due to differing conditions during formation, all of which are not yet fully understood. Hail storms can cause a tremendous amount of damage, particularly to crops.

halite *See* rock salt.

halo A ring (of rings) or light seen around the Sun or Moon. It is caused by the refraction of the light by ice crystals or water droplets in high thin clouds. It is usually white but when it is well developed it may be faintly coloured, shading from red on the inner edge to blue on the outer.

halophyte †A plant that can tolerate a relatively high salt content in its environment. Examples include salt-marsh vegetation such as *Spartina* and *Salicornia*, which may be covered regularly at high tide.

hamada (hammada) An extensive flat rocky surface in a desert from which sand has been removed by the wind. Hamadas are found in the Sahara and Gobi Desert. *See also* erg, reg. [Arabic]

hamlet A small settlement consisting of a group of houses that is too small to be termed a village. In the UK, the term is also applied to a village without a church that forms part of a parish centred on another village.

hanging valley A tributary valley that lies above the main valley and is separated from it by a steep slope down which the stream may flow as a waterfall or series of rapids. Hanging valleys are common in areas that have been glaciated. When spurs were truncated during glaciation the stream valleys lying between them were cut back and left hanging above the level of the main valley.

Hanging valleys may also be formed during the retreat of a coastline under rapid erosion. This is illustrated by the cliff face of the Seven Sisters in Sussex.

harbour A stretch of water where ships can shelter to obtain protection from storms. Harbours may be natural features such as sheltered bays or inlets, e.g. Sydney Harbour. Some harbours are man-made, usually by the building of extensive groynes and breakwaters, e.g. Tema Harbour in Ghana.

Mohs scale

hardness	reference mineral
1 (softest)	talc
2	rock salt or gypsum
3	calcite
4	fluorite
5	apatite
6	orthoclase feldspar
7	quartz
8	topaz
9	corundum
10 (hardest)	diamond

hardness scale (Mohs' scale) A scale devised by the mineralogist Friedrich Mohs in 1812 as a guide to the hardness of minerals. This determination of hardness is one of the most important tests in the identification of minerals. Ten test minerals are arranged in order of increasing hardness and given numerical hardness values. An unknown mineral is classified by seeing which test minerals it scratches or is scratched by; for example, if a mineral scratches quartz but not topaz it has a hardness between 7 and 8.

hardpan A hard compact layer at the surface of the soil or at a lower horizon within the soil profile. It may consist of concentrations of particles transported by soil solution or chemical compounds precipitated from solution. *See also* claypan, ironpan, duricrust.

hard water Water that contains calcium, iron, and magnesium salts in solution. Hard water forms scale in kettles and hot water pipes and does not easily form a lather when used with soap. Water that runs off limestone and chalk tends to be hard as it contains salts derived from the solution of the rock.

hardwood A broad-leaved tree that provides relatively heavy and hard-wearing close-grained timber, or the wood obtained from this tree. Both tropical (e.g. teak) and temperate (e.g. beech) species of hardwoods exist. Much temperate hardwood forest has been replaced by faster-growing softwoods. *Compare* softwood.

hariq The burning of wild grasses to fertilize the ground just before a crop is sown. It is practised in the Sudan where the wild grass has been allowed to grow as part of the rotation pattern. [Arabic]

harmattan A W African wind blowing SE from the Sahara towards the coast. It is hot and often dust-laden, particularly inland, but its dryness helps to relieve the stifling humidity as it nears the coast. In January it may reach as far S as 5°S but in July it rarely reaches 15°N.

Hawaiian eruption A volcanic eruption in which basic lavas of low viscosity and little explosive activity are extruded in large quantities to form large gently-sloping shield volcanoes, such as the Mauna Loa, Hawaii. *See also* shield volcano.

hazards, natural *See* natural hazards.

haze In meteorology, an obscurity of the lower atmosphere that limits visibility to under 2 km but over 1 km. It is normally formed by water particles that have condensed around nuclei in the atmosphere, but may also be a result of particles of smoke, dust, or salt in the air. The term is also used for other phenomena that limit visibility (e.g. heat haze).

headland A promontory of land that projects into the sea. The headland often possesses steep rocky cliff faces, e.g. Flamborough Head on the Humberside coast is faced with almost vertical chalk cliffs.

head of navigation The farthest point up a river that can be reached by ocean-going ships. Over the centuries, as ships have increased in size, the head of navigation has moved downstream. Settlements were sited at the original highest navigable point, e.g. Norwich on the River Yare and Bristol on the River Avon.

headwall (backwall) The steep rock wall at the back of a cirque. *See* cirque.

headward erosion The erosion at the source of a stream in an upstream direction. Headward erosion frequently results from spring sapping, in which water undercuts the slope from which the spring flows. This leads to the development of subsurface pipes and tunnels, which eventually become exposed as streams as they cave in. As the source of a stream recedes under headward erosion it may breach the watershed and capture streams from the other side. *See also* river capture.

headwater The upstream reach of a stream that is close to its source. The term headwaters is commonly used when referring to the streams within the drainage basin of a river that contribute to its source.

heartland The central area of a country or continent. The term was used originally by the British geographer Sir Halford Mackinder (1904) to describe the area of the inhabited world inaccessible to sea power and therefore removed from coastal political power. He envisaged a 'world island' consisting of Eurasia and Africa, the core of which was the heartland.

heat balance The state of equilibrium that exists on Earth between incoming radiation, mainly from the Sun, and outgoing reradiation and reflection from the Earth. It is a worldwide average balance as low-latitude areas receive more radiation than they lose, while the opposite occurs nearer the poles, the heat necessary to preserve the balance being transferred by the movement of air masses and ocean currents.

heath (heathland) An open uncultivated lowland area in temperate regions. It usually has podzol soils (poor acidic soils) overlying sandstones or acidic sands and gravels. It is characterized by such vegetation as heather (*Calluna*) and related species of heath (*Erica*), gorse (*Ulex*), and scattered trees (e.g. birches). Damp heaths (i.e. those occupying hollows or on peaty soils) have sphagnum mosses.

heat island An urban area where temperatures tend to be higher than those of the surrounding countryside. This is the result of several factors, including escaping heat from buildings, reflection and radiation from concrete, tarmac, and bricks, and heat emitted by motor vehicles circulating in the urban area.

heat wave A continuous spell during which temperatures are considerably higher than the average for the time and place. The term has no official definition in meteorology.

Heaviside layer *See* E-region.

heavy industry The production of goods which, when compared with other products, are heavy and normally bulky. The most important of the heavy industries is the manufacture of iron and steel. Industries that use large quantities of iron and steel, such as shipbuilding and some branches of engineering, including locomotive and boiler making, are also regarded as heavy industries.

hectare Symbol: ha A metric unit of land area, equal to 10 000 square metres.

helm wind A local easterly or northeasterly wind affecting the Eden Valley of Cumbria to the W of the Pennines. It is a strong gusty wind that blows down the flanks of Crossfell forming an inversion of temperature at a height of about 3000 m; this produces an unusual cloud effect with a banner cloud (the helm) above Crossfell and a line of cloud parallel to the ridge a few kilometres to the W (the helm bar).

hemisphere Half a sphere. A plane passing through the centre of the Earth will divide it into two hemispheres. The plane of the equator divides the Earth into the N and S hemispheres. The plane passing through 30°W longitude and 150°E longitude roughly divides the Earth into the W hemisphere (containing the Americas, or the New World) and the E hemisphere (containing Europe, Asia, and Africa, or the Old World).

hemp One of a number of plants that produce strong coarse fibres. *True hemp* (*Cannabis sativa*) is an annual plant, which grows to a height of 2–5 m. It is grown commercially in central and E Europe, N Italy, central China, and Korea for its fibre, which is obtained from the stems and used for making rope, twine, and matting. The plant also produces the narcotic drugs hashish and marijuana. Other plants producing hemp are known as *tropical* (or *hard*) *hemps* and the fibre is obtained from their leaves. These include *abaca* (or *Manila*) *hemp*, which is grown mainly in the E Philippines, Central America, and Colombia; *sisal hemp* grown in E Africa, Brazil, and Haiti, and the closely related *henequen*, which is grown in Mexico and Cuba. Henequen and sisal have thorny sword-shaped leaves 1–2 m long, which absorb water during the dry season.

Hercynian The period of mountain building that took place during the Carboniferous and Permian periods in N Europe. It gave rise to the Variscan fold belt, which runs from SW Ireland to the Sudeten Mountains in E Europe. [Derived from the Harz Mountains in Germany]

heritage coast A stretch of coastline in the UK that has been identified as being subject to demands on land use which require careful planning. The aim is not only to protect the landscape but to provide for the conflicting interests of farmers, conservationists, and visitors. Such areas include parts of the East Sussex coast and the South Glamorgan coast.

heterotroph An organism that derives its energy from organic material, in the form of greenstuff or animal matter. †Heterotrophs therefore need a biotic environment, and belong to the higher trophic levels in an ecosystem. They include consumers such as herbivores and carnivores, and decomposers such as bacteria and fungi. *Compare* autotroph.

hierarchy of central places *See* central place hierarchy.

high In meteorology, any area in which atmospheric pressure is higher than in the surrounding areas. The term is used in a more general sense than an anticyclone, which is centred on a closed isobar. *See also* anticyclone, ridge of high pressure.

highland A region that is at a higher elevation than adjacent areas.

high-rise estate A planned housing estate consisting of high-density multistorey housing blocks. In the UK, high-rise estates have been constructed chiefly by local authorities. Many were built between 1955 and 1970 with the intention of speeding up slum clearance and urban renewal. It was believed that high-rise development

left more land for recreation and landscaping and, by making possible the rehousing of population at high densities, it alleviated land shortage in the cities. Since about 1970 the increased awareness of social disadvantages has virtually ended the building of such estates.

high-water mark (high water) The highest level that the sea reaches on an incoming or flood tide. The high-water mark varies throughout the year and is marked on British Ordnance Survey (OS) maps as the high-water mark medium tides (HWMMT).

hill A rounded elevation of the Earth's surface, which is of lower altitude than a mountain and with less steeply inclined sides. Examples include the Chiltern Hills and Cotswold Hills in England.

hill shading (plastic shading) A method of showing relief on maps by shading some of the slopes. Normally the shading is superimposed as if the area is being illuminated from the NW so that slopes facing S and E are in shadow and therefore shaded. Sometimes vertical lighting is used with steeper slopes having darker shadows. The method is often used in conjunction with contour lines and layer tinting and gives the effect of an illuminated relief model.

hill station A settlement or resort at a high altitude in the tropics. Because of its altitude the station has lower temperatures than the plains during hot seasons. In India, hill stations (e.g. Simla) were established by the British.

hinterland The area that has close economic, social, and cultural ties with a central place, such as a port, city, or town. In the case of a port it is the region inland from which it receives trade or to which it sends goods, e.g. the hinterland of Rotterdam extends to Switzerland. Cities and other settlements have trade areas surrounding them, e.g. York serves as a centre for shopping and other services for people from neighbouring parts of North Yorkshire, West Yorkshire, and Humberside.

histogram A type of graph that represents frequency distributions by means of rectangles or bars. The widths of the bars are the class intervals, the heights are the frequency. Histograms are frequently used, for example, to show the incidence and amounts of rainfall over periods of time. †*See also* frequency curve.

historical geography The branch of geography concerned with the past. The two main aspects of this are the reconstruction of past environments at a particular point in time and the study of the sequence of changes that take place with the passage of time at a place.

historical geology *See* stratigraphy.

hoar frost A form of frost that occurs when the dew point of the atmosphere is below freezing point. If the ground or objects are cooled below this dew point tiny crystals (spicules) of ice form directly from the atmospheric water vapour and are deposited. It is white and in the USA is often called 'white dew'.

hogback (hog's back) A narrow steep ridge in which the strata are either tilted vertically or dip very steeply, resulting in both the front and back slopes being equally steep. Resistant rock with this structure may be left upstanding as a ridge when less resistant rock around it is removed by denudation. The Hog's Back near Guildford in Surrey is a well-defined ridge of this type.

holding Agricultural land that is being cultivated by an owner, tenant, or manager as one unit. Sometimes a number of farms are worked as one holding. In the UK units of only a few hectares are regarded as smallholdings and not as farms.

holiday resort A place that is frequented by people on holiday. Many resorts have coastal locations, e.g. Scarborough, Nice, Atlantic City. However, a number of inland towns attract tourists because, for example, of their leisure facilities (e.g. Las Vegas), scenic surroundings (e.g. Windermere in the Lake District), or historic interest.

holistic †Denoting a philosophical outlook or attitude that maintains that small units tend to combine to form 'wholes' greater than the sum of the constituent parts. A holistic view of a problem aims to see it as a whole, to synthesize rather than analyse. Geography is said to have this holistic approach.

Holocene (Recent) The most recent geological epoch of the Quaternary period, covering the 10 000 years or so from the end of the Pleistocene epoch to the present day. A rise in sea level resulted in the English Channel being flooded, thus separating Britain from the rest of the continent of Europe. The deposits from this period contain most of the evidence of early man. It is also sometimes known as the Postglacial epoch.

homestead 1. A type of rural settlement consisting of dispersed farms or houses as distinguished from a nucleated hamlet or village.
2. A house or estate and the land that adjoins it, especially a farm.
3. In the USA, a farm occupied by the owner and his family, especially a farm of 160 acres originally granted to a settler under the 1862 Homestead Act.

homolographic projection A type of map projection in which the area within two adjoining parallels and meridians is equal to any other area similarly enclosed, i.e. the projection is equal area and preserves the ratios between an area on the map and the corresponding area on the globe. An example is Bonne's projection. *See* Bonne's projection. *See also* equal-area projection.

homoseismal line A line on a map joining points that are affected simultaneously by an earthquake shock. The line may be either roughly circular or elliptical, depending on whether the source of the earthquake was a point or a line such as a fault.

honeycomb weathering The weathering of rock resulting in a surface of small sharp ridges separating deep pits that closely resembles a honeycomb. The process frequently occurs in rocks along a coastline where water-filled hollows are enlarged by physical and chemical weathering to produce the characteristic pits and ridges.

hook 1. A curved spit of land resembling a fish-hook that projects from a coastline into the sea, e.g. the Hook of Holland in the Netherlands.
2. A sharp bend or elbow in the course of a river or in the line of a valley.

horizon 1. *celestial horizon* (true horizon, astronomical horizon) A great circle on the celestial sphere, the plane of which passes through the centre of the Earth and is parallel to the geographical horizon.
2. *geographical horizon* (apparent horizon, sensible horizon, visible horizon) The boundary of the Earth's surface visible from a certain point, i.e. the line at which land (or sea) and sky appear to meet. The distance of the horizon is determined by the height of observer. Obstacles that interrupt the view do not form part of the horizon.

horizon, soil *See* soil horizon.

horn *See* pyramidal peak.

horse latitudes Zones of high atmospheric pressure occurring over the oceans in latitudes about 30°–35°N and 30°–35°S, between the belts of the trade winds and the westerlies. They are areas of descending air and are characterized by comparatively dry quiet stable conditions. The name may originate in the practice, in the days of sail, of throwing some of the cargo of horses overboard when fodder ran

short during long periods when ships were becalmed in these zones between Europe and America.

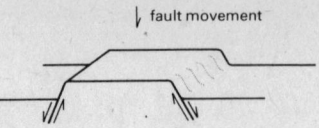

Horst

horst An elevated block of rock bounded by faults. It is left upstanding either through uplift of the block between the faults or through subsidence of the surrounding land. The horst may be denuded to the level of the surrounding land so that only the structural form remains. Examples include the Vosges of E France, the Black Forest of S Germany, and the Sierra Nevada of the USA. *See also* block faulting, graben. [German]

horticulture The cultivation of flowers, fruit, or vegetables in small plots using intensive methods of farming. The most intensive form of horticulture is probably the cultivation of crops such as tomatoes or cucumbers under glass in a controlled environment. Less intensive forms include the cultivation of daffodils in small fields in Cornwall and the Scilly Isles and fruit growing around Evesham and in the Vale of Kent.

hot spring (thermal spring) A spring of hot water that flows out of the ground. Unlike a geyser, the water is not forced out under great pressure and flows continuously instead of intermittently. In many places hot springs are associated with past or present volcanic activity, e.g. in Iceland and New Zealand. They are also found in nonvolcanic areas, e.g. at Bath, Avon. The water that flows from hot springs often contains a large proportion of dissolved minerals, which may be deposited as basins or terraces around the hot springs.

huerta A small intensively cultivated plot using irrigation water, found in the Mediterranean coastlands and the Ebro valley of Spain. The plots are used for crops such as tomatoes, cotton, oranges, and vegetables. Huertas can yield two or more crops a year and are crisscrossed by numerous irrigation channels. *See also* vega. [Spanish]

human geography The geographical study of phenomena of the Earth's surface which are due, or relate directly, to the activities of man. It is the study of man's reciprocal relationship with his environment. The main branches of human geography are economic geography, historical geography, political geography, population geography, regional geography, social geography, and urban geography.

humic acid An organic acid that is formed when water passes slowly through a mass of decaying vegetation (humus). †The acid is effective in the chemical decomposition of rocks and formation of soils. This efficiency is aided by the action of the humus in retaining the acid so that minerals are subject to decomposition over long periods. Mosses and lichens on rock surfaces may develop small saucer-shaped depressions under them that are the result of weathering by humic acid.

humidity The amount of water vapour present in the atmosphere. This may be expressed in several different ways. *See* absolute humidity, relative humidity, saturation.

hummock A low rounded mound of earth, rock, or ice. Hummocks are small-scale features and may be formed by numerous different processes. Hummocky ground is a common feature in areas that were formerly glaciated.

humus The dead organic content of the soil. Humus comprises both vegetable (e.g. plant litter) and animal matter, which has been decomposed beyond recognition to a dark-coloured structureless material. It is characteristic of

the A horizon of many soil profiles. †*Humification* – the formation of humus, mainly through bacteriological activity – requires aerobic conditions in soils. Humus may be discrete, i.e. in the form of fibrous masses partially decomposed and incompletely mixed with the mineral fraction, or intimate, i.e. in the form of tiny particles inseparably bound to minerals in a clay–humus complex. This colloidal state improves the soil structure by binding loose light sands and dividing heavy clays, thus improving the water retaining capacity of these textures. Humus also increases soil fertility by providing essential nutrients such as nitrogen and trace elements. Different vegetation generates humus of different properties.
Compare litter. *See also* moder, mor, mull.

hurricane 1. A wind that reaches force 12 on the Beaufort wind scale, i.e. it has a velocity in excess of 32.7 m per second.
2. A tropical cyclone occurring around the Caribbean Sea and Gulf of Mexico. It usually develops in the Atlantic to the E of this area and tracks W across the West Indies and the E coast of Central America, often turning to the NE to affect the Gulf and Atlantic coasts of North America. Hurricanes are most frequent in September and October and their passage often results in severe damage due to the high winds and the destructive seas generated by them. *See also* cyclone, typhoon, willy-willy.

hydration A weathering process in which water is taken up by the minerals of a rock. This may cause considerable expansion of the minerals. The stresses caused in the rock by alternate wetting and drying may lead to its disintegration. Hydration is particularly effective in clays where it causes the break-up (slaking) of the clay. Certain minerals change in character when subject to hydration; for example, anhydrite may be altered to gypsum and the feldspar constituent of granite may be altered to clay minerals.

hydraulic force The erosive force of moving water, which depends for its efficiency upon the volume of water involved, the velocity of the water, and the nature of the surface on which the force is exerted. The load carried by the water is not involved in producing this force. Hydraulic force is most effective on highly jointed and creviced rock where the water impounds and compresses an air pocket. The pressure exerted in this way may gradually break up the rock. Hydraulic force is an important component of both fluvial and marine erosion.

hydraulic geometry †The study of the relationship between the cross section of a stream channel and variations in the volume and velocity of the water. Channel width, depth, and the slope of the river bed, together with the roughness of the bed and banks, help to determine the stream velocity. The channel cross section is important in its effect upon the movement of load and water, i.e. a wide shallow cross section is less efficient than a narrower deeper one.

hydrocarbon A compound of hydrogen and carbon. Different types of hydrocarbons form the economically important deposits of peat, coal, petroleum, and natural gas.

hydroelectric power (HEP) Electric power generated from the energy of falling water. The water drives turbines, which in turn drive electricity generators. Conditions suitable for a power station site include high annual precipitation; a river supplied with precipitation from a wide catchment area; an existing lake, or a valley narrow enough to dam in order to create a reservoir to guarantee regular water supply during dry periods; a good drop in height between lake and power station (known as the head of water); a relatively frost-free climate; and a location not too far from large-scale electricity demand. Pump-storage power stations use electricity available from elsewhere at periods of off-peak demand to pump water back up into the reservoir, which is then available for generating electricity during times of peak demand.

hydrogenation of coal The manufacture of oil from coal with the use of hydrogen. In normal circumstances oil drilled from the ground is cheaper and more plentiful than oil produced from coal. The process was used by Germany during World War II because of a shortage of crude oil.

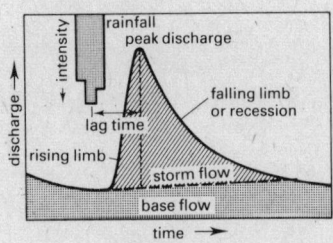

Components of a hydrograph

hydrograph †A graph that shows the variation in the level, velocity, or discharge of a body of water such as a stream with time. It usually shows both the level of baseflow, which originates chiefly from the groundwater supply, and the stormflow (quickflow), which is the storm runoff. The peak of discharge occurs after the most intense rainfall has ceased – the interval is the basin lag.

hydrography The science that investigates the waters of the Earth's surface (oceans, seas, lakes, rivers, etc.) with particular reference to their physical properties. This includes the study of such features as temperature, salinity, tides and currents, and the surveying and mapping of the water bodies.

hydrological cycle (hydrologic cycle, water cycle) The cyclic movement of water between the atmosphere, the land, and the sea. Water is released into the atmosphere as water vapour

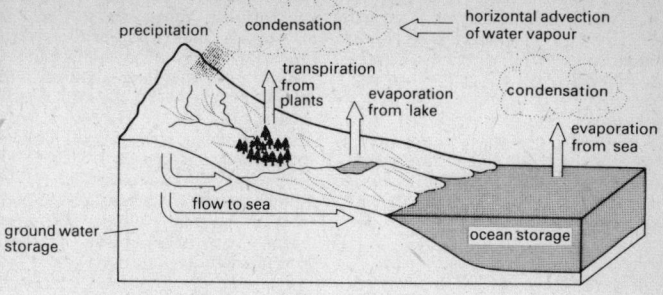

Hydrological cycle

through evaporation from the oceans, rivers, and lakes, and through evapotranspiration from plants and the ground surface. Within the atmosphere water vapour condenses to form clouds and is returned to the land and to its water bodies as precipitation (e.g. rain, snow, hail). This water may run off the land in rivers and streams into lakes and the oceans or move underground as ground water.

hydrology The study of water on the Earth, including its chemical and physical properties, occurrence, distribution, and circulation on the surface and below ground.

hydrolysis †A form of chemical weathering involving a chemical reaction between water and rock. In the reaction a mineral salt combines with the hydrogen ions in the water to produce an acid and a base. The process of hydrolysis is particularly important in the weathering of feldspars to form clay; for example, in granite, the feldspar is broken down to form kaolinite (china clay).

hydrophyte †A plant that grows in fresh-water environments and saturated soils. Examples include the floating water lilies, submerged hornworts, and emergent bulrushes. Hydrophytes show adaptations to their habitats such as large leaf area, projecting stomata, and thin cuticles for efficient transpiration, and short root systems. *See also* hygrophyte.

hydrosere †A form of sere (complete plant succession) that develops in freshwater habitats. A typical hydroseral sequence with its associated communities is:
open water (floating or submerged hydrophytes) → swamp (reeds) → fen (herbs) → carr (trees, e.g. birch and alder).
Such a sequence is often manifest as a zonation from deep water to dry land, and is matched by vertical evidence of historical vegetational change. Hydroseres are ultimately taken over by forests. *Compare* xerosere. *See also* plant succession, sere.

hydrosphere The total water surrounding the Earth. This includes not only the surface water (in solid or liquid form) of oceans, seas, lakes, rivers, ice caps, etc., but also atmospheric water and water below the Earth's surface (e.g. within the soil and in underground streams).

hygrometer An instrument that measures the relative humidity of the atmosphere. The most common type uses a human hair, which expands or contracts as the relative humidity increases or decreases, while another type uses a lithium chloride wire, the resistance of which varies with the changing humidity. A continuous record, a *hygrogram*, showing changes of relative humidity of the atmosphere over a given period of time is obtained from a *hygrograph*, which is a hygrometer fitted with an inked pen resting on a rotating drum. *See also* psychrometer.

hygrophyte †A plant adapted to a plentiful and regular moisture supply, but not to fully waterlogged conditions. Examples of hygrophytic vegetation include tropical rainforest and humidity-loving ferns and mosses. *See also* hydrophyte.

hypabyssal rock An igneous rock that has cooled at a moderate rate at a moderate depth, usually forming a small intrusion such as a sill or a dyke. Hypabyssal rocks are intermediate between the extruded volcanic rocks and the deep plutonic rocks. Many different types are known; some have developed large crystals of one mineral set in a background of much smaller even-sized crystals, while other types are even grained. *See also* extrusive rock, plutonic rock.

hypsometer A scientific instrument that is used to calculate altitude or atmospheric pressure by measuring the temperature at which water boils. As the boiling point varies with air pressure this can be calculated from the temperature at which boiling occurs. An equation is then used to calculate the altitude. The accuracy of this method is not very high.

I

ice The solid state of water. It is less dense (916.8 kg m⁻³) than water (999.84 kg m⁻³) at 0°C. As a result water expands on freezing and ice floats in water. It can be formed by the freezing of water, by sublimation (in which water vapour is condensed directly into ice crystals), and by the compression of snow.

ice age A time in the Earth's history when ice spread towards the equator covering large parts of the continents. The most recent of these periods, the

Pleistocene or Quaternary glaciation, is popularly known as the *Ice Age*. During it most of Europe, North America, parts of N Asia, and S South America were covered with sheets of ice. There were four main glacial phases in the Pleistocene, separated by interglacial periods, the last phase ending about 10 000 to 15 000 years ago. The movement of the ice sheets created characteristic landforms that can be seen in most of the mountainous areas of the British Isles, such as Snowdonia.

There are several theories concerning the causes of ice ages; these include a change in the Earth's orbit, a change in the position of the poles, variations in the Sun's radiation, and the sunspot cycle.

Between about 1550 and 1850 there was a marked fall in temperatures throughout most of the N hemisphere; this period is known as the *Little Ice Age*.

iceberg A large mass of ice floating in the sea. Icebergs are formed when blocks of ice break away from a glacier or ice sheet where it protrudes into the sea in a process known as calving. They then drift under the influence of winds and currents. Those originating from glaciers are generally irregular in shape; those from ice sheets tend to be tabular. The main sources of icebergs are Greenland in the N hemisphere and Antarctica in the S hemisphere, where the icebergs are characteristically tabular and may extend up to 80–100 km in length. The great proportion of an iceberg's bulk lies beneath the surface of the water and it is a considerable hazard to shipping.

ice cap A permanent and often isolated mass of ice, which is smaller than an ice sheet or an ice field. Glacial and periglacial processes are active at the ice cap and its margins. Examples include the ice cap on the island of Spitsbergen and the local ice cap that formerly covered the Scottish Highlands during past geological periods.

ice-contact terrace *See* kame terrace.

ice-cored moraine †A ridge of ice or bank of snow that has been covered with debris. Such features are commonly found on the surface of existing glaciers or at their margins. On the eventual melting of the ice or snow that is at the core of the moraine, the feature collapses.

ice field 1. A large continuous area of sea ice, which is subject to pronounced seasonal variations in size. In winter, the ice field that covers a proportion of the Arctic Ocean is comprised of numerous floating ice floes that have been compressed together.
2. A large mass of ice that permanently covers an area of land, e.g. the Columbia Icefield in Jasper National Park, Alberta, Canada.

ice floe A sheet of floating sea ice that is detached from the ice field. It may be distinguished from an iceberg by having a greater horizontal extent than vertical. Ice floes are numerous in the Arctic Ocean; at the onset of the Arctic winter the ice floes become frozen together to form much broader areas of floating ice.

Icelandic low A fluctuating area of generally low pressure situated in the subpolar region of the N Atlantic between Greenland and Iceland. It is not a permanent area of low pressure but is on the track of depressions moving W to E across the N Atlantic, which makes its annual average pressure relatively low.

ice lense †A body of ice that lies within a soil layer in periglacial areas. When water freezes in the soil to form these features the forces of expansion that are produced cause the disruption of the soil layer above. This is the process known as frost heaving or congeliturbation.

ice sheet A large continuous tract of ice and snow of great thickness, that covers vast areas. Many developed during the Quaternary glaciation to cover large expanses of North America and Europe. Today the only major ice sheets are the Greenland ice sheet, covering about 1.6 million km², and the Antarctic ice sheet, covering about 13 million km².

ice shelf An extension of an ice sheet beyond a coastline so that it floats upon the sea. For example, the Ross and Filchner ice shelves in Antarctica.

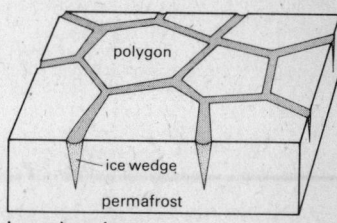
Ice-wedge polygons

ice wedge †A wedge-shaped body of ice tapering downwards into the ground, which occurs in periglacial areas. The intense cold of winters in periglacial areas causes the ground to contract and crack in a similar way to clay when it dries. Meltwater that penetrates these cracks in summer refreezes at depth to prevent the cracks closing and forms ice wedges. With successive seasons the wedges become enlarged and extend deeper into the ground, sometimes to depths of 30 m. The cracks in which ice wedges freeze are generally polygonal in shape, hence *ice-wedge polygons* are formed. Ice wedges may become fossilized through infilling by sands and gravels.

igneous rock Rock that has solidified from molten magma. Igneous rock may be either extrusive or intrusive. Extrusive rocks are those that are poured out onto the surface of the Earth as magma before cooling and are generally glassy or fine-crystalled. Intrusive rocks cool and solidify within the crust of the Earth; these may be either plutonic, which cool deep within the crust and usually have large crystals, or hypabyssal, which cool at intermediate depths and contain moderate-sized crystals.

Igneous rocks may also be classified according to their chemical composition. For example, one classification is based on silica content and has the following divisions: acid rocks (over 66% silica), intermediate rocks (55–66% silica), basic rocks (45–55% silica), and ultrabasic rocks (less than 45% silica). *See also* metamorphic rock, sedimentary rock.

illuviation †The deposition in a soil of fine material such as clays, bases, and colloidal organic matter that has been washed downwards from above. Such accumulations often form pans in the B horizon of a soil profile, constituting an *illuvial horizon*. *Compare* eluviation.

immigrant 1. (*Population Geography*) A person who has voluntarily moved from another country to the country in which he is now a settled resident. The voluntary movement may have been dictated by economic or social circumstances.
2. (*Biogeography*) A plant or animal that by natural means has entered an area or habitat from which it was previously absent. †The immigrant may remain only temporarily in the new area (before either withdrawing or becoming extinct) or may become established in the community of adoption, in which case the plant or animal's geographical range is extended.
Compare introduction.

impermeable Denoting a rock that is nonporous and does not allow water to sink into it. For example, granite is impermeable, although the presence of cracks and fissures in the rock may make it pervious. *Compare* impervious.

impervious Denoting a rock through which water cannot easily soak. It can apply to porous but fine-grained sediments such as clays and to nonporous rocks such as massive unjointed granite. *Compare* impermeable.

imports Goods and services that are obtained from another country. Most British imports are in the form of goods, particularly food and raw materials such as wheat and iron ore, which cannot be supplied from the country's own resources. Imports of services include banking, insurance, and shipping. The UK is a net exporter of these services, which are described as 'invisible' imports and exports to distinguish them from visible goods.

improved land Land that has had its value increased either because it has been reclaimed from waste for agriculture or because it has had its yield potential increased by such methods as drainage or levelling. For example, reclamation of the Veluwe region of the Netherlands has resulted in the improvement of barren heathland, which now produces crops. In the UK meadows that are ploughed and planted with ley grasses become improved grassland.

incised meander †A river meander that has been cut deeply into bedrock. This may be caused by the continued downcutting of a river while it is developing meanders or by rejuvenation of a river in which meanders are already established. Two forms of incised meander are recognized. These are intrenched meanders, in which the valley sides are steep and symmetrical resulting chiefly from vertical erosion, and ingrown meanders, in which the sides are assymmetrical and result chiefly from lateral erosion by the river. *See* ingrown meander, intrenched meander.

inconsequent drainage †A drainage pattern that is not related to the present structure of the rock over which it flows. The rivers are therefore described as being discordant. Inconsequent drainage is either superimposed or antecedent. In a superimposed pattern the rivers have cut down through the rock upon which they originally developed and now lie upon older rocks. The streams have thus superimposed their pattern onto these old rocks. In antecedent drainage the streams cut down during a period when the land was uplifted, and at a faster rate than that of uplift. In this way their pattern is not directly related to the present structure.

Indian corn *See* maize.

Indian summer A period of fine warm weather that occurs with some regularity after the summer season has ended. In the UK this occurs in early September and can be recognized in roughly four out of every five years, while in North America it occurs later, in October and early November. The term was first used in New England for this warm spell during which time the Indians were supposed to be making their preparations for winter. The term is used generally and has no precise meteorological definition.

indigenous Describing an organism that belongs in, or is native to, a certain locality or habitat, i.e. it is not introduced. The description may be applied to species or communities of plants and animals, and to populations. It is sometimes applied to rocks and minerals that have remained in the locations in which they were formed. *Compare* exotic, introduction.

industrial complex An area that is used for industrial purposes, usually by a number of firms. Whereas an industrial region is very large, e.g. the Ruhr, an industrial complex is much smaller, e.g. Europoort in the Netherlands. In some instances one raw material source such as an oil refinery will stimulate the growth of a number of industries close to the refinery forming an industrial complex, e.g. the industries along Southampton Water using by-products from the Fawley refinery.

industrial inertia The tendency for industries and individual firms to remain in a particular location after the causes which determined the original choice of that location have diminished or disappeared. For exam-

ple, the Sheffield steel industry survives even though the original iron ore is worked out and other factors such as local water power have lost their relevance. The proximity of a number of steel-using firms enables the industry to retain its importance today. *See also* industrial momentum.

industrial linkage †The connections that exist between different types of industries with the objective of achieving economies and lower costs. There are four common types of linkage: *vertical linkage* involves firms that are contributing to different stages of an operation, e.g. metal processing. *Horizontal linkage* involves firms that are producing components that at some stage will be assembled into a particular product, e.g. a car or a TV set. A third type, called *diagonal linkage*, involves firms such as machine-tool makers that provide a product or service for a variety of other firms in the area. A fourth type involves firms with 'common roots', such as dependence on local skilled labour and other industries in the region, e.g. the nail and screw industry of the West Midlands.

industrial location theory †*See* location theory.

industrial momentum The tendency of an industry in a specific locality to maintain or increase its importance even though the conditions that favoured its establishment in that place have altered or disappeared. For example, the production of chinaware and porcelain in the Potteries first developed using local clay and the coal of the N Staffordshire coalfield. Clay and other raw materials are now brought in from other parts of the UK or overseas and most furnaces use electricity but skilled local labour and the growth of ancillary industries offer continuing locational advantages. *See also* geographical momentum, industrial inertia.

industrial town A town in which a large number of the workforce is employed in industry. Occasionally towns are dominated by one industry, e.g. the Ford car works at Dagenham in Essex, or have a particular specialism, such as the boot and shoe industry of Northampton. However, most industrial towns contain a number of different industries, e.g. Nottingham has engineering, chemicals, textiles, and tobacco manufacture.

industry Economic activity that is concerned with the production of goods, extraction of minerals, or the provision of services. In a narrow sense industry is confined to the production of goods, i.e. manufacturing industry. In a wider sense it is used to describe the service industries such as tourism, banking, and transport, as well as coal-mining, oil-drilling, building, and contracting.

infant mortality rate (infantile mortality rate) †A measure of mortality based on the deaths of very young children. The rate is usually expressed as the number of deaths of infants under one year old per 1000 live births in any given year.

infield That part of a farm, usually in a hilly district, that is close to the farmhouse. The outlying area, or outfield, is likely to consist of rough grazing or moorland with occasional arable patches. The infield is normally fenced, manured, and cultivated. This system of farming was to be found in Scotland and Ireland until recent times.

infiltration The seepage of water into the soil. †Soil characteristics such as the structure, texture, presence of a vegetation layer, and existing moisture content, and the nature, especially intensity, of the precipitation influence the rate of percolation. The maximum rate at which rainfall can be absorbed by a soil in a given condition is known as the *infiltration capacity* (usually expressed as mm per hour). Once this is exceeded the soil quickly becomes saturated and overland flow takes place.

ingrown meander †A type of incised meander in which one valley side is steeper than the other. The steeper side is often concave in shape while the less steep side is convex. The assymmetrical cross section of an ingrown meander may be caused by the development of meandering in a river at the same time as it continues downcutting into bedrock. The river erodes laterally on the outside of meander bends, undercutting the banks to produce steep concave-shaped slopes. On the inside of meander bends deposition takes place resulting in the gentler convex slope. *Compare* intrenched meander. *See* incised meander.

inland basin (interior basin) A depression in the surface of the Earth that is entirely surrounded by higher land; for example, the Great Basin in the USA. *See* basin.

inland drainage *See* internal drainage.

inland sea A large isolated expanse of water, which lies in an inland basin and has no link with the oceans. Such bodies of water are larger than lakes and may be the focal point of several rivers that drain into them. Examples include the Caspian Sea and the Dead Sea.

inlet An inland opening of the coastline or opening into the shore of a lake. Inlets vary widely in their size and nature and include fiords, which are narrow, deep, and rocky, and estuaries, which are usually wide, silted, and low lying.

inlier An outcrop of rock that is entirely surrounded by rocks of a younger age than itself. This is usually the result of folding, faulting, and erosion. For example, inliers may form when the crest of an anticline is breached by streams eroding along it and older rocks are exposed. *Compare* outlier.

inner city An area near the centre of a city that is densely populated and contains dilapidated housing, often in multiple occupation. The inner city is characterized by recent loss of population, a declining economic and industrial base with the loss of jobs outstripping that of population loss, and rates of unemployment significantly higher than for the city as a whole. The inner city is often a reception area for immigrants who are employed in low-wage occupations. Since the 1960s an 'inner-city crisis' has been identified in the UK, which has resulted in measures such as the establishment of enterprise zones, which are designed to revive and modernize the economic base. *See also* enterprise zone.

input-output analysis †The measurement of the relationships between the level of production of a particular set of goods and services (output) and the raw materials, labour, and other manufacturing inputs required to produce at that level of output. By identifying input-output coefficients it is possible to trace through the effects of a set of output requirements on the production pattern of the economy. This system of analysis was invented by the Soviet economist R. H. Kantorovitch and developed by W. Leontief in the 1950s.

inselberg A prominent isolated steep-sided mass of rock, which rises abruptly from a plain and is characteristic of savanna landscapes. †Inselbergs may be the remnants of plateaus that were eroded and cut back during pediment formation until only the residual hill was left. It has also been suggested that inselbergs

represent the remains of resistant cores of rock that were exposed when weathered debris was removed by denudation. Many resistant crystalline rock inselbergs are found on the plateaus of central Africa. [German: island mountain]

insolation The radiant energy that reaches the surface of the Earth from the Sun. The Sun emits a wide variety of energy waves from very short x-rays, through the visible spectrum, to the longer infrared rays, and while only about 1 part in 2 000 000 000 of the total of the Sun's radiation reaches the Earth, it is essential for the maintenance of life. The amount of radiation reaching the upper layers of the atmosphere varies very little and is known as the solar constant, but differences in atmospheric conditions and geographical position result in great variations in the amount of insolation received in different parts of the world. Within the atmosphere some of the radiation is absorbed, some is 'scattered' by dust particles and water vapour, and some is reflected back into space by clouds and dust, leaving a worldwide average of about 45% of the solar constant actually reaching the Earth's surface. Insolation is also reduced when the angle of the Sun's rays to the surface is low and when daylight hours are short, thus the greatest amount is received at places where the skies are generally clear, the angle of the Sun is high, and the days are relatively long, i.e. around the tropics. On average throughout the year the amount of insolation received by the surface decreases from the equator towards the poles and while equatorial areas have little variation throughout the year there is great variation near the poles. *See* solar constant. [From *in*coming *sol*ar *radiation*]

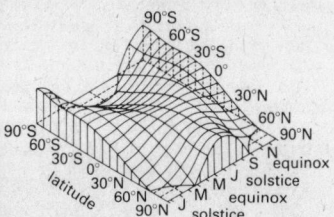

Global insolation variations with latitude and season, assuming no atmosphere

instability, atmospheric The condition that occurs in the atmosphere when an air mass has a higher temperature than that of the air surrounding it. Such an air mass tends to rise, expand, and cool. This is one of the commonest causes of rainfall, which is often heavy and accompanied by thunder.

installed capacity The output capability of a factory or power plant. For example, the coal-fired Drax power station near Selby in North Yorkshire has an installed capacity of 1980 MW. Only at peak demand times is this capacity likely to be fully used.

insular climate The climatic conditions experienced by island and coastal areas where the sea rather than a land mass is the major influence. The main effect of this marine influence is a small seasonal temperature range (i.e. equable conditions). This is in direct contrast to the conditions found in a continental climate. *Compare* continental climate. *See also* maritime climate.

intake The point at which air or a liquid is channelled into a plant or a mine. For example, surface intakes allow fresh air to be drawn through an underground working. Some factories have intakes where water from a lake or river is piped into the works for cooling or cleaning purposes.

intensity of rainfall A measure of how heavily rain has fallen. This can be described generally (e.g. light drizzle, torrential downpour). More precisely, the hourly intensity of rainfall can be measured by dividing the total rainfall by the number of hours of rain. A daily intensity covering a longer period of time can be obtained by dividing the total rainfall over a set period by the number of raindays. These calculations are important as the distribution and intensity of rainfall are of more interest than simple rainfall totals, especially to farmers.

intensive cultivation Methods of farming in which large amounts of capital and/or labour are applied per unit of land. In advanced societies machinery, modern irrigation techniques, fertilizers, and other capital intensive methods are used, e.g. for market gardening S of Amsterdam. In some Third World countries, such as Indonesia, intensive agriculture involves the use of a large labour force rather than capital to produce rice and vegetables.

intensive livestock rearing The application of large amounts of capital and/or labour per unit of land for the purpose of rearing animals. Capital is normally required, not only for buildings and equipment but also for the associated cultivation of fodder crops. Much of Danish agriculture is of this type, involving the use of specialist breeds (e.g. the Landrace pig), mechanical farming, and the purchase of additional feedstuff.

interception †The capture of drops of rain by the leaves, branches, and stems of plants. The interception of the rainfall by the vegetation cover prevents some of it from reaching the ground. However, above a certain level or duration of rainfall the amount that can be held by the plant cover is exceeded and the water will reach the ground either by dripping off the leaves (throughfall) or by flowing down stems and branches (stemflow). Some of the rainfall that is intercepted by the vegetation will be evaporated directly back to the atmosphere. The amount of interception that takes place depends on the character of the vegetation and the duration of the rainfall.

interchange A road junction that is designed so that streams of traffic from different directions do not converge at the same level, enabling traffic flow to be continuous. For example, N of Bristol an interchange allows traffic from the M4 to join the M5 or vice versa. A more complex interchange is Spaghetti Junction near Birmingham.

interfluve The ridge of higher ground that lies between two rivers in the same drainage system.

interglacial period The time interval between two glacial phases in the Pleistocene glaciation. During this time the climate became milder and the ice sheets retreated or disappeared. These periods are represented by soil, lake, and river deposits sandwiched between layers of glacial drift. The deposits contain plant remains, which provide valuable information on the climate and allow the comparison of conditions in the interglacial periods. *See also* ice age.

interior drainage *See* internal drainage.

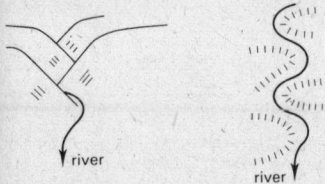

river

river

Interlocking spurs

interlocking spur A ridge projecting from one side of a river valley that interlocks between two corresponding ridges on the opposite side of the valley. As seen from above the spurs of a valley interlock rather like serrated teeth with the river winding between the spurs. It is characteristic of the upper course of a river where the stream is easily diverted and tends to flow around obstacles. *See also* truncated spur.

intermediate rock An igneous rock containing 55–66% silica and less than 10% free quartz. It contains some feldspars, which can be either sodic plagioclase or alkali feldspar or both. Examples of intermediate rocks include andesite, diorite, and syenite. *See also* acid rock, basic rock, ultrabasic rock.

intermittent saturation, zone of A layer of soil that lies just below the surface layer of soil and which will contain ground water after rainfall, but which will tend to dry out during periods of drought. In this way the zone contains water only intermittently and the frequency of rainfall

will determine the periods during which the zone is saturated.

intermittent spring A spring that flows at irregular intervals; the term is also applied to springs that flow continually but with varying discharges. Such springs may be the result of variations in precipitation that lead to fluctuations in the level of the water table. The tapping of ground water through bores and wells may also cause fluctuations in the water table with its consequent effect on spring discharge.

intermont (intermontane) Denoting a feature that lies between mountains. For example, a basin of deposition, such as the Great Basin of the USA, which lies between the mountain ranges of the Sierra Nevada and the Wasatch, or a plateau, such as the Bolivian Plateau, which lies between the E and W ranges of the Andes in South America.

internal drainage (inland drainage, interior drainage) A drainage system in which the waters of rivers empty into an inland sea or lake without reaching the ocean. For example, the Volga River has its mouth on the shores of the Caspian Sea.

internal migration *See* migration.

International Date Line A line that roughly follows the 180° meridian, with some deviations to avoid land masses and groups of islands. The date immediately E of the line is one day earlier than to the W. This situation occurs because of the accumulated time change of 1 hour for each 15° of longitude W and E of the Greenwich Meridian; i.e. 180°W of Greenwich is 12 hours slow but 180°E is 12 hours fast. For example, when it is noon on Wednesday at Greenwich it is midnight at the 180° meridian on either Wednesday/Thursday or Tuesday/Wednesday. To adjust for this, travellers in aircraft and ships crossing the International Date Line from W to E repeat a day (e.g. Wednesday is followed by Wednesday) and travellers

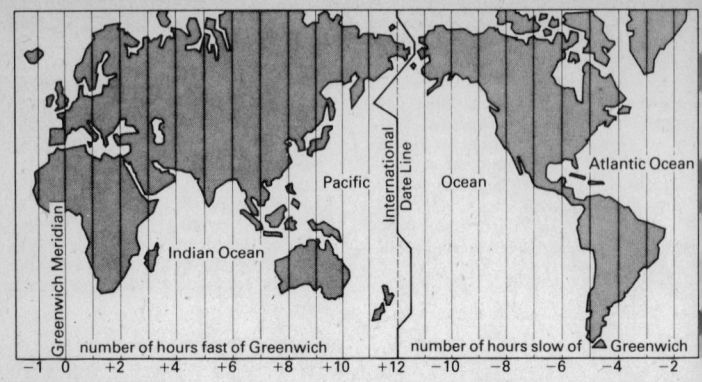

International Date Line

crossing it from E to W lose a day (e.g. Wednesday is followed by Friday).

international migration *See* migration.

International Million Map (IMW, International Map of the World) A map series on the scale 1:1 000 000. It has been drawn by a number of countries following approved specifications. The series is incomplete since many countries have not carried out routine mapping of their territory and some sheets are provisional or in need of revision. The projection used is a modified form of the polyconic with the sheets normally covering 4° of latitude and 6° of longitude.

intertropical convergence zone (ITCZ) A broad zone of relatively low atmospheric pressure situated in equatorial areas between the northeast and southeast trade winds. It is not static and tends to move N and S with the apparent seasonal movement of the overhead Sun. As it is a zone of convergence the winds are light and variable and there is a considerable amount of convectional rainfall. The zone is narrower and more sharply defined over the land than over the oceans, but the only time conditions approach those of a front is when the converging air masses have very differ-

ent humidities; for example in W Africa where the air from the north blows off the desert and from the south off the ocean. The zone is also one source of tropical storms (cyclones), when intense localized areas of low pressure develop and move N or S away from the equator.

intertropical trough *See* equatorial trough.

intrazonal soil A mature soil that differs from the zonal soil developed under the prevailing climatic conditions as a result of local factors such as relief or parent material. Intrazonal soils include hydromorphic soils (associated with areas of poor drainage), calcimorphic soils (which are developed on calcareous parent material), and halomorphic or saline soils (which are developed in saline conditions). *Compare* azonal soil, zonal soil.

intrenched meander (entrenched meander) †A form of incised meander in which both of the steeply sloping sides are of equal gradient. This produces a symmetrical valley cross profile. An intrenched meander is formed when a meandering river is rejuvenated and renewed downcutting into the bedrock takes place. It indicates that vertical erosion by the river is dominant, unlike in ingrown meanders, which are largely a result of lateral erosion by the river. *Compare* ingrown meander. *See* incised meander.

introduction In biogeography, a plant or animal that has been introduced by man into a locality to which it is not native. Important food plants, such as the potato in Europe, and commercial crops, such as rubber trees in Malaysia, have been past introductions. Some pests, such as the rabbit in New Zealand, have spread after deliberate or accidental introduction. *Compare* immigrant.

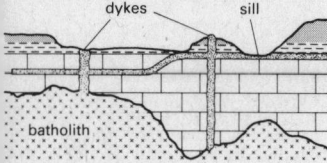

Igneous intrusions

intrusion A solidified mass of igneous rock formed from molten magma that has forced its way into and through pre-existing rocks, especially along lines of weakness, and then cooled. *See also* batholith, boss, dyke, sill, stock.

intrusive rock Igneous rock that has solidified below the Earth's surface. Granite is the most common intrusive rock. *Compare* extrusive rock. *See* hypabyssal rock, plutonic rock. *See also* intrusion.

inversion of temperature An increase in temperature with height in the atmosphere. This is a reversal of the normal vertical temperature distribution in the atmosphere. An inversion may occur either near ground level or at high altitude. A *surface (ground) inversion* is usually formed when rapid radiation takes place at night, cooling the ground, which chills the air above it. The effect is intensified in valleys and hollows into which the cold air drains and collects. A similar effect is obtained when a warm air mass moves over a cold surface and the lower layer of the air is chilled. *High-*

altitude inversions are commonly formed at fronts where either a warm air mass rides over a colder one (warm front) or a cold air mass forces its way under a warmer one (cold front). In an anticyclone subsidence within the air mass may produce an inversion.

inverted relief Relief of land that is opposite to the underlying structure of the rocks, i.e. synclines (downfolds) form high ground and anticlines (upfolds) form low ground. This results from prolonged denudation. During folding the rocks of an anticline may become fractured leaving it as a zone of weakness. Streams subsequently flowing along the crest of the anticline exploit the weaknesses in the rock and rapidly erode it. Eventually the crest of the anticline may be completely denuded, so leaving the anticline as a lowland when compared to the syncline. Snowdon in Wales is an example of a synclinal mountain.

invisible exports Those items, such as services, provided for foreign countries that produce income but do not involve the transfer of goods. In the UK balance of payments account invisible exports include net payments for shipping services, commissions for banking and insurance, and the interest on national capital invested abroad.

involution 1. (*Geomorphology*) †A lobe of one material projecting either upwards or downwards into another material. This results from the effects of frost heaving and other disturbances in the upper layers of the soil in periglacial environments. The presence of involutions in deposits indicates that the area was formerly subjected to periglacial conditions. **2.** (*Geology*) †The refolding of nappes resulting in extremely complex fold structures. *See* nappe.

ionosphere That part of the Earth's atmosphere extending upwards above the stratopause from an altitude of about 60 km. It is so called because incoming solar radiation ionizes the gases within it forming ionized layers, which are indicated by the letters D to G (e.g. E layer). The layers reflect radio signals and other electromagnetic waves back to Earth. The aurora also occur within it.

ironpan A thin compact layer within the soil consisting of an accumulation of sesquioxides, notably of iron. †In humid climates (e.g. in the British Isles) ironpans are characteristic of the B horizons of podzols and are formed by the redeposition of iron after leaching from above. *See also* hardpan.

irrigation The artificial application of water to the land in order to grow crops or to improve crop yields. Some of the most primitive forms of irrigation, such as the Archimedes screw, sakiyeh, and shaduf, are still in use. Where large amounts of capital are available concrete dams and channels are built. A flexible system, which is increasing in popularity, is the use of mobile overhead pipes, which can be moved from field to field. *See also* basin irrigation.

isallobar A line drawn on a weather chart joining places that have experienced the same change in atmospheric pressure over a set period of time. The pattern produced shows regions of increasing and decreasing pressure and the movements of pressure systems, and is therefore used in forecasting.

isanomal (isanomalous line) A line drawn on a map or chart joining places having the same anomaly (or departure from the mean or normal) in any meteorological or climatic element. One of the commonest uses of isanomals is to show the difference between the average temperature of a place or region and the average temperature of all places lying along the same latitude. Such a map gives a very clear picture of the influence on temperature of the distribution of land and sea, ocean currents, prevailing winds, etc. For example, in winter a

place on the W coast of Europe will show a high positive anomaly (i.e. it will be much warmer than the average for the latitude) while a place on the same latitude in the interior of the USSR will show a high negative anomaly (i.e. it will be much cooler than the average for the latitude).

island A mass of land that is surrounded by water and is smaller than a continent. Greenland, with an area of about 2.2 million km², is the world's largest island.

island arc *See* arc.

isobar A line drawn on a map or chart joining places having the same atmospheric pressure. †On weather charts, which are used in forecasting, isobars are plotted to show the distribution of pressure at a particular time and are usually drawn at intervals of 2 or 4 millibars (mb). On a larger scale, on maps and charts used for climatological study, isobars are used to show the average pressure distribution over large areas for a given period of time. As the weather stations at which the recordings are taken are frequently at different altitudes corrections are made to the pressure readings to make comparisons between them valid. A common level is taken, usually mean sea level, and the pressure values that result are those that would be found were the stations at mean sea level.

isobath A line on a chart or map joining all places of the same depth below the surface of an ocean or a lake. It is the equivalent of a contour line on an ordinary relief map and similarly indicates the relief features on the floor of the ocean or lake.

isocline A fold in which the layers of rock have been so highly compressed that the limbs of the fold are nearly parallel. Folds of this type are closely packed and look rather like a concertina. Examples can be seen in the uplands of Scotland.

isodapane †A line joining places with the same total transport costs, both for moving raw materials for processing and for moving finished goods to the market. This concept was introduced by Alfred Weber (1909) as part of his theory of industrial location. It can be shown diagrammatically by drawing transport cost contours around a raw material source and similar contours around a market point. The isodapane can be plotted at those contour intersecting points where the total cost values are the same.

isogonic †A line on a map that connects places having an equal magnetic declination or variation from true north.

isogram *See* isopleth.

isohaline †A line on a map joining points in the oceans that have equal salinity.

isohel A line on a map joining places having equal amounts of sunshine over a certain period.

isohyet A line on a map joining places with equal amounts of rainfall over a certain period.

isoline *See* isopleth.

isopleth (isogram, isoline) A line on a map joining places with the same value for a certain element. This is the collective term for the various types of lines representing specific values that are drawn on a map. [From Greek]

isoseismal line A line on a map joining places that have experienced equal intensity of shock from an earthquake. These lines form irregular curves around the origin of the earthquake.

isostasy †The principle that variations in the height of the Earth's surface are compensated for by the underlying distributions of mass and hence are in a state of balance. For example, continental areas of a greater elevation project down more deeply into the denser but weaker asthenosphere than the thinner and denser oceanic crust. As a result of erosion or deposition the state of balance is put out of equilibrium and has to be compensated for by movements of the Earth's crust and underlying asthenosphere. For example, the growth of continental ice sheets causes a downward deflection of the Earth's crust with a corresponding outflow of asthenospheric material from beneath it. When the ice melts the asthenospheric material flows back below the continental area and the crust is uplifted. If a mountain area is eroded and the eroded material subsequently deposited along the margins of a continent there will be a rise in the area eroded and subsidence in the area of deposition.

isotherm A line drawn on a map or chart joining places having the same air temperature. Isotherms may be used to show the temperature distribution over an area at a particular point in time, but are more commonly used to show average temperatures over a period of time. The most common isothermal maps are those constructed to show the distribution of mean monthly temperatures for January and July, a comparison of which reveals the seasonal variation of temperature. The temperatures plotted on most of these maps are reduced to mean sea level to eliminate the variations caused by altitude.

isothermal layer Any layer of the atmosphere in which the temperature remains constant with height. The term was formerly applied to the stratosphere in which such conditions were believed to exist.

isthmus A narrow bridge of land with water on either side that connects two larger bodies of land, e.g. the Isthmus of Panama, which connects North and South America and separates the waters of the Pacific Ocean and Caribbean Sea on either side of it.

ITCZ *See* intertropical convergence zone.

J

jebel (djebel, jabal) A mountain or range of mountains found in arid lands. The word is frequently incorporated into Arabic place names. [Arabic]

jet stream †A very strong steady westerly wind blowing at high altitudes (about 12 000 m) just below the tropopause. It is usually confined to a narrow band 150–500 km wide and a few kilometres deep and its speed averages 50–60 knots, although 200 knots has been recorded. The highest speeds always occur during winter. There are two main jet streams, one associated with the polar front and the other in subtropical latitudes between 20° and 30°N and S. The *polar front jet stream* is very irregular in its location and is commonly discontinuous, whereas the *subtropical jet stream* is fairly persistent in location for any given season. Similar smaller features can be observed in other areas. [The term was introduced by the Swedish-born US meteorologist Carl-Gustaf Rossby in 1947]

joint A crack in a mass of rock. Sedimentary rocks are commonly divided into blocks by joints both parallel to and at right angles to the bedding. The joints can be fractures occurring when the rock is either compressed or torn apart under tension but, unlike faulting, with little displacement or movement. In igneous rock joints are produced by cooling, for example, the characteristic columnar jointing found in basalt. Where exposed on the surface the joints are susceptible to weathering and to being opened up by plant roots. The joints are used by quarrymen to cut out rectangular blocks of stone.

jökull A small local ice cap or a mountain that is covered with ice. For example, Vatnajökull in Iceland. [Icelandic]

journey to work The daily movement of people to and from work that results from the separation of residential from industrial and business districts in towns and cities. The concentration of workplaces in and near the city centre while suburbs occur on the fringes of cities generates a centripetal morning movement and a centrifugal evening movement. The movement of industries to the edges of cities has created countercurrents to the main flow.

jungle Any equatorial forest with luxuriant vegetation, such as monsoon forest. The term usually implies tangled undergrowth and wild animals.

Jurassic The middle geological period of the Mesozoic era, extending between the end of the Triassic period, about 195 million years ago, and the beginning of the Cretaceous period, about 136 million years ago. The Jurassic rocks were laid down chiefly in shallow-water conditions. It is divided into the Lower, Middle, and Upper Jurassic. This period saw the start of the mountain building that gave rise to the Alps in Europe. The first mammals and earliest-known birds appeared and some forms of Jurassic plants remain in existence today, e.g. ferns, conifers, and ginkgoes. [Named by the German geographer Alexander von Humboldt in 1795 after the Jura Mountains in France and Switzerland]

jute The fibre of two species of tropical plants (*Corchorus capsularis* and *Corchorus olitorius*) which, because of its cheapness and coarseness, is used for making such products as sacking, twine, and the backing material for carpets and lino. The plant grows to a height of about 3 m and the inner bark produces fibres of about the same length. Most of the world's jute comes from the lower Ganges-Brahmaputra valley in Bangladesh and India and bales of the fibre are exported through Calcutta. [From Sanskrit *jūta*: braid of hair]

K

Kainozoic *See* Cenozoic.

kame An isolated mound or low ridge of water-sorted sands and gravels that has been deposited by meltwater at the margin of a stagnant or decaying ice sheet or glacier. The sorting action of the water leaves the deposit with distinct layers or beds. A *kame delta* may be formed where meltwater streams deposit their load when flowing from a stagnant or retreating glacier or ice sheet into an ice-dammed lake. Kames may also be formed as accumulations of debris in a crevasse or moulin in the ice of a glacier or ice sheet. When the ice melts the debris is let down to ground level to form a *moulin kame*. Kames are found, for example, in central Ireland. *See also* kame terrace. [Scottish]

kame terrace (ice-contact terrace) A ridge of sand and gravel deposited by meltwater flowing between a decaying or stagnant glacier and the confining valley sides. Kame terraces are typically flat-topped and steep-sided, and have distinct layers as a result of water sorting.

kampong A cluster of buildings forming a small village in SE Asia, especially Malaysia. It is surrounded by intensively cultivated land. [Malay]

kaoliang A small grain of the millet group, which is grown in N China and Inner Mongolia. It can be used to produce flour or as an animal feed. *See also* millet.

kaolin (china clay) A fine white clay that easily crumbles to powder, resulting from the alteration of feldspars in granite. This alteration is caused either by weathering (hydrolysis) or by the action of ascending gases associated with igneous activity. There are extensive deposits in SW England, which are associated with the St Austell granite and formed by the action of ascending gases. Dartmoor, Bodmin Moor, and St Austell in Cornwall are

the chief producing areas. Kaolin is quarried with high-pressure water jets. It is used in the manufacture of fine porcelain and china, and as a filler in the manufacture of paper, rubber, and paint. [Named after the Kaoling mountains in China from where supplies for Europe were first obtained]

karaburan A strong hot northeasterly wind that occurs in the Tarim Basin of central Asia in summer when the continental interior is very hot. It is frequently strong enough to produce duststorms, which darken the sky. The dust carried by this wind is one of the major sources of the great loess deposits of this area of Asia.

karren †An exposed limestone surface that is dissected by grooves and furrows as a result of limestone solution. Many limestone pavements exhibit this characteristic. *Rillerkarren* are grooves separated by sharp ridges and *rundkarren* are grooves with rounded separations. [German]

karst A region of well-jointed Carboniferous Limestone in which carbonation is the dominant weathering process, producing characteristic features such as sinkholes, limestone pavements with clints and grikes, caves containing stalactites and stalagmites, steep-sided gorges, and underground drainage. Examples of such limestone regions include the Karst region of Yugoslavia; the Causse region of SW France; and Malham in Yorkshire, the Peak District in Derbyshire, and the Mendip Hills in the British Isles. The process by which karst scenery is formed is *karstification*. †It has been suggested that karst regions are characterized by their processes rather than their forms. In tropical regions karstic processes may lead to the formation of cockpit karst. It is also possible to find features resembling those of karst areas in rocks other than limestone (e.g. granite); such features are known as pseudokarst.
[Named after the Karst region of Yugoslavia – an extensive limestone area where most landforms have been produced by solution]

katafront

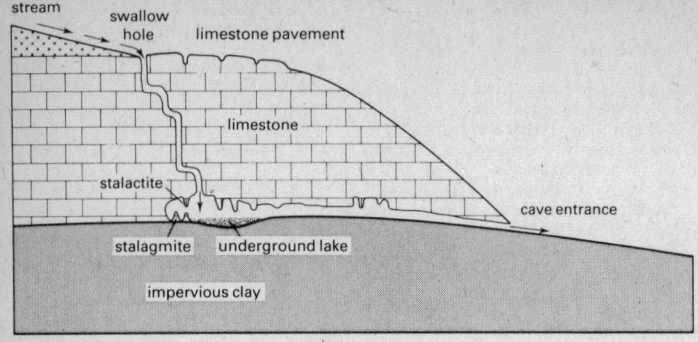

Cross section of features of karst topography

katafront †A front at which the warm air is tending to sink relative to the cold air. As a result the weather changes are generally more subdued than at the more normal anafront where the warm air tends to rise. At a kata-warm front there is no extensive medium- and high-level cloud because of the inversion formed at the base of the subsiding air, most of the cloud present being thick low-level stratocumulus. Rainfall is light. Similarly, at a kata-cold front low stratocumulus cloud from which little precipitation falls is found. Changes in temperature, humidity, pressure, and winds are slow. *Compare* anafront. *See* front. [The term was coined by the Norwegian meteorologist Tor Bergeron in 1937]

katabatic wind A cold downslope wind caused by the gravitational movement of cold dense air near to the Earth's surface. Air chilled by night-time radiation under calm clear conditions or by contact with a glacier or ice cap is denser than the surrounding air and, if the ground surface is sloping, will tend to flow downhill. Its direction of movement is usually controlled by the valleys into which it collects. The strongest katabatic winds are those that blow down from an ice cap, for example those off the Greenland and Antarctic ice caps. *Compare* anabatic wind. *See* glacier breeze, mountain wind, nevados.

kegelkarst †*See* polygonal karst. [German]

kelvin Symbol: K The SI base unit of thermodynamic temperature. †It is equal to 1/273.16 of the temperature of the triple point (i.e. the point at which the gas, liquid, and solid phases are in equilibrium) of water. One kelvin is equal to one degree on the Celsius scale. Zero Kelvin (0 K) is absolute zero (i.e. −273°C).

kettle hole A hollow in the ground surface that is found in deposits of glacial drift. Kettle holes commonly contain small lakes. They originate where blocks of stagnating ice become covered by deposits of drift. The ice slowly melts beneath the deposits, which eventually collapse leaving a depression. This may be filled with water or by sediments. †Kettle holes are often found in groups forming a distinctive topography of hollows and ridges known as *kettle moraine*. For example, the Kettle Moraine region of Wisconsin, USA.

Kew barometer A mercury barometer that allows the atmospheric pressure to be read directly from the meniscus of a column of mercury contained in a graduated tube. *See* barometer.

key village (key settlement) A village that has been chosen as the location for future growth and development in an area. †A number of development plans in the UK have been based on the idea that the concentration of services and people into a number of key villages makes the best use of limited resources. Such policies often rely on the use of threshold analysis to determine the level and type of services a settlement can support.

khamsin A hot dry southerly wind, very similar to the N African sirocco, that blows across Egypt and the SE Mediterranean area from the Sahara. It precedes depressions moving eastwards through the Mediterranean Basin and occurs for a period of about 50 days between April and June. Duststorms are frequent, but in areas where the wind has to cross a stretch of ocean the air may be quite humid. [Arabic: fifty]

kibbutz (*plural* kibbutzim) A collective agricultural settlement in Israel. The ownership of the property and the profits are communal and the work on the settlement is shared among the inhabitants. Unlike the collectives of the USSR and China, the kibbutzim are voluntary organizations which can determine their own programmes. Some have introduced light industries to supplement the farm income. Members are free to leave the kibbutz at any time. The degree of family private life permitted varies. There are about 250 kibbutzim in Israel with a total population of about 100 000. *See also* collective farming. [Hebrew]

kilogram (kilogramme) Symbol: kg The SI base unit of mass, equal to the mass of the international prototype of the kilogram, which is a piece of platinum–iridium kept at Sèvres in France.

kilometre (kilometer) Symbol: km An SI unit of length, equal to 1000 metres.

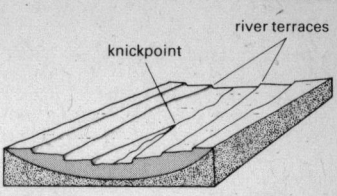

Knickpoint

knickpoint (rejuvenation head) †A marked break of slope in the long profile of a river. This steepening may have been caused by the lowering of the base level of erosion that occurs in rejuvenation. The knickpoint in a rejuvenated river marks the point at which the old river profile meets the newly graded profile. A knickpoint may also be formed by faulting or by the presence of a band of more resistant rock. Knickpoints will recede upstream under the headward erosion by the river.

knob and kettle topography (knob and basin) In the USA, a glaciated area of terminal moraine that is characterized by low hills (knobs) separating shallow hollows (kettles). It is formed during the retreat of an ice front, which leaves irregular mounds of moraine and blocks of ice that thaw to leave the hollows.

knoll A small rounded hill or mound, which is often crowned with trees.

koembang A föhn-type wind that occurs in Java, Indonesia. As with most such winds it can cause damage to young crops such as tobacco. [Indonesian]

kolkhoz A collective farm in the USSR, the land of which is owned by the government and leased to the workers of the farm. Farm production and policy are determined by national and regional plans and the farm is run by a committee. Collective farms own their machinery and production in excess of the agreed quota can be sold at market prices. Members of the kolkhoz may farm private plots, keep a few animals, and sell the produce on the open market. *See also* collective farming, sovkhoz. [Russian]

kopje (koppie) In S Africa, a small isolated hill that rises above a relatively
flat plain. †A kopje is a form of
inselberg and may be formed during
the development of pediments in semiarid areas. If the rock is well jointed
the feature may take on a castellated
form that is known as a *castle kopje*.
[Afrikaans]

Köppen climatic classification †A climatic classification first proposed in
1918 by the German climatologist W.
Köppen and subsequently modified
several times. In the classification the
Earth is divided into temperature
zones based on mean monthly temperatures. The five major zones are:
(1) Tropical, all months over 20°C.
(2) Subtropical, 4–11 months over
20°C and 1–8 months between 10°C
and 20°C.
(3) Temperate, 4–12 months between
10°C and 20°C.
(4) Cold, 1–4 months between 10°C
and 20°C and 8–11 months below
10°C.
(5) Polar, all months below 10°C.
The temperature zones are subdivided
on the basis of seasonal incidence and
amounts of annual rainfall.

kraal 1. A native hut village in S
Africa.
2. An enclosure for livestock in S
Africa.
[Afrikaans]

L

laccolith (laccolite) A large dome-like
mass of igneous rock that was
intruded along a bedding plane in a
sedimentary sequence of rocks, causing
the overlying sediments to become
arched above it. In well-developed laccoliths the base tends to be relatively
flat, so that the resulting intrusion has
a lens shape, but they are frequently
more irregularly shaped. *See also*
cedar-tree laccolith.

lacustrine Of or relating to a lake. For
example, lacustrine sediments, which

are deposits laid down within a lake.
See also lacustrine delta.

lacustrine delta A delta formed where
a river flows into a lake, projecting
from the shore of the lake out into
the water. An example of this feature
is found in Switzerland where the
River Rhône flows into Lake Geneva.

ladang Shifting agriculture as it is practised in Indonesia. *See also* shifting
cultivation. [Indonesian]

lag The delayed effect seen in the relationship between the date of maximum insolation at a climatic station
and the date of the mean maximum
temperature recorded there. In the N
hemisphere maximum insolation
occurs at the summer solstice (21
June) when the North Pole of the axis
is tilted towards the Sun, yet the
warmest months are July or August.
The lag is greater in temperate maritime climates because the neighbouring seas retain summer warmth longer
than would land areas in the same
location.

lagoon A shallow stretch of water that
is separated from the open sea by a
barrier (e.g. sand, shingle, coral) such
as a spit or bay-mouth bar. For example, the Fleet in Dorset is a lagoon
that is separated from the sea by the
barrier of Chesil Beach. Lagoons are
frequently formed between coral reefs
and the shore. The term is also
applied to the stretch of water
enclosed within a ring- or horseshoe-
shaped atoll. Gradually a lagoon may
be filled with sediment, or it may
slowly decrease in size as the barrier
migrates inland under the influence of
wave action.

lake A body of water that lies in a
hollow in the Earth's surface and is
entirely surrounded by land. It is
unconnected with the sea except by
rivers. In arid and semiarid areas
lakes may have no outlet to the sea
and form the focus of an area of
inland drainage. As a result the water
of the lake becomes increasingly

saline. Some inland seas (e.g. the Caspian Sea, Dead Sea) are large lakes.

lake rampart A ridge of material along a lake shore, formed as a result of the pressure exerted by the expansion of ice when the lake is frozen. The feature is therefore confined to areas that experience cold winter temperatures. For example, the Great Lakes of North America are surrounded by lake ramparts.

lalang Thick coarse grass (*Imperator cylindrica*) that colonizes areas of Malaysian rainforest that have been cleared or disturbed by man. Lalang is thus an indication of abandoned sites of shifting cultivation. [Malay]

Lambert's zenithal equal-area projection A map projection constructed as if a flat piece of paper touches the globe at one point (either the North Pole or the equator), onto which the details of the Earth's surface are projected. Shapes are compressed towards the edges but the projection is equal area and all bearings are true compass directions from the centre of the map. The projection is used in atlases to show the hemispheres.

laminar flow 1. †A smooth non-turbulent movement of a fluid (e.g. water, air) in which parallel layers with different relative velocities slide over each other with no mixing. This form of flow is rare in natural streams and air flows. *Compare* turbulence.
2. †The movement in glaciers that occurs when ice fractures and shears under pressure and is thrust along glide planes.

land breeze A cool wind that blows from the land to the sea (or a large lake) during the night. It occurs when night-time radiation chills the land, and the air in contact with the ground surface. The atmospheric pressure over the land is raised relative to that over the sea resulting in a pressure gradient and the flow of air towards the sea. It is most common in tropical areas but also occurs in higher latitudes, particularly when the weather is calm. The land breeze is not as strong as the sea breeze, which blows off the sea during the day. *Compare* sea breeze.

land bridge 1. (*Geology*) A tract of land, either in the past or present, that has linked two continents otherwise separated by sea. For example, the Isthmus of Panama, which links the continents of North and South America.
2. (*Zoogeography*) A tract of land that formerly linked two continents and enabled animals to become dispersed from one continent to another.

land classification †The grouping of land into different classes according to its physical characteristics. Various classification systems have been devised to measure land potential, i.e. its suitability for cultivation or other uses. In the UK the Soil Survey has categorized good, medium, and poor quality land. The United States Soil Conservation Service has carried out a land capability classification, which deals particularly with erosion hazards.

landes A sandy lowland area bordered by sand dunes, lagoons, and belts of

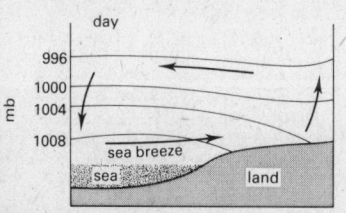
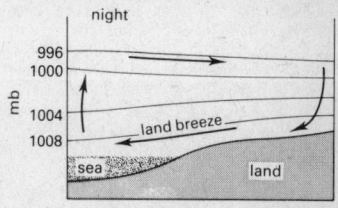

Land and sea breezes

137

pine trees. The term is applied especially to the Landes region of SW France, bordering on the Bay of Biscay, but is also applied to such areas elsewhere. [French]

landform Any feature of the Earth's surface that possesses a particular shape and form. Geomorphology is concerned with the study of landforms and their relation to each other. The precise measurement of the dimensions of landforms is an important aspect of practical geomorphology.

land-locked state A country surrounded on all sides by land and therefore with no access to the sea. Land-locked states have historically faced problems in trading and communications. Examples include Czechoslovakia, Switzerland, Mali, and Zimbabwe.

landscape The visual sum of all the landforms in an area together with the vegetation and the land use that is employed. †Many efforts have been made to assess the quality of a landscape in a quantitative manner. These concern the use of landscape-evaluation techniques, which attempt to assess a landscape with reference to its performing a particular function. *See also* landscape evaluation.

landscape evaluation †An attempt to judge the quality of the landscape. The evaluation of landscape is a relatively recent development and is particularly difficult because of the degree of subjectivity that can be involved. There are two broad approaches: a largely subjective qualitative approach, which relies on the evaluation of landscape quality by the individual, and secondly, the quantitative approach, where landscape quality is determined by the presence of specific features in the landscape. Such an approach was used in the Coventry, Solihull, Warwickshire sub-regional study.

landslide (landslip) A form of mass movement in which rock and debris moves rapidly downslope under the influence of gravity as a result of failure along a shear plane. The movement is often lubricated by water and tends to occur suddenly. It may result from undercutting of the slope, for example the marine erosion of a sea cliff. Other causes include earth tremors and heavy rainfall, which may saturate the slope and make it unstable. The landslide leaves a basin-shaped scar on the slope and a mass of debris at the foot, often projecting forward as a tongue. A landslide may take a number of forms that include slumping and rotational slips. *See* mass movement. *See also* rotational slip.

land use classification The classification of land according to the use to which it is put. In the UK, the First Land Utilization Survey, carried out in the 1930s by L. D. Stamp, identified six areas of land use – arable, heath and rough pasture, orchards and nurseries, meadowland, forest and woodland, and urban areas. The second survey begun in 1960 and conducted by Alice Coleman, was based on 13 major classes of land use.

lapiés An exposed limestone pavement that is dissected by grooves and furrows. These furrows are the result of carbonation that is guided by the jointing of the rock. The term is synonymous with karren, the German term for a limestone pavement: *See* limestone pavement. [French]

lapilli Fragments of cinder that are thrown out by a volcano during an eruption. The particles tend to be small, generally between 2 and 64 mm in diameter. [Italian]

lapse rate The rate of decrease of temperature with altitude expressed in °C per 100 m. †Within the atmosphere as a whole, up to the level of the tropopause, the average decrease of temperature with altitude is about 0.6°C per 100 m and this is known as the *environmental lapse rate* (ELR). However, differing conditions within the atmosphere can produce a wide range of varying lapse rates; for example, if there is an inversion of temperature the lapse rate will have a negative value. Variations also occur if 'parcels' of air are forced to rise either by localized heating or by orographic means. Such variations are very important in terms of weather production as they determine whether the rising air is stable or unstable. *See also* dry adiabatic lapse rate, saturated adiabatic lapse rate.

latent heat The energy transferred when a substance changes state, e.g. from solid to liquid, liquid to gas, gas to liquid. In meteorology, the most important changes in state that occur are those from water to water vapour (evaporation) and from water vapour to water (condensation) within the atmosphere. In evaporation, the transfer of heat is from the atmosphere to the water therefore the temperature of the atmosphere is reduced, while in condensation the transfer of heat is from the water vapour to the atmosphere therefore the temperature of the atmosphere is increased.

lateral erosion The horizontal erosion carried out by a river on its banks. This acts chiefly on the outside (concave) bank of a meander bend where it causes undercutting of the bank and its eventual collapse. As a result each meander (and thus the valley) is widened and at the same time migrates slowly downstream. The higher land bordering the river is gradually eroded back leaving a low cliff known as a bluff bordering the meander belt.

lateral moraine A ridge of angular rocky debris that lies on the surface of a glacier close to and parallel with its junction with the valley sides. The debris consists chiefly of material that has fallen onto the glacier from the valley sides as a result of freeze-thaw weathering. Lateral moraine may form distinct ridges running the length of the glacier. The Aletsch Glacier in Switzerland displays lateral moraine. *See also* moraine.

laterite A layer of deposits resulting from the deep weathering of rocks in humid tropical climates. The high temperatures and rainfall cause rapid removal of silica and alkalis from the weathered zone leaving a concentration of sesquioxides of iron and aluminium in the form of a thick red clay. The clay is plastic when wet but hardens once exposed to the atmosphere. It is used for building purposes, blocks being cut while it is in the plastic state and then dried. For example, Angkor Wat in Kampuchea was built of laterite. Exposed layers of laterite may occur as a capping on uplifted areas where the overlying material has been removed.

lateritic soil *See* latosol.

laterization †A weathering process in which iron and aluminium oxides become concentrated in the upper layers of the soil. This process takes place in tropical and subtropical regions with abundant rainfall and a marked dry season. Rocks that are low in iron minerals but with a high proportion of aluminium are converted by laterization to form bauxite. *See also* laterite.

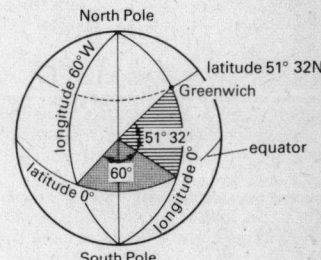

Latitude and longitude

latitude The distance of a point N or S of the equator, measured as an angle from the centre of the Earth. The equator is latitude 0° and the poles are latitudes 90°N and 90°S. Lines joining places with the same latitude are called parallels. *See also* longitude, parallel.

latosol (lateritic soil) A zonal soil formed in lowland regions of the humid tropics (e.g. equatorial Africa) over laterite. †Latosols often support dense forests, but once these are removed the reduction of shade and humus accelerates laterization, the soil soon loses fertility and friability, and becomes agriculturally poor.
See also laterite.

Laurasia The supercontinent of the N hemisphere that is thought to have existed about 200 million years ago following the break-up of Pangaea. It fragmented to form the present-day landmasses of North America, Greenland, Europe, and Asia (N of the Himalayas). *See also* continental drift, Gondwanaland, Pangaea.

lava Molten or partially molten magma that is extruded from a volcano or volcanic fissure out onto the surface of the Earth, where it cools and solidifies. Lava can be acid (i.e. rich in silica), intermediate, or basic in composition. Acid lavas are very viscous, have a high melting point, and do not flow very far; basic lavas are fluid, have a lower melting point, and can cover great distances. The decomposition of lava can give very fertile soil. *See also* aa, block lava, pahoehoe, pillow lava.

law of diminishing returns †An economic law stating that at a certain level of production additional units of input will yield proportionately smaller units of output and the additional cost incurred will be greater than the additional revenue received. For example, the addition of more fertilizer to a field of corn will at first increase the yield per hectare considerably, but a point will be reached at which the additional yield will have a lower value than the cost of the additional fertilizer which is applied.

layer tinting The use of colour on a map between pairs of contours so that the distribution of high and low relief can be quickly appreciated. Each level is given a different tint or colour, usually ranging from green for low-lying land through a sequence to brown, red, purple, and sometimes white for the highest relief features.

leaching †The downward movement of material in solution or colloidal suspension within the soil profile. The material includes organic matter; compounds of nitrogen, calcium, magnesium, sodium, and potassium; and clays. It may be redeposited lower in the soil profile or washed right out of the soil system. High rainfall, freely draining soil, and the presence of acid humus encourage this process. Strong leaching leads to increased acidification of the soil and the development of an eluvial horizon from which material has been removed. If the material is redeposited lower in the profile an illuvial horizon is formed. *See also* eluviation.

lead A bluish-grey metal that is rarely found in its native state but most commonly occurs as the sulphide ore galena. This ore occurs in veins in igneous rocks or as the result of metamorphism. The primary producing countries are the USA, Canada, and Australia. About one third of the world's output of lead is used for making antimonial lead electrodes for batteries; other uses include sheeting and piping, ammunition, foil, and in alloys such as pewter, solder, and bronzes. Lead compounds are used in glass making and tetraethyl lead is used as an additive in petrol to prevent 'knocking'. Lead is poisonous and its use is very strictly controlled.

leat A trench that channels water for use in factories, water mills, mines, or homes. *See also* flume.

Lebensraum †Land that is claimed by a nation on the grounds that it is necessary for the nation's continuance or growth. The concept was employed by Nazi geopoliticians justifying German claims to expanded frontiers. [German: living space]

lee (leeward) The side away from the direction in which the wind is blowing, for example the side of a hill or mountain or a stretch of water sheltered from the wind. *Compare* windward.

lee depression †A low-pressure system formed in the atmosphere downwind of a mountain barrier. It occurs where an eddy is set up in the airstream, producing a low-pressure system having many of the weather characteristics associated with a depression. It is common in mountainous areas, two of the best examples being found to the south of the Maritime Alps in the W Mediterranean area, and to the east of the Canadian Rockies.

lee wave †A standing wave form in the airflow that is generated on the leeward (downwind) side of a mountain. This is formed when a strong or steady wind flow crosses the relief barrier. The wave form remains more or less stationary to the barrier. A lenticular cloud may develop outlining the upward motion of the wave. *See* lenticular cloud.

lenticular cloud (lenticularis) †A lens-shaped cloud found mainly over or to the lee of mountains and hills and associated with lee waves or eddies in the airstream (e.g. helm cloud). The cloud appears to be stationary over the crest of the rising air current, but is actually continuously forming at the upwind end and dispersing at the downwind end.

levante (levanter, llevante, llevantades) An easterly wind experienced in SE Spain and the Strait of Gibraltar. It blows when a depression moves eastwards across the W Mediterranean area and is usually fairly strong and humid, bringing heavy rain. A banner cloud often forms at the summit of the Rock of Gibraltar when this wind blows. It is sometimes known as the solano. *See also* banner cloud. [Spanish]

leveche A hot dry southerly wind blowing from N Africa over S Spain and occurring when a depression is passing through the W Mediterranean area. It is a sirocco-type wind and frequently carries dust from the Sahara as far as Spain. [Spanish]

levee (levée) A bank of alluvium bordering a river in its lower course that is built up by deposition and lies above the level of the floodplain. Levees originate during floods when water overflows the banks of the river. The coarsest materials of the load carried by the river are deposited first, near the river channel, and these deposits are gradually built up to form the levees. Finer material (silt and clay) is spread over the rest of the flooded area. In time the levees may be large enough to support the river at a higher level than the floodplain. Levees are found, for example, along the Mississippi River in the USA. If the levees are breached then a considerable area of floodplain may be flooded. [French]

level An instrument used in surveying to determine height. It consists of a spirit level mounted on a telescope, which swings in a horizontal plane. It is used in conjunction with graduated levelling staves. *See* Abney level, dumpy level.

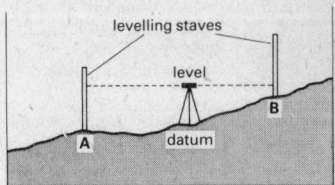

Levelling

levelling †A method of surveying used to determine height. The instruments used are a level (e.g. Abney level, dumpy level) and graduated levelling staves. The level is set up between two, usually equidistant, levelling staves, one (A) at a point of known height and one at a point (B) for

141

which the height is required. A sight is made onto A, known as a *backsight*, and a reading is taken. A sight is then made onto B – the *foresight* – and a reading is again taken. The difference between the two readings gives the difference in height between A and B. By using equidistant staves errors are cancelled out as are differences of altitude caused by refraction or the Earth's curvature.

levelling staff (*plural* levelling staves) †A graduated staff used in levelling. It is usually 4 m in height with heights marked clearly so that they can be read off from the level to the nearest millimetre. A small bubble within the staff enables it to be positioned vertically. *See* levelling.

ley Land that has been under grass or clover for a period of years. The land is ploughed and then seeded to produce pasture, which may be used as grazing for animals.

liana (liane) A climbing plant that twines around and up trees, especially in equatorial rainforests, in search of light. The stems of many lianas are thick and woody, and may damage other vegetation. *See also* rattan. [French]

libeccio (libecchio) A gusty westerly or southwesterly wind that blows across Corsica mainly in summer. It cools the W coast but is frequently hot and dry when it reaches the E coast. When it blows in winter it can bring rare snow showers or rain to the mountains of the island. [Italian]

light industry Industry in which the weights of materials handled are relatively light. Light industries generally provide fewer environmental hazards than heavy industries and can therefore be located close to residential areas. Industrial estates with light industries such as printing, furniture making, and precision-tool making are often located on the edge of towns.

lightning A visible discharge of static electricity in the form of a flash of light that occurs in thunderstorms. It neutralizes the electrical tensions that build up between oppositely charged areas within clouds, or between a cloud and the ground. Forked lightning is the most common form taken by the discharge; others include ball lightning (a persistent drifting spherical mass of glowing air) and St Elmo's fire. *See* thunderstorm.

lignite A type of coal that contains a high proportion of water and oxygen. It is brown or brownish-black in colour and crumbles when it is exposed to air long enough to dry out. It is formed mainly of wood, bark, and leaves. The only British example is found at Bovey Tracey, Devon, but important deposits are found in Germany and North America. Compact lignite is called jet and is found in the UK in the jet shales of Yorkshire.

limestone A dense sedimentary rock made up chiefly of calcium carbonate (calcite, $CaCO_3$). Organic limestones (e.g. chalk) are made up of the calcareous remains of marine plants and animals. Precipitated limestones include evaporites and oolites. Clastic limestones consist of rocks made up of fragments of pre-existing limestones. Pure limestone can be white or pale grey, occasionally black, but is often coloured by impurities such as iron and manganese. Because of its chemical composition it is easily weathered by the action of water containing carbon dioxide resulting in many characteristic features, such as underground drainage, of limestone areas. *See also* karst.

limestone pavement A platform of bare limestone with a surface of irregularly shaped blocks (clints) separated by fissures (grikes) formed by carbonation. Limestone pavements develop where the rock strata are horizontally bedded and well-jointed. Carbonation is concentrated along the joints in the rock forming the fissures. *See* clint, grike. *See also* karst.

limon A fine-grained deposit, which is of similar wind-blown origin to loess, and is often treated as synonymous, but occurs in more humid conditions (e.g. N Europe). It includes brick-earths. See loess. [French]

limonite ($2Fe_2O_33H_2O$) A yellowish-brown ore of iron, containing 60% iron. It results from the alteration of other iron minerals, such as iron pyrites. It can form rusty-looking deposits on the surface of basalt that has been weathered and is formed from the iron compounds in the rock. It is also found as a cement between the grains in sandstones, giving them a brown colour, and in association with bauxite. Three types of limonite are found: bog iron ore, pea iron ore, and ochre, the last being used as paint pigment.

limpo (campo limpo) A type of tropical grassland (savanna) in Brazil with very few trees. See savanna. See also campo. [Portuguese]

linear village See street village.

line squall A cold front formed when a cold air mass replaces a much warmer one producing a narrow band of squalls, which may extend for several hundred kilometres. The warm air rises rapidly over the cold and low dark clouds are formed with heavy rain or hail and violent gusts of wind. As the band is narrow and usually moves rapidly, the changes in the weather associated with this feature can be very sudden.

links In Scotland and NE England, an area of near level or gently undulating sandy land near to the seashore. Sand dunes and coarse grasses are often associated with such land.

linseed The oil produced from the seed of the flax plant. The oil is used in paints, varnishes, and linoleum, and the residue is compressed into a cake as a cattle feed. See also flax.

lithogenesis The first stage of mountain building during which sediments accumulate in a gradually sinking geo-syncline. These eventually become compacted to form rock. See also orogenesis.

lithology The study of the physical and chemical characteristics of rocks, including colour, grain size, composition, and texture. It is sometimes restricted to sedimentary rocks.

lithosol A shallow stony azonal soil consisting chiefly of unweathered or partly weathered rock fragments, e.g. scree. Little soil development takes place and organic matter accumulates in the soil horizon giving it a dark appearance. Compare regosol.

lithosphere The outer portions of the Earth, including the crust and the upper mantle. It lies above the weaker and less dense asthenosphere. The term is also used, but less frequently, to refer to the crust of the Earth alone. See also asthenosphere.

litre Symbol: l A metric unit of volume now defined as 10^{-3} metre³ (i.e. 1000 cm³).

litter The vegetable matter, mainly fallen leaves and twigs, that loosely covers the surface of the soil, sometimes to a depth of many centimetres. Plant litter is only slightly decomposed so that its source (e.g. beech leaves) remains recognizable. It is converted into humus by further decomposition. †Different types of litter confer different properties on the soil. For example, pine needles are physically tougher and chemically more acidic than oak leaves. Compare humus.

Little Ice Age See ice age.

littoral Denoting the shore of a sea, lake, or river. The littoral zone is that part of the shore that lies between the highest high-water mark and the low-water mark along a coast. The eulittoral zone is the zone that extends between the high-water mark and the limit of attached plants in the sea, which is usually at a depth of between 40 and 60 m.

littoral drift *See* longshore drift.

livestock The animals and birds that are kept on a farm for profit or to provide food for the family. The main types of livestock to be found on farms include cattle, pigs, sheep, horses, goats, and poultry.

llano The tropical grassland (savanna) of the N part of South America, especially the Guiana Highlands and Orinoco Basin. *See* savanna. [Spanish: flat ground]

load The material transported by ice, wind, or water (e.g. rivers, seas). A river carries some of its load in solution (in which it is dissolved in the water), some as suspension load (in which small particles are held up within the water), and the rest as bedload. The bedload moves by sliding and rolling along the river bed under the influence of gravity and by saltation, in which particles bounce along the stream bed. When the river is in flood the velocity is increased and it can transport larger loads, even boulders. The load is deposited when the river is no longer able to carry it; for example, when the velocity is reduced on entering a lake.
†The total amount of load that can be transported by a river is known as its stream capacity. The stream competence is indicated by the size of the largest particle in the load.
See also bedload, saltation, suspension load.

loam soil A soil that consists of a mixture of sand (less than 52%), silt (28–50%), and clay particles (7–27%). Loam soils are generally the best agricultural soils, combining the free movement of water and air between the coarse grains with the nutrient-retaining properties of the clay colloids. *See also* soil texture.

local authority In the UK, a body that provides services to the population at a local level. Largely established in the 19th century, local authorities underwent reorganization in 1974 (England and Wales) and in 1975 (Scotland). In England, a new hierarchy of metropolitan counties and districts and counties and districts was created; in Wales, new counties and districts; and in Scotland, regions and districts. The local authorities are responsible for providing a variety of services, including housing, roads and public transport, leisure, planning, and environmental services, although not all authorities provide all these services.

local climate The climate of a small area that is distinct from the climates of surrounding areas.

localization of industry †The concentration of an industry in a particular region or regions. For example, shipbuilding is concentrated on certain river estuaries but not on others. This concentration can only be explained by a careful analysis of such factors as the historical background to the development and the economies to be gained by shipbuilders selecting certain sites and rejecting others.

local plan In the UK, a land-use plan designed to give detail to the broad policies described by a structure plan (a plan that usually covers a county). There are three types of local plan: district plans, action area plans, and subject plans. District plans affect fairly large areas planned for comprehensive development. Action area plans affect smaller areas undergoing intensive change. Subject plans give detailed treatment to a particular aspect of planning (e.g. the development of minerals). *See also* structure plan.

local time (apparent time) The time of day at a place as indicated by the position of the Sun. *Local noon* occurs when the centre of the Sun appears to cross the meridian of the place, i.e. when the Sun reaches the highest point in its apparent diurnal movement across the sky. Local time varies from Greenwich Mean Time (GMT) by 4 minutes per degree of longitude from the Greenwich Meridian (0°); it is earlier in the day to the E of Greenwich. *Compare* standard time.

locational rent †The concept, first proposed by the British economist David Ricardo (1817), that the rent paid for a piece of land is a surplus earned by the owner on nonmarginal land. Farmers and other tenants will pay for the use of land that will bring them returns greater than their costs. The larger the profit margin, the more rent can be charged because the demand for the land will also be high. For example, high rents can be charged for shops in London's Oxford Street because this location attracts many customers and profits are high.

location of industry The spatial distribution of industrial activity, i.e. the siting of manufacturing plants. For example, a map showing the location of industry in the London area would show certain areas of concentration such as the Lea Valley, lower Thames Valley, and W London.

location quotient †The degree of concentration of an industry in a particular area compared with the distribution of industries as a whole. The location quotient for an industry in a particular region is obtained by using the formula:

$$LQ = (Ax/Ex)/(An/En)$$

where LQ is the location quotient, Ax is the number of people employed in industry A in area x, Ex is the number of people employed in all manufacturing in area x, An is the number of people employed nationally in industry A, and En is the number of people employed nationally in all manufacturing industry. A location quotient of more than 1 shows that the region has more than its share of a particular industry. Conversely, a value less than 1 indicates that it has less than its share.

location theory †A body of theory that attempts to explain and predict the location of economic activities. One of the earliest attempts to explain

agricultural land-use patterns was the von Thünen model (1826). Many models to explain the location of industry have been formulated. Earlier industrial location theories sought to find the best or optimum location with maximum profits. In practice maximum profit theories have been difficult to formulate and separate theories have been established for least costs and maximum revenues. The two best-known theories are the least-cost theory developed by A. Weber (1909) and the maximum revenue theory of A. Lösch (1940). Weber focused particularly on transport costs and how these could be minimized in the choice of a location. Although many of his assumptions such as perfect competition are unreal, his ideas highlight the importance of transport costs and the possible different concentrations of industries around materials, labour, and markets. Lösch made an analysis of the size of market areas and attempted to identify the optimum market area for firms in competing industries.

More recent theories are behavioural models in which the search is for areas of profitability and satisfactory locations. These recognize that in reality decision-makers rarely have all the facts or the ability necessary to make decisions that minimize costs or maximize profits. *See* market-area analysis, von Thünen model, Weber's model.

loch In Scotland, a lake or a long narrow sea inlet. Loch Tay and Loch Ness are examples of lakes in Scotland; Loch Fyne is an example of a sea inlet.

lodgement till Material that is deposited directly from moving ice (e.g. a glacier or ice sheet). It is frequently compacted by the weight of the ice above it. Meltwater is not involved in the process and as a result the material remains unsorted and unstratified. †Large fragments of rock in the till are aligned so that their long axes point along the direction in which the ice flows. *See also* till.

loess (löss) A fine-grained deposit, yellowish or grey in colour, consisting of wind-borne dust. It probably originated as wind-blown dust from desert surfaces or vegetation-free areas at the margins of ice sheets. Loess covers large areas in N central Europe, the Mississippi Plains (USA), and E China. Deposits accumulate to a depth of tens of metres. It forms a fertile soil, especially when irrigated. †Loess is usually stone-free and unstratified. It is coherent (i.e. it binds together) but not cemented, and therefore is porous. Sometimes it may be calcareous with shell fragments. *Compare* limon. [German]

logan stone *See* rocking stone.

longitude The distance of a point E or W of the Greenwich (prime) meridian, which is longitude 0°, measured as an angle from the centre of the Earth. Longitude is measured in degrees E or W of Greenwich up to 180°. Lines of longitude are known as meridians. At the equator the lines of longitude are about 111 km apart. The distance becomes progressively shorter N and S of the equator until it is zero at the poles. *See illustration at* latitude.

longitudinal coast *See* concordant coast.

longitudinal dune *See* seif dune.

longitudinal valley A valley that lies parallel to the grain or structural trend of an area. For example, in the Dalmatian region of Yugoslavia a series of folds runs parallel with the Adriatic Sea. Within these folds are numerous longitudinal valleys, many of which are flooded.

long profile The profile of a river channel that extends from its source to its mouth. Ideally the long profile is a smooth concave curve. In reality it is usually broken by knickpoints (as a result of past base level changes), stretches of rapids, lakes, hard rock bands, etc. A smooth profile from source to mouth may indicate that the river has reached a state of grade, in

which there is a balance between erosion and deposition. The river is then said to possess a graded long profile.

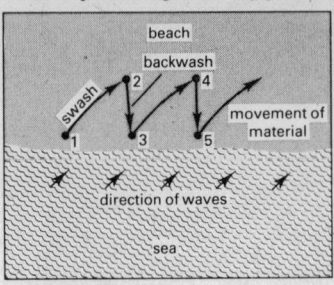

Longshore drift

longshore drift (littoral drift) The movement of material (sand or shingle) along the length of a beach as a result of waves breaking at an oblique angle upon the shore. The swash of the waves moves material diagonally up the beach. The retreating water or backwash of the waves washes some of that material back towards the sea and at right angles to the beach. The overall effect is that there is a net movement of material along the shore. Longshore currents aid this process. The prevailing winds will indicate in which direction the movement will occur. In S England the drift is to the E because the prevailing winds blow from the SW. Groynes are often constructed to halt the movement of material. *See also* groyne, backwash, swash.

long wave *See* Rossby wave.

lopolith An igneous intrusion that lies between rock strata and is saucer-shaped, forming a large shallow basin at the surface. The feature is concordant because it lies between and not across the layers of rock. The Bushveld in South Africa is underlain by a large lopolith.

Lorenz curve †A graphical technique that is used to compare an uneven distribution with a hypothetical even one. The Lorenz curve may be used, for example, for studying industrial diversification in a country. The curve is drawn by expressing the frequency in each category (e.g. numbers of people employed in each manufacturing industry in a place) as a percentage of the total frequency (e.g. total number of people employed in industry in a place) and plotting the result graphically in the form of a cumulative frequency curve. An even distribution results in the curve being a straight line at 45° to the horizontal. The more uneven the distribution the more concave will be the curve.

lough In Ireland, a lake or sea inlet. For example, Lough Neagh in Northern Ireland. [Irish]

low, atmospheric Any area in the atmosphere where the atmospheric pressure is lower than that of the surrounding area (e.g. a depression). On a weather chart such areas are usually identified by closed isobars with values decreasing towards the centre.

lowland An area of land of low elevation. It may be low relative to adjacent lands with a higher elevation (e.g. the Scottish Lowlands), or it may be an area of land that lies close to sea level (e.g. the Fens).

loxodrome See rhumb line.

lucerne See alfalfa.

L wave See earthquake.

M

macchia See maquis. [Italian]

mackerel sky A cloud pattern formed by some cirrocumulus (Cc) and altocumulus (Ac) clouds that resembles the pattern of scales on a mackerel.

macroclimate The general climate of a large area (e.g. Mediterranean climate). Compare microclimate.

maelstrom A whirlpool or strong tidal eddy. It may be the result of turbulence within a tidal current or it may

develop at a point where different currents converge. The most famous example is the Maelstrom, which is of tidal origin, between two of the Lofoten Islands off the NW coast of Norway. [Dutch]

maestrale A cold northerly or northwesterly wind that blows in winter over N Italy. It is the Italian equivalent of the mistral of S France. [Italian]

maestro A northwesterly wind that blows most commonly in the Ionian and Adriatic Sea areas and around Corsica and Sardinia. It occurs as the winds swing around behind a depression that is passing eastwards through the Mediterranean basin.

magma Molten rock material charged with gases and at a very high temperature lying beneath the Earth's surface. On solidification and crystallization it forms igneous rock. When magma is extruded out onto the surface of the Earth it forms lava; if it solidifies within the crust it either forms hypabyssal or plutonic rocks.

magnesite (magnesium carbonate, $MgCO_3$) The most common ore of magnesium. It is usually white in colour but can be grey, yellow, or brown. It results from either the weathering of magnesium-rich rocks or the replacement of dolomite and limestone under the action of volcanic gases. Magnesite is found in India, Australia, the USA, and Canada, and is used in the production of carbon dioxide and magnesium salts, and in the manufacture of fire bricks, crucibles, and fertilizers.

magnetic declination (magnetic variation) The angle between true north and magnetic north, measured on a horizontal compass on the Earth's surface. This varies with time and with position on the Earth's surface. In 1982 declination at Greenwich was 4°57′W, decreasing about 12′ a year.

magnetic field, Earth's See geomagnetic field.

magnetic pole One of the two poles in the Earth's magnetic field. At these two points a freely pivoted compass needle will point vertically downwards. The magnetic poles do not coincide with the geographical poles: the north magnetic pole is located in Canada, about 1300 km away from the North Pole, and the south magnetic pole is located in S Victoria Land in Antarctica, about 2700 km away from the South Pole. The magnetic poles are not fixed and their locations slowly change with time. *See also* geomagnetic field.

magnetic reversal *See* geomagnetic field.

magnetic storm A worldwide but temporary disturbance of the Earth's magnetic field that appears to be associated with the occurrence of solar flares and sun spots. The aurora is also usually highly visible and extensive during such a storm, but whether this is a cause or an effect is not fully understood. The main effects are the disruption of radio and telegraphic communications and magnetic surveys.

magnetic variation *See* magnetic declination.

magnetite (lodestone, Fe_3O_4) A black magnetic iron ore, containing 72% iron. It is found in small quantities in most igneous rocks. Workable deposits of magnetite are found in the Urals in the USSR, in the E USA, and at Kiruna and Gällivare in N Sweden.

magnetosphere The area surrounding the Earth in which the Earth's magnetic field exists. It extends to about 60 000 km on the side of the Earth facing the Sun but has a much greater extent on the opposite side.

magnitude of an earthquake The amount of energy released by an earthquake. Several methods have been developed to assess this energy but the method used today is the Richter scale. *See also* Richter scale.

maize (Indian corn) A grass cereal that produces large edible seeds. It is known as corn in the USA. Maize requires a growing season of at least 140 days and 200 mm of rain in the summer months. The plant grows to a height of 3–4 m and produces seed cobs on the main stem. It forms a staple food in South America and many other parts of the world. From the seeds a number of food products, such as corn flakes, popcorn, corn oil, corn beer, and cornflour, are obtained. In the developed world, e.g. the USA, which is the world's chief producer of maize, the most important use of the seed is as an animal feed. A garden variety known as sweet corn is cultivated as a vegetable. *See also* corn.

mallee A scrub vegetation that grows in parts of SW and SE Australia under desert conditions. Mallee consists mainly of evergreen eucalyptus shrubs, about 2–3 m high, forming dense low thickets.

Malthusianism †The ideas on population of the British economist Thomas Malthus (1766–1834), which were put forward in his *Essay on the Principle of Population* (1798) and later developed. He argued that population growth is a function of the growth of the food supply: if the food supply was increased then the population would also grow. He propounded the idea that population, if unchecked, tends to increase at a geometrical rate (i.e. 1, 2, 4, 8, ...), while food production increases only at an arithmetical rate (i.e. 1, 2, 3, 4, ...) over the same time interval. He also held that population would increase to the limits of the means of subsistence and is only checked by war, famine, and pestilence. Population increase would result in misery and lower standards of living.

Today *Neo-Malthusians* believe that only through the limitation of births can the size of world population be sufficiently controlled to make possible the elimination of misery and raising of the standards of living.

mandate The power given to a country to govern or supervise another nation not previously under its control. After World War I the League of Nations designated the principal victors, including the UK and France, to become mandatory powers over the former colonial possessions of Germany and the Turkish empire, until the territories were ready for self-government. The mandated territories included Syria, Tanganyika (now part of Tanzania), and South West Africa (now Namibia). On the creation of the United Nations (1945) the remaining mandated territories became trust territories. *See also* trust territory.

manganese A light pinky-grey metal that does not occur in the native state. The most important ores are oxides that occur in two ways: in sediments associated with iron compounds, and in association with manganese-bearing rocks that have been altered by weathering to produce manganese nodules. The most important ore is manganite, containing 62.5% manganese, which occurs in association with bauxite. Manganese nodules, which contain about 24% manganese together with other metals such as iron and nickel, occur on the deep ocean floor. Most of the output of manganese (about 95%) is used in the production of hard steels; it is also used in the manufacture of chlorine, paint and varnish, and in disinfectants. The chief producers are the USSR, India, South Africa, Ghana, and Morocco.

mango showers Heavy showers, often accompanied by thunder, that occur in India between March and May, before the breaking of the main monsoon. The name arises from the fact that at this time the mangos are beginning to ripen. *See also* blossom showers.

mangrove A type of swampy vegetation that grows along many tropical coastlines, notably in Asia. It consists of a number of broad-leaved shrubs and trees that are adapted to the intertidal environment. These adaptations include arching aerial roots to anchor the trees and special breathing roots with 'knees', which operate at low tide. Mangrove requires calm muddy conditions and shallow water. It grows especially along rivers, in deltas, and where protected inside the surf zone. †The tangled mangrove vegetation shows a zonation from the shoreline outwards and progresses through seral succession. The mangroves reduce the local salinity to brackish levels and by preventing erosion and trapping sediments they extend the land out to sea.

mantle That portion of the interior of the Earth lying beneath the crust and above the core. The evidence obtained from the recording of earthquake waves suggests that the mantle is made up of ultrabasic rock. Its upper limit is the Mohorovičić discontinuity, which lies at an average depth of 35 km below the surface of the Earth, and it descends as far as the Gutenberg discontinuity, which lies at about 2900 km.

manufacturing industry Those industries that process materials or assemble components to produce finished products. Manufacturing industry is distinct from other types such as service industries or the tourist industry because goods and materials are produced.

map The representation on a flat surface (i.e. plane surface) of all or part of the Earth's surface, or some other celestial body such as the Moon. The representation is drawn to a specific scale and map projection and shows distinctive aspects of the surface such as relief features, settlements, and routes. Examples of maps designed to show specific features include geological maps, which show the nature of the rocks beneath the Earth's surface.

map projection A method by which the curved surface of the Earth, or a part of that surface, is represented on a flat surface (i.e. plane surface) as a map. Mathematical formulae are used to construct a graticule on the map corresponding to the intersecting parallels and meridians on the Earth. A number of map projections have been devised; the three main groups are cylindrical, conical, and zenithal (azimuthal). It is impossible to construct a projection of a large part of the Earth's surface without some distortions of shapes, relative areas, or directions. The map projection chosen for an area thus depends on which of these three features needs to be accurate, the extent of the area, and the use the map is intended for. Projections of smaller countries such as the UK can be extremely accurate. *See* conical projection, cylindrical projection, zenithal projection.

maquis (macchia) A scrub vegetation that occurs on acidic soils in the Mediterranean region (e.g. S France). Maquis consists of a dense mass of low-growing evergreen shrubs (e.g. laurel, myrtle, and rosemary), herbaceous plants, and occasional taller trees (e.g. wild olive), that can withstand the hot dry summer. †It is regarded as a plagioclimax vegetation determined by man's clearances of the original evergreen oaks, regeneration of which is suppressed by fire and the grazing of goats.
Compare chaparral. [French]

marble A limestone that has been completely recrystallized by metamorphic processes involving pressure and heat. The resulting rock is hard, patterned, and shiny, with no trace of its original fossil content. Examples include the white marble from Carrara in Italy, which is used for statues. Some marbles are formed by the mixing of limestone with igneous rock, e.g. the green Connemara marble from Ireland. The term marble is sometimes used commercially for polished limestones but this is incorrect.

mares' tails High thin cirrus (Ci) clouds in the form of delicate white filaments that have a hair-like appearance. They often indicate high winds in the upper atmosphere.

marginal land Land that is sufficiently fertile to yield a return that just covers the costs of production. Marginal land will be abandoned if the price obtained for its products falls or the cost of production increases. In the UK some hill farms are situated on marginal land, which is only grazed when prices are sufficiently high for the farmers to make a profit. At other times the land is abandoned.

margin of production The point at which the costs of production are just covered by the profits obtained from the sale of the product. Any additional costs would not be matched by additional revenue. At this point a firm is producing goods without a net profit.

marine transgression †*See* transgression.

maritime climate (oceanic climate) A climatic type in which the ocean is the dominant influence; it is therefore found on islands and near coasts, especially those having prevailing onshore winds. Such an area, which can occur in any latitude, is equable (i.e. it has small diurnal and seasonal temperature ranges) because the sea heats up and cools down much more slowly than the land. It also tends to be humid with a considerable amount of cloud cover and precipitation, the air passing over it having picked up moisture from the ocean. *Compare* continental climate. *See also* insular climate.

market The place at which there is a demand for commodities, e.g. a market town such as Norwich. The term is also used in a more generalized sense to indicate the demand for a commodity, e.g. the world cocoa market, or the market for antique silver in the USA.

market area analysis †A concept of industrial location introduced by the German economist August Lösch in which the best location was that which maximized profits. He assumed:

(1) an isotropic surface (i.e. a uniform land surface); (2) a uniformly distributed population; (3) a uniform transport cost surface; and (4) the existence of perfect competition. An industry (e.g. a brewery) would serve the area around it with price increasing away from it. The market area for the brewery would thus be a circle with demand greatest at the centre. If other breweries, with their own circular market areas, locate outside the market area of the original brewery there will eventually be a hexagonal pattern of market areas as the circles become packed together. See also location theory.

market gardening The intensive cultivation of vegetables, soft fruit, or flowers for sale. This form of cultivation is highly specialized and involves considerable capital and/or labour expenditure per unit of land. Where capital is not available, e.g. in countries of the Third World, market gardening is labour intensive. In the UK market gardening on a large scale is found in the Fens and Bedfordshire. In North America the term used is 'truck farming'.

market town A town in which a regular or permanent market is held, especially a town that serves an agricultural area. †The concept of a market town has been reinterpreted in urban geography in recognition of the functions that the town performs for the surrounding countryside. A market place is distinguished by its role as the central place for a tributary area.

marl 1. A type of soil that is a natural mixture of clay and calcium carbonate. It is sometimes used as a fertilizer and for making bricks.
2. A mudstone containing a high proportion of calcareous material.

marsh A poorly drained low-lying tract of land that possesses distinctive flora. The marsh may be subject to occasional flooding and the water table is nearly always very close to the surface. Rushes, reeds, and certain mosses are characteristic of marshland. A marsh may be drained to provide agricultural land as has occurred, for example, in the Fens of East Anglia. See also salt marsh.

massif An upland region that is formed by a mountainous plateau with well-defined boundaries. The Massif Central of France is an example of this form of upland. [French]

mass movement (mass wasting) The movement of material down a slope under the influence of gravity. The movement is often aided by lubrication by water. †Mass movements may be slow or rapid, and take the form of flows, slides, or falls. Slow mass movements include soil creep. Examples of sudden and rapid mass movements include landslides and rock falls. A flow of material involves internal deformation in which the particles at the contact surface of the flow move more slowly down, due to friction with the ground surface, than those particles above; these include earthflows, mudflows, and solifluction. Slides and falls act along a shear plane within the rock and all the particles move at the same speed.
See also creep, earthflow, landslide, rockfall, solifluction.

mass production The manufacture of a commodity on a very large scale. This is normally achieved in a factory by conveyor-belt techniques or the continuous use of automatic equipment. The high volume of output provides economies of scale which may cheapen the commodity. For example, mass-production techniques have reduced the comparative prices of TV sets and cars in recent years.

material index The measurement of the loss of weight during the processing of raw materials when compared with the weight of the finished product. The formula is:
material index = weight of raw material inputs/weight of finished product
The higher the index is above 1 the greater the significance of the cost of

moving the raw material to the factory.

maximum thermometer A thermometer that records the maximum temperature achieved over a specified period of time. The most common type consists of an ordinary mercury thermometer with a metallic marker resting above the mercury column. The marker can be pushed upwards by the mercury but cannot slide down again as the temperature drops and the mercury recedes. *See also* minimum thermometer.

meadow soil An intrazonal hydromorphic soil that occurs in low-lying areas such as floodplains. It has a humus-rich dark brown A horizon that overlies a mottled gleyed zone. Lush grassland, often managed for a hay crop, provides the surface humus. The gleying results from a high water table or recurrent flooding. *See also* gleying.

mean (arithmetic mean) †In statistics, the average value. It is obtained by the formula:

$$\bar{x} = \Sigma x / n$$

where $\bar{x}$ is the mean, n is the number of values being considered, and Σx is the sum of all the values. *See also* median, mode.

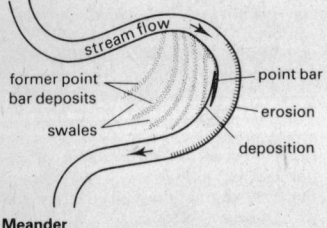

former point bar deposits — point bar
swales — erosion
deposition

Meander

meander A pronounced curve or loop in the course of a river channel. The flow of water in a meandering stretch of river is not even. The current is fastest on the outside of the meander bend and this area is subjected to the greatest erosion. Concentrated erosion on the outside of meander bends results in the formation of a narrow neck. The river may eventually break through the neck in times of flood bypassing the meander bend and forming a cutoff, the ends of which rapidly silt up forming an oxbow lake. The current is slowest on the inside of the bend and material is deposited forming point-bar deposits and a gentle slip-off slope. As the outsides of the meander bends are eroded and move downstream, the insides of the bends become filled by sedimentation and also move downstream. This results in a continuous downstream migration of the meanders.

†A number of theories of meander development have been suggested. One theory is that a local feature leads to the development of a meander, which then influences the development of a series of meanders downvalley. A second theory is that the presence of pools and riffles (alternating deeps and shallows) in the river's course leads to their development. The size and wavelength of a meander is related to the river's discharge; wavelength is normally about ten times the channel width.

See also oxbow, †pools and riffles.

meander belt †That part of a valley floor over which a river will shift its channel. The boundaries of a meander belt are marked by the concave banks of successive meanders. Beyond the meander belt there may be low river bluffs or river terraces.

meandering valley †A sinuous valley with pronounced curves. The slopes on the inside of these curves tend to be less steep and frequently convex whereas the slopes on the outside of the bends are steeper and may be concave in shape. The river that occupies the valley is likely to have developed meanders on a much smaller scale over the alluvial deposits of the valley floor.

mean sea level The average level of the surface of the sea. It is calculated from data which takes into account the changes in tide that have occurred over a long period of time. In the UK this calculation is based on Newlyn in Cornwall (the Ordnance Datum).

mechanical weathering (physical weathering) The break-up of rock by processes of weathering that do not involve chemical change. Mechanical weathering is responsible for the disintegration of rocks, as opposed to decomposition, which results from chemical weathering. The physical stresses and strains causing mechanical weathering may be the result of expansion and contraction of grains or crystals in the rock itself or the result of changing external pressures. These processes include thermal expansion, which results in exfoliation; freeze-thaw action; salt-crystal growth, in which the growth of crystals creates pressures in the rock; pressure release (unloading), in which the removal of surface rocks leads to the splitting of rocks; and animal and plant activity. *See also* exfoliation, freeze-thaw, granular disintegration.

medial moraine The angular weathered debris that lies in the form of a ridge running lengthwise along a glacier. It results from the convergence of two or more lateral moraines at the point where two or more valley glaciers meet. For example, the moraine formed by the confluence of the Mittel-Aletsch and the Ober-Aletsch glaciers in Switzerland. *See also* moraine.

median †In statistics, the central value or class in a series of ranked values. *See also* mean, mode.

medical geography The study of the distribution of human disease and causes of death, together with the study of the environmental factors that affect human health.

Mediterranean climate (warm temperate western margin climate) The broad climatic type that occurs on the W coasts of the continents between latitudes 30° and 40°N and S. The major influence in these areas is the seasonal shift of the wind belts, with the trade winds blowing off the land in summer and the westerlies blowing off the sea in winter. Thus summers tend to be hot and dry while in winter depressions move in from the west bringing mild moist conditions. The natural vegetation is adapted to the hot dry summers and consists mainly of drought-resistant shrubs and small trees, e.g. the maquis of S France and the chaparral of the SW USA. The crops grown are also those requiring generally warm conditions, e.g. grapes, olives, citrus fruits, and flowers (for the perfume industry).

The largest area having this climate is the Mediterranean basin, and while there are great variations in temperature and rainfall within such a large area, the whole basin satisfies the basic criteria of summer drought and winter rainfall. Other areas having this climatic type are central California, central Chile, SW Australia, and Cape Province, South Africa.

megalopolis An urbanized region that contains several large metropolitan areas. For example, the wholly urbanized region extending from New York to Washington in the USA. [From Greek: 'great' and 'city']

meltwater Water that originates from melting snow or ice (e.g. a glacier). †It frequently carries large loads of debris. Material that has been deposited by meltwater differs from that that has been laid down by the ice in that it is sorted into layers and often rounded. The meltwater may flow above, within, or below the ice. It has been suggested that most of the erosion in glaciated areas is carried out by meltwater rather than the glacier ice. The term fluvioglacial is usually applied to the effects or features resulting from meltwater in glaciated areas.

mental map †The images of space and preferences held by an individual or a group. These will vary between individuals and between groups according to the information flows received from external stimuli and to individual characteristics such as sex, age, socio-

Mercator projection

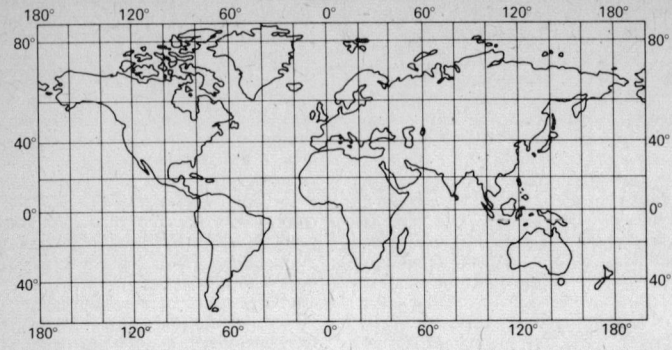

Mercator map projection

economic status, ethnicity, and individual personality factors. People often act according to their mental maps rather than to more objective information.

Mercator projection A cylindrical map projection. The equator is drawn true to scale but all other parallels of latitude are drawn the same length and therefore exaggeration increases away from the equator. This is balanced by exaggeration of scale along the meridians of longitude equal to that along the parallels at any point, i.e. the map is stretched N–S to balance the E–W stretch of the parallels. Scale from a given point is the same in all directions and shapes over small areas are shown correctly. However, scale and area are considerably exaggerated in high latitudes. Compass bearings are accurately shown, and plotted as straight lines drawn between two points. As a result the projection is widely used for navigation; it is also used for world maps. *See also* transverse Mercator projection. [Named after Gerardus Mercator, a Flemish cartographer who first used the projection in 1569]

mercury (quicksilver) A silver-white metal that is liquid at normal temperature and pressure. It is obtained chiefly from its ore cinnabar, though it is also found in the native state.

Cinnabar deposits are formed by volcanic activity and are being deposited in hot springs today. Mercury is used chiefly as a liquid electrode in the electrolytic production of chlorine and caustic soda; it is also used in the manufacture of drugs, paints, thermometers, and barometers. Important deposits are found in Spain, Italy, and the USA.

mere A small shallow lake or pool. The term is applied especially to waterfilled hollows in Cheshire that result from subsidence of the ground due to the extraction of salt deposits.

meridian A line of longitude passing from the North to the South Pole and forming half of a great circle. From the Greenwich meridian, sometimes called the prime meridian, all other meridians are measured. *See also* longitude.

meridional heat flux The general flow or transfer of heat from tropical latitudes towards the poles. This is achieved in two main ways: by the N and S movement of air masses (meridional flow) and by the mass movement of warm waters in ocean currents (e.g. the Gulf Stream).

mesa A flat-topped and steep-sided tableland that develops in semiarid areas on horizontal or very gently dipping strata. Mesas are commonly topped by a resistant cap rock and are the remains of plateaus that have been denuded. Mesas are similar in shape to buttes but occur on a larger scale. Examples are found in Arizona, USA. *See also* butte. [Spanish]

meseta An extensive mesa or tableland with steep well-defined sides and a flat top. The term is applied to the plateau of central Spain. [Spanish]

mesopause A temperature discontinuity in the ionosphere at an altitude of 80 –90 km, below which, in the mesosphere, temperatures decrease with altitude and above which, in the thermosphere, temperatures increase with altitude. The temperature at the mesopause is approximately –100°C.

mesophyte †A plant that requires a moderate amount of water, but cannot withstand saturation or drought. Temperate vegetation that grows in areas of reliable rainfall is mesophytic.

mesosphere A layer in the atmosphere extending between the stratopause (at an altitude of about 50 km) and the mesopause (at an altitude of 80–90 km). Within the mesosphere the temperature decreases with altitude from about 0°C at the stratopause to about –100°C at the mesopause.

Mesozoic The geological era that extended from about 225 million years ago to about 65 million years ago. It followed the Palaeozoic era and preceded the Cenozoic era. It comprises the Triassic, Jurassic, and Cretaceous periods. The era was characterized by the great number of giant reptiles and by the appearance of the first mammals at the beginning of the Triassic period.

mesquite A spiny shrub or small tree of the genus *Prosopis* that grows in arid parts of the SW USA (e.g. Texas) and Mexico. The plants grow in dense thickets. Their deep roots aid resistance to drought. [Spanish]

mestizo In South America, the offspring of a Spanish American and an American Indian. The term is sometimes used elsewhere for a person of mixed parentage. [Spanish: mixed]

metamorphic aureole The zone of altered or metamorphosed rock that surrounds an igneous intrusion resulting from contact with high temperatures, pressure, and gases from the intrusion.

metamorphic rock Rock formed when pre-existing sedimentary or igneous rock is altered as a result of changes in physical or chemical conditions. This process of *metamorphism* may be through intense stress or pressure caused by earth movements, increased temperature caused by volcanic activity, or the action of gases and liquids of magmatic origin.

There are three main types of metamorphism:

(1) *contact (thermal) metamorphism*, which results from the intrusion of a mass of molten rock.

(2) *dynamic (regional) metamorphism*, which takes place when rock layers undergo strong structural deformation during mountain building.

(3) *dislocation metamorphism*, which occurs when pre-existing rocks undergo localized deformation along a fault plane or thrust plane.

The mineral composition and the structure of the rock can be altered; for example, under stress some minerals such as mica take on a parallel arrangement. Examples of metamorphic rocks include marble (metamorphosed limestone), quartzite (metamorphosed sandstone), and slate (metamorphosed shale). *See also* igneous rock, sedimentary rock.

meteor A small fragment of rock and metal travelling through space. When drawn into the Earth's gravitational field it becomes white-hot through friction as it falls through the atmosphere, and is seen as a *shooting star*. If it reaches the Earth's surface it is called a meteorite, but most meteors burn up by the time they reach 75 km above the Earth. Some 8,000 million meteors enter the Earth's atmosphere every day, many no larger than sand grains. Meteor showers occur on certain dates in the year, probably originating as debris in comet orbits.

meteorite A mass of rock or mineral that enters the Earth's atmosphere from outer space and falls to the surface of the Earth without being burnt up. About 500 fall to Earth every year but only about 10 are ever recovered. Meteorites are composed chiefly of either iron or stone, or a mixture of both.

meteorite crater An indentation in the Earth's crust caused by the impact of a meteorite. Much of the mass of a meteorite may be burnt up as the object falls through the atmosphere, but large meteorites have struck the Earth and have produced large steep-sided craters. For example, the Great Meteor Crater in Arizona, USA, which is approximatcly 5 km in circumference and 174 m deep and was formed by the impact of a meteorite about 50 000 years ago.

meteorological satellites Spacecraft placed into orbit around the Earth and equipped with instruments for recording meteorological data. They are playing an increasingly important role in weather observation and forecasting. In 1960 the first US weather satellite – Tiros I – was launched. This was followed closely by the Russian Cosmos and Meteor series. Today many nations operate meteorological satellites enabling virtually the entire surface of the Earth to be covered. The data sent back to Earth includes photographs of cloud patterns and development, temperature (using radiometers that measure infrared radiation), and upper wind statistics. Satellites are also used for the collection of data from instrumented balloons, ocean buoys, and unmanned weather stations in remote areas, and they speed up the communication of data around the world, allowing more accurate weather forecasts to be made.

meteorology The scientific study of the atmosphere and the physical processes at work within it, including pressure, winds, temperature, precipitation, clouds, sunshine, etc. The worldwide collection and study of data relating to these processes over a long period of time has enabled meteorologists to discern recurring patterns of weather development, which are the basis of weather forecasting. The forecaster uses a weather (or synoptic) chart plotted from data obtained simultaneously at a great number of weather stations and from this is able to predict the likely weather conditions for a limited time ahead.

metre Symbol: m The SI base unit of length. It was first introduced in France in 1799 when it was assumed to be one ten-millionth of the surface distance from the equator to the pole (in fact 0.02% too short). It is now precisely defined as the length equal to 1 650 763.73 wavelengths in a vacuum corresponding to the transition between the levels $2p_{10}$ and $5d_5$ of the krypton–86 atom.

metropolis The chief town or city of a country or region. For example, London may be described as the metropolis of the UK.

mica A group of silicate minerals that have perfect cleavage, so that they split easily into thin plates. It is a common mineral in igneous and metamorphic rocks and is responsible for the characteristic shiny appearance of many metamorphic rocks such as schists. The most common types are biotite, which is dark in colour, and muscovite, which is light.

microclimate The climate of the immediate surroundings of a particular object of study, e.g. a plant or building. The size of the object determines the area under consideration. The microclimates of urban areas have received much attention in recent years. *Compare* macroclimate.

mid-latitude Denoting those areas of the world lying between latitudes 30° and 60°N and S. In meteorological and climatic terms, it is the area within which the major factor is the meeting of polar and tropical air masses producing frontal activity.

midnight sun A phenomenon observed in latitudes higher than 66½°N and S where, during summer, the Sun does not sink below the horizon. This results from the tilt of the Earth on its axis, each hemisphere being inclined towards the Sun during its summer. The duration of the phenomenon increases towards the poles, where it may be observed for six months of each year.

mid-ocean ridge A ridge, often of great length, rising above the deep-sea plain. Such ridges occur where the plates of the Earth's surface are moving apart, allowing material to well up from the interior. The two largest are the Mid-Atlantic Ridge, which extends from N of Iceland (about 70°N) to Bouvet Island (about 55°S), with only one appreciable gap near the equator, and the Chagos-St Paul Ridge, which runs down the whole length of the central Indian Ocean.

migration The movement of an individual or group from one place of residence to settle in another, either permanently or semipermanently. Together with fertility and mortality, migration is one of the chief elements determining the population change of an area. *International migration* is the movement of population from one country to another (e.g. from the UK to Australia). *Internal migration* is the movement of population within a country or land unit. It results in a change in the distribution of population but does not effect the total numbers. It includes inter-regional movement (e.g. from NE England to the Southwest Peninsula); inter-urban movement (e.g. Belfast to London); rural-urban (e.g. from the Scottish Highlands to Edinburgh); rural-rural (e.g. from central Wales to the Severn plain); and intra-urban movements (e.g. from Southwark to Barnet in London). *See also* fertility, mortality.

mile, nautical *See* nautical mile.

millet One of the cereal grasses. There are many varieties of millet grown in Asia, N and E Africa, and S Europe. The plant grows to over 1 m in height and produces a head containing large numbers of small seeds. These can be made into flour for human consumption or used as a fodder crop. *See also* kaoliang.

millibar Symbol: mb The unit of pressure used in meteorology. It is equal to one thousandth of a bar, i.e. 100 pascals.

million city A very large city with a population of more than one million.

mineral A naturally occurring substance having a definite chemical composition and atomic structure, and formed by inorganic processes. About 2000 minerals are known. Rocks are made up of various combinations of minerals. Some minerals form a series in which there is gradual replacement of one element by another, e.g. the feldspars. The most common mineral in rocks is quartz.
The term is also used for any naturally occurring material that is mined and of economic value.

mineralogy The scientific study of minerals. This includes the study of crystal form and atomic structure; the chemistry of minerals, and the chemical tests for their identification; the physical properties, including colour, hardness, taste, and feel; and the optical properties that are used in the identification of minerals when rocks are examined in thin sections under a polarizing microscope.

mineral spring A spring of water that contains a high proportion of dissolved mineral salts such as iron compounds, magnesium chloride, and sodium chloride. Medicinal properties have frequently been attributed to such waters; as a result spa towns (e.g. Bath, Harrogate) have frequently developed around mineral springs. *See* spring.

minimum thermometer A thermometer that records the minimum temperature achieved over a specified period of time. The most common type consists of an ordinary alcohol thermometer, fixed horizontally, with a metallic marker within the column of alcohol. The surface tension of the meniscus of the alcohol moves the marker down the tube as the temperature falls, and leaves it at the lowest point reached when the temperature rises again. Thus the position of the marker indicates the minimum temperature reached in any given period of time. *See also* maximum thermometer.

mining The extraction of metallic ores and minerals of economic value from the Earth. In the UK the extraction of rock, sand, gravel, and other building materials is termed quarrying, whereas coal, lead, gold, diamonds, and tin are mined, whether or not the working is on the surface or underground, e.g. opencast coal mining.

minority A group of people that is different from the larger group of population of which it forms a part as a result of factors such as race, politics, religion, culture, and language. For example, ethnic minorities of Asians and West Indians in the UK.

Miocene The geological epoch of the Tertiary period following the Oligocene epoch, which ended about 22.5 million years ago, and preceding the Pliocene epoch. It extended for about 17 million years. During this time the Himalayas, the W chains of North and South America, the Carpathians, and the Pyrenees, as well as the Alps, were formed. In Europe, the mountains are characterized by nappe structures. In the British Isles the seas continued to withdraw, resulting in an absence of Miocene rocks. The effects of the Alpine mountain building were felt in the S coast area of England

and in the Isle of Wight, where the rocks were uplifted and tilted but there was no metamorphism. Miocene mammals included elephants, deer, pigs, monkeys, and possibly hominids.

mirage An optical illusion that distorts images of objects, caused by the refraction (bending) of rays of light as they pass through layers of air of differing densities. The most common type, an *inferior mirage*, occurs when the layer of air near the ground is intensely heated, making it less dense than the air above it. This causes the rays of light from the sky to be bent upwards, producing an image of the sky near the ground, which gives the illusion of a sheet of water. As intense ground heating is required to produce this effect it is most common in hot desert areas. It may be observed in higher latitudes on very hot days over surfaces that are good conductors of heat (e.g. metalled roads).
In high latitudes, especially over ice caps, the opposite may occur. A dense layer of very cold air at the surface and warmer air above causes the rays of light to bend downwards and produce a *superior mirage*, the effect being to bring close the images of distant objects.

misfit river (underfit river) A river that is too small in proportion to the size of the valley through which it flows.
†This may result from:
(1) The capture of a river's headwaters, which reduces its volume.
(2) A climatic change, for example one from a pluvial climate to a more arid one. Many rivers in the British Isles are misfits as a result of climatic change. At the end of the Pleistocene glaciation when the glaciers thawed large volumes of meltwater were available to erode valleys.
(3) The glaciation of a valley or the erosion of a valley by an overflow from an ice-dammed lake.

mist A reduction of visibility within the lower atmosphere to 1–2 km caused by condensation producing water droplets within the lower layers of the atmosphere. It is intermediate between fog and haze.

mistral A cold dry and usually strong northerly or northwesterly wind experienced in S France, especially in the Rhône delta area. It occurs when there is high pressure over France and low pressure over the Mediterranean (i.e. mainly in winter). This pressure gradient causes a southerly movement of air, which is strengthened when funnelled down the Rhône valley. It tends to be strongest when the pressure gradient is steepest, i.e. when a deep depression is moving E through the Mediterranean Basin. It frequently reaches gale force and in the Rhône Valley houses and crops must be protected from it by wind breaks of trees oriented E–W. [French]

mixed farming Farming that includes the growing of arable crops and the keeping of livestock. In many cases part of the arable land is used for growing fodder crops for the livestock. In the corn-growing area of the US Midwest, cattle and pigs are reared and fed on part of the corn crop. Cash crops such as malting barley and potatoes are grown on mixed farms in the UK and W Europe.

mode †In statistics, the value or class that occurs most often (i.e. has the greatest frequency). *See also* mean, median.

model 1. A scaled-down representation of a landscape or a feature of the landscape. It is possible to construct working models that show processes within a laboratory, for example, a river in a laboratory flume.
2. †An idealized representation of reality. This may be in the form of a diagram, an analogue, or an equation. Early models in human geography included the von Thünen model of agricultural land use and the central-place models of Christaller and Lösch. In meteorology, mathematical models of the atmosphere are used, for example, for weather forecasting.

moder A form of humus that is intermediate between acidic mor and neutral mull. Moder shows some mixing with the coarser mineral particles in the soil, but retains a stratified appearance. *Compare* mor, mull. *See* humus.

mofette A small hole or vent in the Earth's crust from which carbon dioxide, nitrogen, and other gases are emitted. Mofettes occur in regions where volcanic activity is virtually extinct; for example, in the Auvergne region of central France. [French]

Mohorovičić discontinuity (Moho) A boundary surface within the interior of the Earth that marks the base of the crust and the upper limit of the mantle. It lies at an average depth of 35 km below the Earth's surface and marks a change in the speed of earthquake waves. [Named after the Yugoslav scientist A. Mohorovičić who discovered it in 1909]

Mohs' scale *See* hardness scale.

Mollweide projection †A map projection showing the entire Earth's surface within an elliptical frame. The parallels of latitude are represented by straight lines; the central meridian of longitude is also a straight line but the other meridians curve, the curvature increasing progressively towards the edge of the projection. Shapes, directions, and distances are considerably distorted away from the central area, but there is no distortion of area and the projection is suitable for showing distribution patterns. By 'interrupting' the projection over ocean areas and using a central meridian for each continental area the shapes are less distorted.

monadnock A residual hill that rises above the surrounding flat or gently undulating plain. [Coined by W. M. Davis in 1895 after Mount Monadnock in New Hampshire, USA]

monocline A bend in sedimentary rock strata in which there is a sudden increase in the dip of the beds to nearly vertical, followed by a flattening of the strata as they return to the original dip. There is no change in the direction of the dip.

monoculture Cultivation that is limited to a single crop. There are obvious risks to the farmer who depends on one crop for his livelihood. For example, the monoculture of cotton in the S of the USA at the end of the 19th century resulted in considerable hardship when the crop was ruined by the boll weevil.

monolith A large single block of stone. This may be of natural origin or result from the action of man; for example, the standing stones at ancient sites such as Avebury.

monsoon A large-scale seasonal reversal of winds, pressure, and rainfall in the tropics. The largest and best developed monsoonal area in the world is SE Asia. During the summer (April to September) the interior of the continent is intensely heated, creating a low-pressure area into which winds are drawn from over the cooler surrounding oceans. This creates the SW monsoon over India, and as the air from over the ocean is very moist, it results in very heavy rainfall, especially where mountain barriers force the air to rise, producing additional orographic rainfall. During this season, intense tropical cyclones can develop bringing very strong winds and heavy rainfall, often causing widespread flooding and destruction. During the winter (October to March) the continental interior becomes much cooler than the surrounding oceans; the wind direction is thus reversed, blowing from the continental high pressure to the low pressure over the ocean. This creates the NE monsoon over India, which is generally a cool dry wind. However, where the monsoon crosses a stretch of ocean it picks up moisture and deposits some rain on windward coasts (e.g. Sri Lanka). The whole of SE Asia experiences this type of climate although wind directions vary from place to place. For example, in N China it is southeasterly in summer and northwesterly in winter, and in central China it is southerly in summer and northerly in winter. Similar but less complete seasonal changes occur in other areas of the world, including E Africa, N Australia, and the SE USA, and these are often called monsoons locally. [From Arabic *mawsim*: season]

monsoon forest The tall-growing extensive forests of tropical Asia that experience a marked hot dry season. Monsoon forests resemble equatorial rainforests, the main difference being that the trees are deciduous and shed their leaves after the monsoonal rains. This avoids transpiration (water loss) in the even hotter but dry season that follows. Grasses, including bamboo, are commoner in monsoon areas, but epiphytes are less characteristic. *See also* equatorial rainforest.

montaña A large forest on the slopes of the Andes mountains in South America. The term is applied to the densely forested region in Peru on the E flanks of the Andes. [Spanish]

Moon The only satellite of the Earth. It has approximately ¼ of the Earth's diameter, 1/81 of its mass, and has about 1/6 of the Earth's surface gravitation. The Moon revolves round the Earth in an elliptical orbit, completing one orbit in 27.322 days, at a mean distance of 384 400 km keeping more or less the same side facing the Earth. It has virtually no atmosphere, which partially accounts for the extremes of surface temperatures, ranging from −180°C to +110°C. The Soviet spacecraft Luna 9 made the first unmanned soft landing on the Moon in January, 1966; the first manned spacecraft landing was made by the USA with Apollo 11, on 20 July, 1969.
The main influence exerted by the Moon over the Earth is its contribution to the gravitational forces responsible for the tides of the Earth's oceans.

moor (moorland) An open upland with thin peaty soil, which is wet and acidic. It typically supports a vegetation of heather, coarse grass, bracken, and moss. Such areas are usually used for rough grazing; in some areas (e.g. Scotland) moors are used for the rearing of grouse for sport. Extensive tracts of moorland are to be found in Scotland and in England in Bodmin Moor, Dartmoor, and Exmoor.

mor An acid form of humus (with a pH value of 4.5 or less) that develops from tough acidic leaf litter produced by conifers and heath plants in cool moist areas. Mor is characteristic of nutrient-poor soils such as podzols. In such environments soil organisms, especially earthworms, are relatively few, and both decomposition and mixing with the underlying soil is slow. Consequently mor remains layered and sharply differentiated from the underlying soil. *Compare* mull. *See* humus.

moraine The weathered material or glacial till that is transported by moving ice and the landform resulting when this material is deposited. A large proportion of moraine is derived from debris that falls onto the glacier after frost-shattering on exposed rock faces. Some is derived by the plucking action and abrasion at the base of the glacier. The material is characteristically coarse and angular. It may be transported on the surface of the ice (e.g. medial moraine, lateral moraine), within the ice (englacial moraine), or below the ice (subglacial moraine). *See* ground moraine, lateral moraine, medial moraine, push moraine, terminal moraine. [French]

morphogenesis †In geography, the origin of landforms. A *morphogenetic zone* is a region in which under certain climatic conditions particular geomorphic processes will predominate.

morphological region In geomorphology, a region that is characterized by possessing distinctive forms and structure.

morphological system †A system that is defined in terms of its internal geometry; i.e. the number, size, shape, and linkages of its components. *See* system.

morphometry †The precise and objective measurement of landforms. For example, *fluvial* (or *drainage*) *morphometry* involves the quantitative investigation of the geometric properties of rivers and their basins.

mortality The number of deaths within a given period and area. Together with fertility and migration it is one of the three components of population change. It is generally more stable and predictable than fertility. A number of measures are used to express mortality, the most widely used being the crude death rate. *See* crude death rate, †infant mortality rate. *See also* demographic equation.

mortlake *See* oxbow.

moshav (*plural* moshavim) In Israel, a cooperative agricultural settlement in which the farmers own, work, and are responsible for their own land but in which there is also a degree of mutual cooperation. *See also* kibbutz. [Hebrew]

moss forest A tropical forest that is heavily festooned with mosses and liverworts. It occurs in mountainous parts of Sri Lanka, Malaysia, and E Africa. The vegetation develops because of constant high humidity and heavy precipitation, and is almost impenetrable. †Very moist temperate woodlands have similar but less luxuriant bryophyte growth.

mother-of-pearl cloud *See* nacreous cloud.

motorway A road with separate carriageways, four or more lanes, and limited access, that links some major urban areas in the UK. The roads are designed to allow vehicles to maintain high speeds and are limited to certain types of traffic. The motorway network in the UK was started in 1959 with the opening of the M1 from London to the Midlands. *See also* autobahn.

moulin (glacier mill) A rounded and often vertical shaft in a glacier down which meltwater flows under the influence of gravity, sometimes to the bed rock beneath the ice. Moulins are formed by the erosive effect of meltwater and its load flowing down a crack or crevasse in the ice. Some of the load may be deposited in the moulin and exposed when the ice retreats to form a moulin kame. [French: mill]

mountain An upward projection of the Earth's surface that rises to high altitudes. It usually possesses steep slopes of bare rock with sharp ridges and one or more rocky peaks. Mountains are formed in the process of orogenesis. They usually occur in belts or ranges; for example, the Himalayas and the Andes. Mount Everest is the highest mountain in the world with an elevation of 8848 m.

mountain sickness A nausea and shortness of breath that attacks people at high altitudes as a result of breathing the rarefied air. Individuals experience the symptoms at different altitudes.

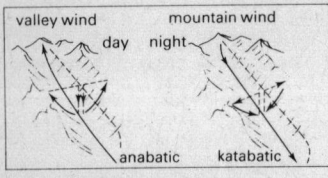

Mountain and valley winds

mountain wind A katabatic wind that blows down a valley axis in mountainous areas at night. It is the night-time reversal of the valley wind. It develops through night-time radiational cooling of the valley slopes. The air in contact with the slopes is cooled and sinks downslope into the valley where it replaces warmer air. The wind is at its strongest just before sunrise when radiational cooling is greatest. *See* katabatic wind. *See also* valley wind.

mouth, river The point at which a stream flows into the sea or an inland lake.

mud A fine-grained clastic sediment in which most of the particles measure less than 0.06 mm. It consists of particles of minerals produced by both physical and chemical weathering. Muds are deposited on lake beds, deep-sea bottoms, and close to shore, where they form mud flats. In the geological record they are represented by mudstones.

mud flat A near-level deposit of silt and mud that is found especially in sheltered environments such as in estuaries and in lagoons. Mud flats are generally submerged at high tide and exposed at low tide and may be vegetated. The vegetation will act as a trap for more sediment and so the mud flat will enlarge. In tropical regions mud flats may be vegetated by mangrove plants.

mud pot A bubbling pool of hot mud that is at or very close to boiling point. It indicates an area of volcanic activity. Mud pots are numerous in Iceland.

mud rain Rain that contains quantities of dust suspended within it. The dust originates chiefly from desert duststorms, although sometimes it may be debris from a volcanic eruption. When the dust is red in colour the rain is locally called blood rain; this is quite common in the islands of the Mediterranean and in S Italy.

mud volcano A vent in the Earth's surface from which hot water and mud are ejected. A small temporary cone of mud may be built around the vent. Such features occur in Iceland, North Island in New Zealand, and Sicily.

mulatto A person who is the offspring of a European and a Negro. [Spanish]

mulga A scrub vegetation of W and central Australian deserts, comprising dense thickets of acacia shrubs, especially *Acacia aneura*, for which mulga is the native name. [Aboriginal]

mull A form of humus with a pH value in the range 4.5–6.5 that develops under grassland and some broad-leaved forests in well-aerated and well-drained soils, such as brown earths and rendzinas. Mull is produced by the relatively rapid decomposition of nutrient-rich plant litter. The presence of soil fauna, especially earthworms, aids the rapid integration of the humus into the soil. Mull is characteristically well integrated with mineral matter, rich in proteins, bases, and nitrogen, and conducive to crumb structure. *Compare* mor. *See* humus.

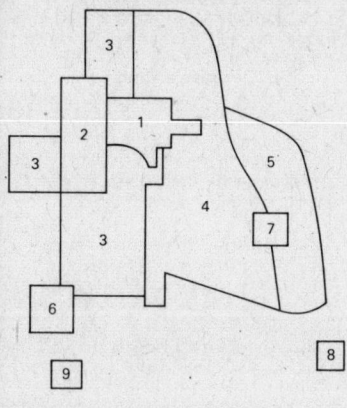

1 CBD

2 wholesale and light manufacturing

3 low-class residential

4 medium-class residential

5 high-class residential

6 heavy manufacturing

7 outlying business district

8 residential suburb

9 industrial suburb

Multiple nuclei model of urban land use

multicropping The growing of a succession of crops during the year on the same plot of land. This is only possible where the growing season is continuous or extends over most of the year and the fertility of the land is maintained. Double cropping of rice is common in S China; in Indonesia and other parts of the tropical world three crops are possible in some areas.

multiple nuclei model of urban land use †A model of urban land use that was proposed by C. D. Harris and E. L. Ullman (1945). It is based upon the assumption that urban functional zones develop through the integration of a number of separate nuclei. These nuclei may be areas of specialization (e.g. a retail area) or may develop as a result of some activities becoming concentrated in particular locations while others repel each other and are scattered. Some activities (e.g. low-class housing) cannot pay the high rents of central areas. The resulting pattern is one of a 'patchwork' of land-use zones. The model is more flexible than the concentric and sector models. *See also* concentric model of urban land use, sector model of urban land use.

muskeg The low-lying and waterlogged hollows characterized by sphagnum moss vegetation that are found in the tundra or permafrost regions of Canada. Stands of tamarack and black spruce trees may surround the hollows. The water table frequently rises above the level of the land. †Muskegs tend also to receive the material that is moved in the mass-wasting process of solifluction. [Algonquian Indian: grassy bog]

Myrdal's model †*See* cumulative causation model.

N

nab *See* naze.

nacreous cloud (mother-of-pearl cloud) A rare type of high cloud with delicate pastel colouring that occurs usu-

nadir

ally in high latitudes at an altitude of about 20–30 km. It may be visible after sunset. Little is known about the formation of nacreous clouds but they appear to be stationary and associated with high-level air currents moving over mountains.

nadir The point on the celestial sphere directly below the observer and exactly opposite the zenith (the point in the heavens directly overhead). The term is used generally to refer to the lowest point. *Compare* zenith.

nappe A large horizontal recumbent fold in rock strata formed through a combination of folding and thrusting. Sometimes the upper limb of the fold is forced over the lower limb along a thrust-fault plane, often for tens of kilometres. This type of structure is formed during mountain building and can be clearly seen in the Alps, where it was first described. [French]

nase *See* naze.

nation A group of people who share certain cultural characteristics and aspirations and who are closely associated politically with a particular area of territory. The group usually forms an independent political unit – a *nation state*. Members of a nation may have shared characteristics, such as race, language, religion, or history, but not necessarily all of these. A nation may, for example, consist of two or more different language groups all having the same nationality; for example, Belgium is a nation of Flemish and French speakers.

national grid The network of lines forming a series of squares that is superimposed on Ordnance Survey (OS) maps. The use of a lettering code and the numbering of the lines provide a reference system enabling locations to be identified throughout the UK. The grid consists of 500 km squares, which are designated by a letter, divided into 100 km squares, which are also designated by a letter, and further subdivided into 1 km

squares, designated by numbers. For example, the national grid reference for HMS *Victory* in Portsmouth Dockyard is SU 628006. *See also* grid.

nationalism A strong feeling of patriotism and national unity that binds a population. Nationalism is revealed in various ways; it may simply take the form of national pride but can be more extreme, such as the driving force behind a struggle for political independence. In the developing countries it is a movement that has been closely associated with anti-colonialism and anti-Europeanism as many colonies have struggled for independence from their former rulers.

nationalization The taking over by the state of land, property, industries, services, or transport facilities previously in private ownership. Between 1945 and 1950 the Labour government nationalized a number of enterprises such as the railways, the coal industry, and the Bank of England. Governments sometimes nationalize foreign-owned concerns, e.g. in 1956 Egypt nationalized the Suez Canal, which had previously been owned by a British-controlled company.

national park An extensive area given special protection to preserve its natural beauty, wildlife, and vegetation for public enjoyment and scientific interest. The first national park to be designated was Yellowstone in 1872 in the USA, where national parks are mainly extensive wild areas. By 1970 34 more national parks had been designated in the USA of which 90% of the total land area was owned by the federal government.

National parks were set up in England and Wales by the National Parks and Access to the Countryside Act (1949). In the UK the land is not nationally owned and recreational activities must coexist alongside established farming and forestry practices, although unsightly quarrying and industrial uses are discouraged. By 1980 there were 10 national parks in England and Wales: Brecon Beacons, Cheviots,

Dartmoor, Exmoor, Lake District, North York Moors, Peak District, Pembrokeshire Coast, Snowdonia, and Yorkshire Dales. Other famous national parks include the Canadian national parks in the Rockies (notably Banff and Jasper), the Kruger park in the Transvaal, and Serengeti, Tanzania.

natural gas *See* gas, natural.

natural grassland *See* grassland.

natural hazards The largely unpredictable sudden changes in the environment, which may have disastrous effects on human activities. These include earthquakes, volcanic eruptions, tsunamis, floods, avalanches, drought, frost, exceptional storms (rain, hail, snow, hurricanes, and tornadoes), diseases of humans, animals, and plants, and swarms of locusts and other pests.

natural increase The relationship between birth and death rates in a particular population. Natural increase may have both positive (birth rate greater than death rate) and negative (death rate exceeds birth rate) values. *See also* crude birth rate, crude death rate.

naturalize To confer the right to citizenship upon a foreigner by the government of his adopted country. The person naturalized then becomes a citizen of that country and subject to all its laws and privileges.

natural region A region on the Earth's surface that displays a distinctive and uniform set of physical characteristics such as climate, relief, or structure. These characteristics mark out one region as being different from an adjacent region. The boundaries of a natural region are difficult to delineate precisely. [The concept of major natural regions was proposed by A. J. Herbertson in 1905]

natural regulation †*See* ecological regulation.

natural resources Those substances, energy sources, and other features of the natural environment that are of value to man. These sources of wealth, which range from soil to fissionable material, can be classified according to the nature of their supply: nonrenewable or wasting resources include fossil fuels, chemicals, and minerals; continuous resources include solar energy; while flow resources are those which may be depleted, sustained, or increased, and include forests and soils.

natural vegetation Vegetation that has developed in an environment untouched by man, and that therefore closely corresponds to the prevailing conditions of soil and climate. Extensive natural vegetation types, such as tropical rainforest, characterize natural regions of the world, and are usually climatic climax communities. Since very little truly natural vegetation survives in temperate regions in particular, the term is generally applied to vegetation only slightly affected by man. *See also* seminatural vegetation, †climax vegetation.

nature reserve An area designated for the preservation of wildlife and plants. Some nature reserves safeguard a single rare species but more usually positive management of the area (e.g. by coppicing, maintenance of water levels, etc.) attempts to protect the entire habitat and its associated flora and fauna. *See also* conservation.

nautical mile (geographical mile) A measure of length, which is based on the length of one minute of arc (i.e. 1/16 of one degree) of latitude. Along the equator this is equal to 1854 m (6082.7 feet), the true geographical mile, but as the Earth is flattened slightly at the poles the length varies elsewhere. An average length of 1853.18 m (6080 feet) has thus been adopted as the standard British nautical mile. The *international nautical mile* is equal to 1852 m (6076.097 feet).

naze (nab, nase, ness) A headland or promontory along a coast; for example, the Naze in Essex on the North Sea coast.

neap tide The smallest amplitude tide (difference between high and low water) recorded at a place. It occurs about twice a month, during the first and last quarters of the Moon, when the Sun, Earth, and Moon are at right angles. At these times the gravitational pulls of the Sun and Moon are opposing, thus high tides are lower than usual and low tides are higher than usual. *Compare* spring tide. *See* tide.

nearest neighbour analysis †A technique that enables geographers to make simple objective comparisons between distributions (e.g. the distribution of settlements). The location of individual points in a distribution are considered in relation to each other. The analysis involves a comparison between the observed spacing of a set of points and the spacing that might be expected in a random distribution pattern. The formula is:

$$RN = 2\bar{d}o\sqrt{n/A}$$

where RN is the nearest neighbour value, $\bar{d}o$ is the observed mean distance (i.e. the measured distance between each point and its nearest neighbour), n is the total number of points in the pattern, and A is the area over which the points are distributed. Values for RN can range in theory from 0 (maximum clustering) to 2.15, which would be produced by a perfectly uniform distribution of points in an area. A value of 1 indicates a random distribution.

neck 1. *See* plug.
2. A narrow projecting tract of land (e.g. an isthmus or peninsula).

negative movement of sea level A rise of the land relative to sea level that may be on a local or worldwide scale. This may either result from an actual rise of the land surface through tectonic movements, or from a fall in sea level such as occurs when water is locked up as ice during an ice age.

Evidence for past negative movements of sea level is seen along coastlines in raised beaches and coastal plains. †The retreat of the sea leads to the rejuvenation of rivers and the formation of knickpoints, river terraces, and incised meanders.
Compare positive movement of sea level.

nehrung (*plural* nehrungen) A long sand spit, especially on the S coast of the Baltic Sea, which is covered with sand dunes and often fixed by coarse marram grass and pine trees. Nehrungen are formed by the deposition of material by longshore drifting and the prevailing southwesterly wind. [German]

neighbourhood unit A small-scale residential area within which a level of services such as shops and a community centre is provided for the inhabitants, with the intention of giving them a sense of identity. The concept was first developed in 1929 by Clarence Perry, an American architect/planner, but its origins have been traced to Classical Greece.

Neogene The upper part of the Tertiary period, i.e. the Miocene and Pliocene epochs.

ness *See* naze.

net reproduction rate (NNR) †A measure of fertility that indicates whether a given population can reproduce itself. It is based on the average number of female children born to every woman of reproductive age in the population, taking into account mortality. The measure recognizes that not all female babies survive to reproductive age and that a number of those who do survive may not have children themselves. If the NNR is equal to 1 the population is stable; a figure of more than 1 indicates a growing population, less than 1 indicates that the population is in decline.

network A system of interconnecting lines. In human geography it refers chiefly to a transport network (e.g. roads, railways) but may also be applied to other line patterns such as boundaries, telephone communications, etc. †Networks can be studied by *network analysis*, which enables them to be measured and planned. Network analysis is based largely on graph theory, a branch of mathematics that produces models of the networks that exist in reality. These models, known as graphs, identify the terminals and junctions of the network and the lines between them. Where two or more lines meet a node or vortex is formed. The lines joining nodes are known as links, arcs, or edges. The graphs may be either planar (two-dimensional) or nonplanar (three-dimensional); examples of nonplanar networks in transport geography include air routes and underground railway systems. Graph theory enables the extent and density of the network to be measured by the size and number of links and nodes. Accessibility can be measured by the examination of the number of links from each node and the connectivity of the network can be established. *See* accessibility, connectivity.

neutral zone A state or piece of territory in which a policy of nonalignment with other powers is followed and military activity is restricted. This may be voluntary, for example, Switzerland will not wage war on any other countries unless it is itself attacked. Neutrality may also be imposed by other powers, for example, Austria was made neutral at the end of World War II. A neutral zone may also be a piece of territory that is demilitarized to act as a buffer in a politically tense area, for example, the 1922 neutral zone between Belgium and Germany.

nevados (nevadas) A katabatic wind that brings cold air from over the snowfields of the Andes down to the higher valleys of Ecuador. *See* katabatic wind. [Spanish]

nevé *See* firn. [French]

new town An urban settlement that is planned and built to provide employment and accommodation so as to relieve the pressures of congestion of existing urban areas. In the UK, the New Towns Act (1946) enabled these to be established; early new towns were built around London and included Stevenage and Hemel Hempstead in Hertfordshire. Since these early new towns the function of the new town has changed and many have been established for a variety of other reasons; for example, to provide economic growth centres in declining areas (e.g. Peterlee, Co Durham). Central Lancashire and Milton Keynes are two recent examples of new towns.

New World The W hemisphere, i.e. the continents of North and South America.

nickel A metal that never occurs in its native state but is found in sulphide and arsenide ores, such as millerite and pentlandite. It is white, unaffected by exposure to air, and capable of being highly polished. It is used in coinage and in the manufacture of steel alloys, such as stainless steel, Invar, and Monel. Because nickel steel is very hard, it is used in the construction of cars and aircraft. One of the main sources of nickel is the nickel-iron sulphide deposits near Sudbury, Ontario, which make Canada the world's primary nickel producer. The Earth's core is believed to contain large quantities of nickel.

nimbostratus (Ns) A thick grey layer cloud that blots out the Sun, and from which more or less continuous rain or snow falls. The base is low and fairly uniform but ragged clouds (known as scud), or trails of rain that do not reach the ground, may be seen below it. It is most commonly associated with the warm front of a depression.

nimbus Denoting clouds from which rain is falling (e.g. cumulonimbus, nimbostratus).

nivation (snow patch erosion) †The weathering and erosional processes resulting from a snow cover, usually in a hollow on the lee side of a mountain or a ridge. Freeze-thaw is the most important process. During the day meltwater from the snow penetrates cracks and fissures in the underlying rock. When this refreezes at night it causes shattering of the rock. This process is most active where the snow cover is thin (i.e. around the edges of the snow patches) as if the layer is thick it tends to protect the underlying material from the freeze-thaw effects. The eroded material is removed by meltwater and solifluction; these processes are most efficient on slopes. Nivation results in the progressive enlargement of a hollow to form a *nivation hollow*, which may eventually develop into a cirque.

noctilucent cloud A rare high cloud form resembling very thin cirrostratus, but with a bluish-white to yellow colour, that may be seen in high latitudes on summer nights. Noctilucent clouds occur at heights of about 80–85 km (around the mesopause). Their composition is not yet fully understood, two theories being that they consist of ice particles or dust.

node 1. One of two points where the plane of the Moon's orbit intersects the plane of the Earth's orbit (the ecliptic).
2. In a network, a point (or vertex) where lines or edges meet, e.g. the main terminal in a railway network. *See* network.

nomad A person who belongs to a tribe or group that lives a roaming or wandering life. The movement from place to place is the result of the need to find pasture for livestock, upon which the nomads are dependent for their subsistence. The rate of movement depends upon the availability of the food supply at each temporary stopping place. True nomads (e.g. the camel nomads in the Sahara and the Lapps of Arctic Scandinavia who herd reindeer) are distinguished from the semi-nomads, who include collectors of food (e.g. Aboriginals) and farmers who practise transhumance.

non-conforming use The use of a particular piece of land that conflicts with the type of use which is shown for the area on a zoning plan. The most frequent example of a non-conforming use is the presence of industrial premises in a zone that is designated as residential.

nonconformity A type of unconformity in which lower igneous and metamorphic rocks are overlain by layered sedimentary rocks. The two types of rock are separated by an erosion surface. *See* unconformity.

nonecumene *See* ecumene.

Norfolk rotation The system of crop rotation that was introduced into Norfolk during the last quarter of the 18th century and later spread to other parts of the UK. Instead of leaving fields fallow for one year in three, the Norfolk four-course rotation of turnips, barley, clover or grass, and wheat enabled fertility to be maintained and output to be increased. Livestock were turned onto the fields in winter to manure the land.

normal distribution (Gaussian distribution) †A symmetrical frequency curve. The mean, the median, and the mode all coincide at the same central point in the distribution. *See* frequency curve.

normal fault A fault in which the fault plane is usually steeply inclined at an angle between 45° and the vertical so that the rocks on one side of the plane are raised, or upthrown, relative to the rocks on the other side, which are downthrown. Normal faults usually occur in groups of parallel faults. In recently active faults, the fault plane forms a cliff-like feature known as a fault scarp. *See also* fault.

norte *See* norther.

northeast trades The trade winds that occur in the N hemisphere. *See* trade winds.

norther A type of polar outbreak or cold wave experienced in North America, which brings a cold dry northerly wind to the S states of the USA. It is caused by the drawing of air southward behind an eastward moving depression. It is extremely dangerous to the crops of the region (e.g. fruit, cotton, tobacco) as it can gust up to 100 km per hour, bring hail storms, and reduce temperatures by up to 20°C in 24 hours. Further S in Mexico and central America the wind is known as the *norte*, and in these tropical and subtropical latitudes the sudden fall in temperature it brings is even more dangerous to both crops and people.

northern lights *See* aurora.

northing Any of the E–W grid lines on a map showing the distance northwards from the point of origin of the grid. The northing coordinates follow the easting coordinates in a map grid reference. *See* grid. *See also* easting.

North Pole *See* pole.

nor'wester 1. In N India, a squall occurring during the hot season (April–June), which may bring heavy rain or hail, the rain being of great importance to the tea crop in Assam.
2. In South Island, New Zealand, a hot dry wind, similar to the föhn, that descends from the Southern Alps to the plains.

notch 1. (wave-cut notch) The part of a sea cliff that has been undercut by wave action at the high-water mark. Hydraulic action by the sea on an area of weakness in the rock, such as a joint, may be largely responsible for the formation of a notch. Eventually the notch may be extended to form a cave.
2. In the USA, a pass or gorge between two mountains.

nuclear power The power obtained from a nuclear reactor when radioactive uranium atoms disintegrate in a fission chain reaction producing enormous quantities of heat. This heat is used to raise steam (in a conventional boiler), which drives electric generators and produces electricity. Nuclear power provided nearly 12% of the electricity used in the UK in 1980.

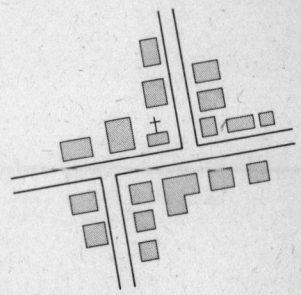

Nucleated settlement

nucleated settlement (nucleation) A form of rural settlement in which farms and other buildings are clustered together, often around some central feature (e.g. a church, green, or crossroads) forming a hamlet or village. Nucleation may have partly evolved as the result of the type of farming practised; for example, cooperative farming such as the medieval open-plan system was most efficiently operated from central nucleated settlements. Other factors encouraging nucleation included the need for defence, the availability of water supply at particular locations, the need for dry-point sites, and the development of modern forms of transport, which created, for example, railway and canal villages. *See also* hamlet, village.

nuée ardente A glowing cloud consisting of steam, other hot gases, and volcanic dust that may be produced during a volcanic eruption. The glow is produced by the incandescent particles of volcanic dust that are carried within it. It may roll down the slopes of a volcano with great destructive force. For example, in 1902, the town of St Pierre on the Caribbean island of Martinique was destroyed by a nuée ardente when Mount Pelée erupted. [French]

nullah The course of a stream with steep and rocky side walls or a drainage channel that only intermittently contains water following heavy rainfall. [Indian]

numerical weather forecasting A modern method of weather forecasting that involves feeding a computer, programmed to analyse weather patterns, with the vast numbers of observations of weather phenomena that are collected each day.

nunatak The rocky peak of a mountain that projects above the level of the ice and snow of an ice sheet. The exposed rock is subject to frost action. Nunataks are most numerous near the margins of ice sheets. They are common in Antarctica and Greenland. [Eskimo]

O

oasis (*plural* oases) A fertile patch in an arid area at which the water table reaches or is very close to the ground surface. The water supports permanent plant growth and settlement. Oases may be found in the deserts of Egypt and Libya. They vary in size from a single spring to areas of several hundred square kilometres.

oats A hardy cereal that can be grown in cooler and wetter climates than wheat or barley. It is used as an animal feed and in the form of oatmeal for porridge and other foodstuffs. In the UK it is grown mainly in the N and W.

oblique fault A type of normal fault in which the fault plane is at an angle between 45° and the vertical but the movement is horizontal rather than vertical. The amount of movement is called the slip. *See* normal fault.

obsequent stream †A stream that flows in the opposite direction to a consequent stream, or in the opposite direction to the dip of the rock strata. *See also* consequent stream. [Coined in 1895 by W. M. Davis]

occlusion (occluded front) A front that forms during the later stages of the evolution of a depression when the more rapidly moving cold front catches up with the warm front and lifts the warm sector above the surface. This is therefore fairly common over W Europe. If the air behind the original cold front was warmer than the air ahead of the original warm front, the cold front will be lifted off the surface forming a *warm occlusion*, but if the overtaking air is colder than the air ahead, the warm front will be lifted, forming a *cold occlusion*. The weather associated with an occlusion is similar to that occurring at the original fronts, but does not last as long and eventually dies out as the air in the warm sector rises and cools. *See* depression.

occupancy rate A measure of the number of persons per residential or housing unit; for example, the number of people per room.

ocean The vast expanse of salt water that covers over 70% of the Earth's surface, or one of the five major expanses into which it is divided, i.e. the Atlantic, Pacific, Indian, Arctic, and Antarctic (or Southern) oceans. The surface waters of the ocean vary greatly in temperature and salinity, being affected by, among other factors, the atmospheric conditions above them. For example, near the equator the surface water temperature may be over 25°C, while near the poles it is frequently below freezing point. However, at depths below about 3500 m (in the abyssal zone) the waters are remarkably uniform, especially in temperature, which is only a few degrees above freezing point. Most of the plant and animal life lives in the sublittoral zone near the coast, and in the pelagic zone in the open ocean. *See* pelagic, sublittoral.

oceanic climate *See* maritime climate.

oceanography The scientific study of all aspects of the ocean, including the nature of the water (salinity, chemical composition, etc.), temperature, movement (tides, currents, waves, etc.), depth, and biology (flora and fauna).

oil palm A species of palm tree (*Elaeis guineensis*). The heavy orange-red clusters of fruit borne by the tree are the source of two vegetable oils: *palm oil*, obtained by boiling the fleshy part of the fruit, and *palm-kernel oil*, obtained from the soft inner part of the kernel. Both oils are used in the manufacture of margarine, soap, and cosmetics. Oil palm trees are grown in plantations in Africa in Nigeria, Zaïre, Sierra Leone, and Cameroon, and in Malaysia and Indonesia.

oil refinery The industrial plant at which crude oil is separated into fractions in a vertical container to produce diesel oil, paraffin, petrol, and refinery gas. The residue is used to produce lubricating oils, paraffin wax, bitumen, and a range of chemicals. The dependence until recently of the UK on overseas oil supplies resulted in the building of oil refineries around the coast close to deep-water anchorages, such as Fawley, Milford Haven, and Teesmouth.

oil sand (tar sand) An open-textured sand or sandstone in which oil has accumulated in the pores in the rock, thus forming an oil reservoir. The largest reserves of oil sands occur in Alberta, Canada, especially in the Athabasca Valley. Other accumulations occur in Venezuela, the USA, the USSR, Madagascar, Albania, Trinidad, and Romania.

oil shale A fine-grained sedimentary rock, normally finely layered, containing organic matter from which oil can be extracted by heating. Free flowing oil is unusual although the rock may contain pockets or small veins of oily

substance. Oil yields of 38 litres per metric ton to 550 litres per metric ton have been recorded. In Scotland the Lower Carboniferous oil shales of the Midland Valley yield an average of 90 litres per metric ton. Oil shales are found throughout the geological record from the Cambrian to the Tertiary, and are thought to have formed in large lakes, shallow marine seas, and small lakes associated with coal-producing swamps.

okta A unit used in meteorology to measure cloud cover. It is equal to a cloud cover of one eighth of the whole sky; eight oktas is therefore equivalent to a total cloud cover.

Oligocene The geological epoch of the Tertiary period following the Eocene epoch and preceding the Miocene epoch. It extended for about 14 million years from about 36 million years ago. During it continental conditions prevailed over the British Isles and France, with sea over what is now Germany and the USSR. A narrow arm of the sea occupied approximately the position of the English Channel. In Britain rocks of this epoch are restricted to the N part of the Isle of Wight and the Hampshire coast.

oligotrophic †Denoting an environment, especially a lake, that is deficient in nutrients and consequently low in productivity. Oligotrophic lakes are frequently dark and peat stained (from slowly decomposing vegetation), acidic, and relatively poor in plant and animal life. *Compare* eutrophic.

olive The fruit of an evergreen tree that grows in Mediterranean regions. The fruit is picked in the autumn and can be pickled or canned. Most of the fruit is pressed to yield a rich oil, which can be used for cooking and as an alternative to butter. Olive oil is also used in soap and cosmetics. Large quantities are grown in S Spain, Tunisia, S Italy, and Greece.

onion weathering *See* exfoliation.

ooze A mud that covers most of the vast abyssal plain of the ocean floor. It consists chiefly of the skeletal remains of minute plants and animals, which have sunk to the ocean floor from the surface waters in which they lived, together with volcanic dust and fine wind-blown material from the land surfaces. The composition of ooze depends on the type of organic remains that are most abundant in the deposit. Each is named after the particular organism (e.g. globigerina ooze). As the organisms are minute, ooze takes a long time to accumulate; estimates of the rates vary from 20 mm per 1000 years for globigerina ooze to 5 mm per 1000 years for red clay. *See* diatom ooze, pteropod ooze, radiolarian ooze. *See also* red clay.

opaco *See* ubac.

opencast mining The extraction of a mineral from just below the ground surface by removing the overburden (surface layers). In the USA the process is called *strip mining*. In the E Midlands of the UK the National Coal Board uses opencast mining methods on the exposed section of the Notts and Derby coalfield. After excavation the land is carefully restored for agricultural use.

open system †*See* system.

opisometer An instrument for measuring distances on a map. There are two types. The first consists of a wheel that drives a pointer on a dial showing distances travelled across the map in centimetres (or inches), which can then be converted into actual distances on the ground. In the second type a wheel moves along an axle; when the distance to be measured has been covered the wheel is put on the scale line and pushed along until it has returned to its starting point at one end of the axle. The point reached on the scale line is the distance travelled.

optimum population †A density of population which, with the given resources and skills, produces the greatest economic welfare (the maximum income per head) or allows the highest standard of living. *See also* overpopulation, underpopulation.

orbit of the Earth The Earth's annual path around the Sun. The orbit is elliptical, with the Sun at one focus. The Earth is nearest to the Sun in December (about 147 000 000 km), and farthest from it in June (about 152 000 000 km). Viewed from above the North Pole, the Earth moves anticlockwise around the orbit, travelling faster nearer the Sun.

order, stream *See* stream order.

Ordnance Datum (OD) The level from which heights on Ordnance Survey (OS) maps are calculated. The level is mean sea level at Newlyn in Cornwall. *See also* datum level.

Ordnance Survey (OS) The official mapping organization in the UK. It is responsible to the government for the survey and mapping of the country, including geodetic and associated scientific work, topographical surveys, and the publication of an extensive range of maps from these surveys. The OS also provides a contract service for those needing special surveys or cartographic work carried out.

Ordovician The geological period that extends from the end of the Cambrian period, about 500 million years ago, to the beginning of the Silurian period, about 440 million years ago, and the system of rocks. The base of this system of rocks is identified by the appearance of particular fossils – the graptolites – which are used to date and correlate the different areas of Ordovician rocks. During this period there was widespread volcanic activity. [Named by Charles Lapworth after the Ordovices, the last of the old Cambrian tribes that had occupied central Wales]

ore The deposit from which a mineral is obtained. An ore contains sufficient useful mineral material to make it profitable to extract it. In some ores several metals are associated together, e.g. silver and mercury are associated with lead and zinc ores. The existence of a second metal in an ore may make its working profitable.

organic weathering The decomposition and disintegration of rocks by plants and animals. This involves both mechanical and physical weathering. For example, mosses and lichens retain rainwater and humic acids in contact with rock surfaces; this enables chemical weathering to break down the rock under the plants. The result of the process can be seen in the shallow depressions that form beneath mosses and lichens. Burrowing animals and the penetration of plant roots can cause the physical weathering of rocks.

orientation The lining up of a map or plane table survey in the field with visible objects so that magnetic north on the map or survey is correctly positioned. Orientation is necessary in plane table surveys when the table is moved to the other end of a fixed base line and when it is moved to a new station. The table is oriented by sighting back to the previous station whose position has already been determined.

origin of a grid The point from which a grid originates and at which all grid numbering commences. For the national grid of the UK a point to the W of Lands End is the point of origin.

origin, seismic See seismic focus.

orogenesis The process by which mountains are formed. Orogenic movements result in the thrusting, folding, and faulting that form the major mountain ranges (e.g. the Andes, Rockies, Alps, and Himalayas). This occurs where two continents collide and the sediments between them are intensely deformed into linear mountain ranges. Volcanic activity and earthquakes are closely associated with orogenesis. Compare epeirogenesis.

orogeny A period of mountain building during which belts of rock are highly deformed. The Caledonian, Variscan, and Alpine orogenies are the chief phases of mountain building to have affected the British Isles since the Caledonian. See orogenesis. See also lithogenesis.

orographic rainfall (relief rainfall) Precipitation that is caused by moisture-laden air being forced to rise over a relief barrier. As the air rises it cools and, if the dew point is reached, condensation occurs forming clouds and precipitation. Orographic rainfall is particularly important in areas where moist prevailing winds off the sea meet a highland coast lying at right angles to them, e.g. British Columbia (Canada) and Scotland. On such coasts, precipitation will increase with altitude on the windward slopes, but once the air has passed over the crest and starts to descend the leeward slopes it begins to warm up, becomes drier, and precipitation is greatly reduced (the rain shadow). In monsoon areas this type of rainfall greatly augments the normal monsoonal rains as occurs, for example, at Cherrapunji, on the windward slopes of the Khasi Hills in Assam, India, which has the highest mean annual rainfall in the world.

orography A branch of physical geography that is concerned with the description of the relief of mountains and mountain ranges.

orthographic projection A map projection in which the globe is viewed from so far away that, in effect, the projecting rays are parallel and perpendicular to the plane of the projection. Only one hemisphere can be drawn and there is expansion at the centre and compression at the margins. Distances, directions, shape, and area are all distorted and apart from showing polar areas, this projection is little used, although it is the most easily recognized since it resembles a view of the globe from afar.

orthomorphic projection (conformal projection) A map projection that preserves the true shape of a small area with minimal distortion. This is obtained by ensuring that meridians and parallels always intersect at right angles. For example, Lambert's conformal conic with two standard parallels, which is used for many French maps.

OS *See* Ordnance Survey.

outcrop The part of a rock formation that occurs at the surface of the Earth. It may be visible or masked by thin layers of soil or debris known as drift.

outlier A mass of rock that is entirely surrounded by rocks that are of an older age. It may be formed by erosion or by structural factors such as folding and faulting. *Compare* inlier.

outport A port than can be reached by larger vessels than the main port that it serves. An outport is usually closer to the open sea than the main port, which probably developed when ships were much smaller than today. Examples include Avonmouth, which is the outport for Bristol, and Cuxhaven, which serves Hamburg.

outwash plain (sandur) An area of fluvioglacial material deposited by meltwater streams emerging from ice sheets and glaciers. It consists of sand, gravel, and clay that has been sorted into layers by the water. The coarser deposits are found closest to the ice and the material decreases progressively in size away from the ice with the finest deposits furthest away. Outwash plains frequently have a hummocky surface resulting from kettle holes. Other fluvioglacial deposits such as kames, kame terraces, and eskers may be associated with the feature. Smaller deposits of outwash material include fans known as *outwash aprons*. Lines of material that have been deposited in valleys are known as valley trains.

overburden The soil or rock that lies above a mineral deposit and which must be removed before opencast mining can take place. In the E Midlands the overburden concealing the coal seams is less than 100 m deep. *See also* opencast mining.

overfishing The depletion of fishing grounds by removing more fish than can be replaced by natural methods. International laws regulating fishing practices, such as the mesh size of nets, have been drawn up but are not observed by all countries. The large herring shoals found in the North Sea 40 years ago have disappeared as a result of overfishing.

overflow channel A water-worn channel formed by the overflow of a lake during a period of high water. Such channels were formed chiefly by the overflow of ice-dammed lakes in the ice ages. The channel may cut across preglacial watersheds. Examples of overflow channels include Newtondale and Forge Valley in E Yorkshire.

overfold A fold in rock strata in which an assymmetrical anticline has been forced over so that one limb of the anticline lies beneath the other.

overgrazing (overstocking) The introduction of more animals to grazing land than the carrying capacity of the land will allow. In these circumstances the pastures cannot be replenished by natural growth and the area may become severely eroded. Overgrazing has, for example, reduced land potential in the Mediterranean basin and savanna regions of W and E Africa.

overland flow (surface flow) The movement of water in films, sheets, or rills across the surface of the ground. It occurs during heavy rainfall when the soil cannot absorb any more water as all pore spaces are full (i.e. the infiltration capacity is exceeded). Overland flow is most noticeable in arid and semiarid areas where there is little vegetation. As rainfall continues in such areas the sheetflow may develop into rills and then into gullies causing considerable erosion.

†R. E. Horton suggested a model of the drainage basin – the *Horton overland flow model* – in which two main types of water supply to the streams are recognized: overland flow and ground-water flow. If the infiltration capacity of the soil is exceeded water will first be retained as puddles on the ground surface by surface retention. These gradually overflow and water flows as sheets until it is concentrated into rills and then stream channels.

overpopulation †An excess of population in an area in relation to the available resources and skills. This may result either from a growth in population or from a decline in the available resources. Overpopulation is accompanied by underemployment, low incomes, and poor living standards. *See also* optimum population, underpopulation.

overspill An excess of population that leaves an urban area (either voluntarily or as a result of redevelopment) as a result of circumstances such as overcrowding and congestion. In the UK, overspill populations have been rehoused in new towns, expanded towns, or in overspill estates in other towns that had agreed to accept the families. This policy was prevalent in the 1950s and 1960s but is now seen as a contributory factor to the decline of the inner cities.

overstocking *See* overgrazing.

oxbow (mortlake) A small crescent-shaped lake in a floodplain that originally formed part of a river meander. Erosion by a river is strongest on the outside of meander bends. As a result the neck of a meander is eroded until it is eventually breached by the river during times of flood leaving the meander bend as a cutoff. The ends of the cutoff soon silt up to form the oxbow lake. Oxbow lakes are found in the floodplains of meandering rivers such as the River Cuckmere in Sussex.

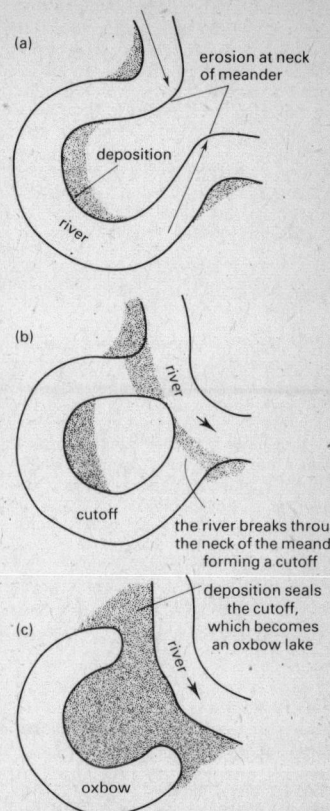

(a) erosion at neck of meander — deposition — river

(b) cutoff — the river breaks through the neck of the meander forming a cutoff

(c) deposition seals the cutoff, which becomes an oxbow lake — river — oxbow

The development of an oxbow lake

oxidation A process of chemical weathering in which atmospheric oxygen reacts with the rock to produce oxides or hydroxides. Oxidation is most effective in rocks that contain large quantities of iron compounds. Decomposition by oxidation in these rocks is shown by the presence of yellow or reddish brown rock, which easily crumbles.

ozone layer (ozonosphere) A zone within the atmosphere, generally between 20 and 50 km, where ozone forms in its greatest concentrations. It results from the splitting of oxygen (O_2) molecules by the ultraviolet radiation from the Sun to form atomic oxygen (O), which then combines with other oxygen (O_2) molecules to give ozone (O_3). The ozone layer prevents most of the potentially damaging ultraviolet radiation from reaching the ground surface and so protects the Earth's life forms.

P

Pacific type of coastline *See* concordant coast.

pack ice A large ice mass that floats on the surface of the sea and consists of numerous ice floes that have been forced together. Ice floes are usually found in polar seas. Areas of pack ice may be of considerable size with little open water to be found within them. When the ice floes have been forced together under great pressure the edges may buckle and warp to form distinctive pressure ridges in the pack ice.

padang The tropical grassland (savanna) of the drier parts of SE Asia. The term is also applied to heathland growing on infertile sandstones in the area. [Malay]

paddy (padi) Unhusked rice; the term rice is not used until the husk has been removed. There are two main forms of paddy, that grown on irrigated fields and upland rice, which is grown without irrigation where the rainfall is sufficiently heavy. Paddy is the staple diet of over one quarter of the world's population. The heaviest cropping regions are the deltas of SE Asia, e.g. the Irrawaddy, Jiangxi, Red River, and the Yangtze. [Malay]

pahoehoe A lava flow that has cooled and solidified with a corded or rope-like surface. This surface structure is due to the escape of gases when the lava is at a high temperature but with-

out explosive activity. The lava solidifies without the jaggedness that is characteristic of lava from which there is an explosive escape of hot gases during cooling. *Compare* aa. [Hawaiian]

paintpot A hole in the ground in an area of volcanic activity out of which hot and highly coloured mud is forced. Geysers and mud pots are usually found associated with such features. The name derives from the bright reds and yellows of the hot mud, which are caused by the presence of certain compounds. Paintpots occur in the Yellowstone National Park of Wyoming, USA.

Palaeocene The earliest geological epoch of the Tertiary period. It extended for about 11.5 million years from the end of the Cretaceous period, about 65 million years ago, to the beginning of the Eocene epoch. The time period represented by this epoch is sometimes considered to be part of the Eocene. In the British Isles it is represented by sedimentary rocks in S England in the London Basin. These sediments are marine in the W and nonmarine in the E. During the epoch mammals became abundant and diversified and by the end of it primates had evolved. [Named by W. P. Schimper in 1847]

palaeoclimatology The study of the climate of a past geological period. The rocks and fossils of the period provide the best evidence for this. For example, the vegetation needed to produce coal deposits required a hot humid climate; the shape of the particles of sand in Old Red Sandstone indicates a desert origin.

Palaeogene The lower part of the Tertiary period, i.e. the Palaeocene, Eocene, and Oligocene epochs.

palaeogeography The study of the geography of geological periods. This includes the mapping of the relative positions of land and sea and the changes in those positions, and the study of topographic features and climate and their changes through geological time. The older the period, the more difficult it is to reconstruct the geography and maps of these times can only be approximate.

palaeomagnetism The study of the changes in the Earth's magnetic field through geological time. Some rocks, for example those containing iron minerals such as magnetite and haematite, have the polarization of the Earth's magnetic field at the time of their formation frozen into their structure. After allowances have been made for later changes such as folding or tilting in the attitude of the bed, the direction of polarization at the time of formation of the rock can be determined. Results have shown that the magnetic poles have moved throughout time and that periodically reversals of the Earth's magnetic field have taken place. Palaeomagnetism has provided important evidence to support the theories of sea-floor spreading and plate tectonics. *See* sea-floor spreading.

palaeontology The study of fossils, including their form and evolution, their place in space and time, and their association with other fossils. It includes the study of both the fossilized remains of the organisms themselves and trace fossils, such as the tracks made by organisms. Palaeontology is used in geology to establish rock sequences, to correlate rocks in different areas, and to determine the environment in which the rock was deposited, for example shallow or deep water. It also provides important evidence on the evolution of organisms and their adaptation to different environments.

Palaeozoic The first of the geological eras, following the Precambrian, extending from about 570 million years ago to about 225 million years ago. It comprises six periods: the Cambrian, Ordovician, and Silurian in the Lower Palaeozoic; and the Devonian, Carboniferous, and Permian in

the Upper Palaeozoic. Two major episodes of mountain building took place during this era: the Caledonian and the Variscan.

palynology †*See* pollen analysis.

pampas The extensive areas of temperate natural grassland in South America that extend across Argentina and parts of Uruguay between the Andes and the Atlantic coast. The E areas are moister, and their original vegetation of pampas grass on rich alluvial soils has been extensively used for grazing beef cattle, or converted to cereal production. In the W, the pampas become desertlike. *See also* grassland. [Spanish]

pampero A type of cold wave or polar outbreak that occurs in parts of Argentina and Uruguay when very cold dry air is drawn in from the S or SW behind an eastward-moving depression. It commonly resembles a line squall with very dark clouds often augmented by dust blown up off the pampas, and sometimes with rain, thunder, and lightning. [Spanish]

pan *See* hardpan.

pancake ice Small, thin, and often near-circular slabs of newly formed ice that float on the sea. The presence of pancake ice indicates a lowering of air temperatures.

pandemic Denoting a disease that is prevalent over a wide geographical area such as a country or continent.

Pangaea The primeval supercontinent that is believed to have broken up about 200 million years ago during the Mesozoic era. The German geologist Alfred Wegener was the first to propose the existence of this supercontinent, which he suggested broke up and drifted apart to form two segments: Laurasia in the N hemisphere and Gondwanaland in the S hemisphere. *See* Gondwanaland, Laurasia. *See also* continental drift, plate tectonics. [Greek: all-earth]

panhandle A narrow strip of a state that projects into another state or between two other states, or extends between another state and the coastline. The term is used chiefly in the USA.

pannage The right to allow pigs to feed in a woodland area. The term is also used for the acorns, beechmast, and litter on which the pigs feed. In the New Forest local farmers (commoners) pay a pannage fee to the Forestry Commission for their pigs to feed in the forest from 25 September to 22 November.

panorama The view of the landscape as seen by an observer. The view, or a section of it, may be used as the basis for a panoramic sketch which puts on paper details visible to the observer that require recording, for example, relief and structural features or urban landmarks.

papagayo A strong cold dry northeasterly wind. It is a southerly continuation of the norther of North America and brings cold clear conditions to the high plateau and Gulf coast of Mexico in winter. *See also* norther. [Spanish]

parabolic dune A curving sand dune in which the horns of the dune point in the direction from which the wind is blowing, i.e. in the opposite direction to the horns of a barchan dune. Parabolic dunes are found, for example, on the W coast of Denmark. *Compare* barchan.

parallel A line of latitude, encircling the Earth parallel to the equator. The equator is the only line of latitude that is a great circle; all other parallels are small circles and become progressively shorter N and S of the equator until they become points at the North and South Poles. *See also* latitude.

parallel drainage A drainage pattern in which the main streams and tributary streams follow virtually parallel courses. This develops where there is a strong structural control in one direction; for example, where strata are gently dipping.

parallel retreat The cutting back of a slope by denudation without a change in the gradient of the slope. Each section of the slope retreats under the forces of weathering and erosion to a new slope that is parallel to that of the original surface. †The process of parallel retreat, as put forward by W. Penck, suggests that slopes may be worn back by denudation rather than worn down, as W. M. Davis proposed. Parallel retreat has also been used to explain the formation of inselbergs and the pediplains that lie between them. *See* slope retreat.

paramo An altitudinal zone that lies between the tree line and the permanent snow line. Such areas are usually bare of trees with only alpine plants among the rock. [Spanish; derived from the paramo, a barren plateau of high elevation in the Andes of South America]

parasitic city †A city within which a certain amount of economic growth occurs but only at the expense of the region within which it is located. The surrounding area is ruthlessly exploited for its natural and agricultural resources. Examples are the early colonial cities of S Africa and the New World. Parasitic cities retard overall regional growth but eventually they may develop into generative cities. *See also* generative city.

parent material The weathered and partially broken rock from which a soil is formed. This may not be the same as the underlying bedrock, for example, it may be a deposit of glacial till or alluvium. Different rock types produce different soils; for example, those formed over limestones are base rich and stone free. In many soil profiles, parent material dominates the C horizon, whereas the B horizon has a mixture of mineral and organic (humus) fractions. †Parent material and climate are the two most influential factors in soil formation. Parent materials are particularly important in areas of varied surface geology, as in the British Isles. They govern the soil's chemical quality through the precise composition of minerals, and also affect physical properties such as soil texture.

parhelion (mock sun) An image of the Sun caused by the refraction of sunlight through the particles of ice that form high clouds. It usually appears at the same elevation as the Sun at an angular distance of 22° from it (the same as for a halo).

parish 1. An ecclesiastical district that has its own priest and church.
2. In England, a subdivision of a county forming the smallest unit of local government in rural areas.

park savanna A type of tropical and subtropical savanna that consists of scattered or clumped trees in a dominantly grassland setting. The patchy tree cover resembles old parkland in the British Isles and reflects the distribution of available soil moisture. *See also* gallery forest, savanna.

part-time farming The combining of farming for part of the year or day with some other occupation because the size of the farm or the climatic conditions make it uneconomical or impractical to farm all the time. For example, the very cold winters in the Midwest of the USA check farming activities and some farmers find work in nearby towns; farmers with smallholdings in Belgium (Flanders) often have part-time jobs in local factories.

pass A routeway through a mountain range that follows the line of a col or a gap. For example, the Khyber Pass through the Safed Koh range between Afghanistan and Pakistan and the Simplon Pass over the Alps between Switzerland and Italy.

pastoral farming The rearing and breeding of animals that feed on grass and other vegetation. There are two main types: nomadic herding to produce food, clothing, shelter, and goods for sale locally, e.g. the Bedouin of the Middle East, and commercial grazing to produce meat, hides, and wool for sale outside the grazing region, e.g. the rearing of beef cattle in N Queensland, Australia.

pasture Land on which grass is grown to provide grazing for livestock. Some pasture fields may be seeded with clover and similar plants instead of grass.

patana A type of vegetation consisting of coarse grassland that grows above the tree line in upland Sri Lanka. Similar high-altitude grassland exists in the Andes (South America) and Himalayas (central Asia). As conditions worsen with increased altitude, xerophytic vegetation becomes more evident. [Sinhalese]

paternoster lake One of a chain or string of usually elongated lakes that lie in a glaciated valley. The lakes are usually separated from each other by dams of moraine or by rock bars. [The name derives from the beaded appearance of the lakes resembling a rosary]

patina †*See* desert varnish.

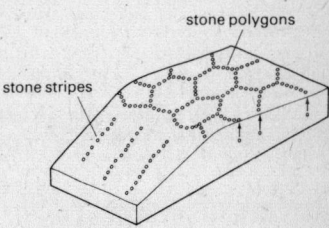

Patterned ground

patterned ground †The sorted and unsorted circles, polygons, nets, and stripes, that are formed under periglacial conditions. The processes by which these features are formed are complex but they are generally the result of freeze-thaw action and solifluction. Relict patterned ground features may be seen in areas that were once subjected to periglacial conditions, for example, in Norfolk. *See*

179

ice wedge, stone polygons, stone stripes. [Coined by A. L. Washburn]

pavement An expanse of near-level bare rock that is the product of denudation. A pavement may be formed by one of a number of different processes such as weathering (e.g. limestone pavement), wind erosion (e.g. hamada), or glacial erosion. *See also* hamada, limestone pavement.

pays In France, a region that possesses distinctive characteristics that distinguish it from neighbouring regions. The characteristics include those of relief, vegetation, and land use.

PCI *See* per capita income.

peak The summit of a mountain. It is generally pointed but may have a more rounded or conical form.

peanut *See* groundnut.

peasant A person living in a rural area who is engaged in agricultural work or cottage industry for subsistence. The term generally implies a low social status. *Peasant farming* is essentially small-scale farming based on family units. The land holdings are usually small. Although a large proportion of the produce may be consumed by the family, peasant farming is not usually subsistence farming; any surpluses grown are generally sold on the open market and cash sales are important, for example, to pay taxes and buy other goods. In tropical countries, for example, peasant farming has diversified into the production of commercial cash crops (e.g. cocoa and tea).

peat The dark soft fibrous or structureless partially decomposed remains of plants forming accumulations over 50 cm in depth. Peat is formed when vegetation dies in waterlogged anaerobic (airless) conditions and the decomposing activity of small organisms is hindered by the lack of free oxygen. Peat thus occurs chiefly in cool moist climates where drainage is poor (e.g. central Ireland) and is normally spongy and saturated. Sphag-

num mosses are often the dominant living and peat-forming species. Peat is cut and dried for fuel and horticultural uses. It is an initial stage in the formation of coal.

pebble A rounded fragment of rock; it is larger than a gravel stone but smaller than a cobble. It is sometimes defined as having a diameter of 4–64 mm. The rounding of the rock fragment may result from either wind or water action. Large numbers of pebbles that accumulate in one place form a *pebble bed*.

pedalfer A mature free-draining zonal soil of humid climates in which leaching has removed base compounds (e.g. calcium) so that the soil becomes generally acidic. Pedalfers are rich in aluminium and iron. Examples include podzols, brown earths, latosols, and prairie soils. *Compare* pedocal. *See also* acid soil.

pediment A gently sloping rock surface at the base of a ridge or mountain in an arid or semiarid area. The pediment surface is often covered with a thin layer of debris. At its upslope end there is a sharp break of slope – the piedmont angle – where it meets the upland. †It is thought that pediments are formed by denudation in arid or semiarid areas. The exact processes responsible for their formation are not known. The various theories include erosion by sheetflood; erosion by the lateral planation of ephemeral (short-lived) streams; parallel retreat of the mountain slope; and weathering that clothes the bedrock with a thin layer of debris.
See also pediplain, piedmont.

pediplain An extensive erosion surface with a concave profile that has been formed by the coalescence of several pediments. †L. C. King and W. Penck have suggested that a pediplain is the final product in an area of parallel retreat (the equivalent of a peneplain in areas of slope decline). Examples of pediplains include the African pediplain in Kenya. The process by which a pediplain is formed is termed *pediplanation*.

pedocal A mature free-draining zonal soil of relatively dry climates that develops where leaching is slight or nonexistent and calcium carbonate accumulates in the soil, e.g. chernozems and chestnut soils. †Pedocals have pH values above neutral (7.0). *Compare* pedalfer.

pedology The scientific study of the characteristics, development, and distribution of soils.

pedon †A column of soil extending from the ground surface down to parent material, used in field investigations of soils. As it is three-dimensional a pedon can complement the conventional two-dimensional soil profile by showing horizontal changes through a soil horizon. *Compare* soil profile.

pelagic Denoting the surface and middle depths of the open ocean, i.e. those areas beyond the sublittoral zone that are negligibly affected by the land. *Compare* sublittoral.

pelagic deposit A deep-water deposit (i.e. ooze) that is derived from the remains of organisms living in pelagic waters. This is only found in the deep waters of the open ocean but there is a broad area, the sublittoral zone, where it is found mixed with deposits derived from the land. *See* ooze.

Peléan eruption (Peléean eruption) A volcanic eruption in which viscous lava accumulates around the vent forming a compact steep-sided cone or dome. Violent eruptions may be accompanied by nuées ardentes. [Named after Mount Pelée on the island of Martinique]

Pele's tears Teardrop-shaped pieces of glassy lava thrown out in volcanic explosions, which have fused and solidified in the air. *See also* lapilli, pyroclast. [Pele is the Hawaiian goddess of volcanoes]

peneplain (peneplane) †An area of low relief forming the final stage of the normal cycle of erosion under humid fluvial conditions. It is a landscape in which almost all hills have been worn down and wide floodplains and gentle interfluves predominate, with only the occasional more resistant peak or monadnock remaining. In practice, no large peneplains exist because processes have been interrupted by changes of climate or tectonic movements. *See also* monadnock. [The term was introduced in 1889 by the physical geographer William Morris Davis; literally: almost a plain]

peninsula A piece of land jutting out into the sea or a lake. [Latin: almost island]

per capita income (PCI) The total income of a group divided by the number of people in that group, i.e. income per head. It is used as one of the indices for comparing standards of living between countries. For example, in Bangladesh per capita income is 85, in Egypt 448, in the UK 4955, and in the USA 8612 (1978 figures in US dollars).

perception *See* environmental perception.

perched block A boulder left resting precariously on top of other rocks after a glacier or ice sheet has melted. Perched blocks are erratics carried from their parent outcrops by glaciers. *See* erratic.

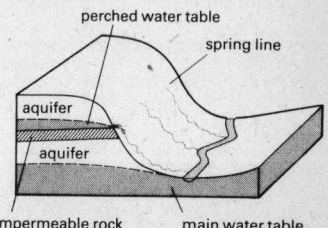

Perched water table

perched water table A body of water formed beneath the ground but above the main water table. It may form wherever a less permeable layer slows down the vertical percolation of water and causes it to pond up, within the soil, regolith, or bedrock.

percolation The downward movement of water through the pores of a rock or soil. The term generally excludes flow through large voids, cracks, or fissures.

perennial irrigation Irrigation methods that provide water for cultivation throughout the year, such as projects based on dams and underground resources that are not subject to seasonal fluctuations. One example of a perennial scheme is the Gezira project in the Sudan, which uses water from the Makwar Dam. *See also* basin irrigation.

periglacial The climate, landforms, and processes typical of cold regions bordering on glacial. The term was originally applied to areas bordering on an ice sheet, but it is now extended to any area with a similar climate (e.g. high mountains at low latitudes). Freezing and thawing are the dominant processes causing, for example, frost shattering, solifluction, needle ice, and erosion by snowmelt. Permafrost is a common feature of arctic periglacial regions but it is by no means ubiquitous. *See* freeze-thaw, frost shattering, permafrost, solifluction.

period An interval of geological time. The corresponding division of rocks is a system. Each period is subdivided into a number of epochs and together a number of periods form an era. *See also* epoch, era.

permafrost Permanently frozen ground formed in very cold (periglacial) regions. The top metre or so may thaw in summer and is known as the active layer, but the ground beneath remains below 0°C all the year round. The frozen ground is a hindrance and a hazard for opencast mining, and melting around buildings can result in subsidence. *See also* active layer.

permeable rock A rock through which water can pass more or less freely from the upper surface to the lower surface. It may be either porous (e.g. sandstone, oolitic limestone) or pervious (e.g. chalk). Water contained within a permeable rock is capable of being extracted by pumping. *See also* pervious rock, porosity.

Permian The final geological period of the Palaeozoic era, extending from the end of the Carboniferous period, about 280 million years ago, to the beginning of the Mesozoic era, about 225 million years ago. Continental conditions extended over much of the British Isles during this period and resulted in the deposition of sandstones, which form part of the New Red Sandstone. Permian rocks are found in the British Isles in County Durham, along the E Devon coast, and in the Midland Valley of Scotland. During this time reptiles became widely established and plants became more advanced. [Named by the British geologist R. I. Murchison in 1841 after the Perm province in Russia]

pervious rock A rock through which water can pass freely as a result of joints, bedding planes, cracks, and fissures in the rock. For example, some limestones (e.g. chalk) and igneous rocks. *See also* permeable rock.

petrochemicals Chemicals that are obtained as by-products of crude oil or natural gas. A large number of goods used today are derived from petrochemicals, e.g. fertilizers, plastics, synthetic fibres such as nylon, pharmaceuticals, and synthetic rubber.

petrology The study of all aspects of rocks, including mineral composition, texture, structure, origin, occurrence, alteration, and relationship with other rocks.

pH A measure of the acidity or alkalinity in a soil or solution. The scale ranges from 0 (completely acid) to 14 (highly alkaline), the median value pH 7 (neutral) being that of pure water. The extremes of acidity (below pH 3) and alkalinity (above pH 11) seldom occur naturally. Most British soils have pH values in the range 4–8.5. The optimum for many crops is pH 6 (i.e. slightly acid). †The pH is given by $\log_{10}(1/[H^+])$, where $[H^+]$ is the hydrogen ion concentration in moles per litre.

phacolith An igneous intrusion in which the magma was injected between rock strata that were being folded. The resulting intrusion usually occupies the crest of an anticline and is lens-shaped with a convex upper surface and a concave lower surface. Phacoliths can occur in sets with each one occupying a successively higher position on the crest of the anticline. Sometimes phacoliths are associated with ore deposits, for example the quartz-gold saddle veins of Bendigo, Australia.

phase A stage in the implementation of a planned programme or in the development of a particular phenomenon. The term is particularly used in planning; for example, a five-year national plan may be divided into a series of phases.

photic zone The upper zone of the ocean within which sufficient sunlight penetrates to allow photosynthesis to take place. This is generally above a depth of about 90 m. *Compare* aphotic zone.

photogrammetry The science of taking measurements and the making of maps from a series of aerial photographs. Elaborate plotting machines are used to eliminate errors and produce detailed and accurate maps. These methods allow the rapid compilation of maps of large areas. Photogrammetry is particularly useful for mapping areas of difficult terrain such as the Canadian Shield.

photo relief The photographing of detailed relief models to provide a realistic representation of hills and valleys. A light is usually placed in the NW so that shadows are cast that highlight the relief features on the model. The completed photograph can be extremely effective.

photosynthesis †The biochemical process by which chlorophyll-bearing organisms use the energy of sunlight to synthesize organic compounds for their own growth. These organisms (chiefly green plants) are known as primary producers; their photosynthetic products are subsequently used by consumers. In green plants, the chlorophyll contained in special receptor cells (chloroplasts) on leaf surfaces absorbs the Sun's energy. The plants use the hydrogen in water (H_2O) and carbon dioxide (CO_2) to form carbohydrate compounds (e.g. sugars and starches), releasing oxygen to the atmosphere in the process. The overall reaction in green plants is complex but can be summarized in the equation:

$$CO_2 + 4H_2O \rightarrow [CH_2O] + 3H_2O + O_2$$

The surface area and position of leaves, besides the duration and intensity of sunlight, govern the rate of photosynthesis.

phreatic water Ground water, especially that below the level of the water table. *Compare* vadose water.

phylloxera An aphid (insect) that lives on the leaves and roots of the grape vine. Between 1870 and 1900 the French wine industry was badly affected by this pest. Vine stocks from the E of the USA were found to be almost immune and the industry has overcome the problem by grafting high-quality French varieties onto resistant stocks from the USA.

physical geography The study of the features of the Earth, which form the environment of man, and their development through time. Contributory sciences include geomorphology (the study of landforms), geology, climatology and meteorology, and pedology (soil science).

physical weathering *See* mechanical weathering.

physiography The study of the surface forms of a region. The word has changed its meaning over the years from covering the whole of physical geography, including geomorphology and climatology, to being restricted to geomorphology.

pictogram (pictograph) A graph based on picture symbols rather than on bars or lines to represent data. For example, the use of sheep symbols to represent a certain number of sheep or symbols of coffins to represent mortality rates.

piedmont A region or landform at the foot of a mountain range stretching between the mountain slopes and a flat lowland. An example is the Appalachian Piedmont, a gently undulating region lying between the Blue Ridge and the Fall Line in the SE USA. †*Piedmont benchlands* are wide platforms that may be formed by discontinuous rejuvenation. A *piedmont alluvial plain*, or bajada, forms in dry lands by the merging of alluvial fans in front of mountain ranges, sometimes downslope from bare rock pediments. *See also* bajada, pediment, platform. [French: mountain foot]

piedmont angle †The junction between a piedmont slope and an upland mass. In arid and semiarid regions the angle may be particularly sharp and mark a change from 30° slopes on the mountain sides to under 10° on the pediment. Some angles seem to be structurally controlled; others may be related to different erosion processes.

piedmont glacier An ice sheet formed at the foot of a mountain range by the coalescence of a number of valley glaciers. †The rate of movement tends to be less than in the neighbouring valley glaciers. An example of this form is the Malaspina Glacier in

Alaska, which is about 35 km wide and 100 km long.

pie graph (divided circle diagram) A graph in the form of a circle divided into sectors. Each sector represents a certain proportion or percentage of the total, with each 1% being represented by an angle of 3.6° on the circle. Pie graphs are used widely in human geography; for example, to show energy consumption in the UK by type of energy used. If more than one circle is used, for example to compare one year with another, each circle can be made proportional in area to the quantity it represents.

pig iron The impure iron that contains silica, carbon, and other impurities obtained from a blast furnace. The molten iron from the furnace is run along channels called 'sows' to which are linked side channels at right angles. These are called 'pigs' and here the iron cools into rectangular blocks.

pilang A close-growing monsoon forest in Indonesia, the dominant tree species of which is a type of acacia. *See* monsoon forest.

pillow lava A volcanic lava that has solidified quickly under water to take on the appearance of a pile of pillows. The sudden chilling of the outer layers of the liquid lava as it is ejected under water or as it flows into the

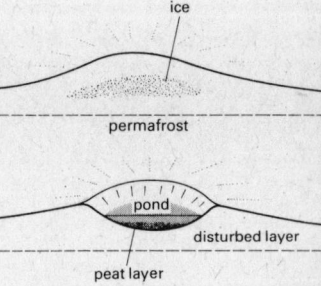

Development of a pingo

water causes a skin to form which fills like a balloon and falls flat under its own weight.

pingo A dome-shaped hill with a core of ice found in periglacial areas, such as Alaska, the Arctic coast of Canada, and Siberia. The dome is formed by the growth of an ice lense below ground caused by the freezing either of local ground water or of water flowing into the site over a period of time. The water expands when it freezes forcing the ground surface up. Expansion cracks on the top of a pingo can allow heat to penetrate in summer, causing melting and collapse. Partially decayed pingos can have craters and crater lakes in the top. [Eskimo]

pipe 1. A vertical shaft or sinkhole, often filled with gravel, found in chalk and other limestones.
2. A mainly horizontal tube found in soil, peat, unconsolidated fine sediments, or earth dams caused by the removal of particles in flowing water. †Pipes are commonly formed where percolation is restricted by a relatively impermeable layer and a perched water table is created; but the most general factors are high velocities of flow (throughflow) and erodible soils or sediments. They may form a subsurface extension of the stream network and drain storm water into the streams. They are found in most climatic regions.
3. A volcanic conduit or tube filled with igneous rock.

pipeline A cylindrical tube used for transporting materials in a fluid form. Apart from transporting crude oil and refined products, gas, and water, pipelines are also used for moving chemicals and materials such as clay in a liquid form. Transport of oil by pipeline is far cheaper than by any other means, but the initial capital outlay is high. A pipeline network carries natural gas from beneath the North Sea to the main centres of population in the UK.

piracy of streams See capture, river.

pitch 1. The direction of dip of the axis of a fold in rock strata.
2. A solid or nearly solid hydrocarbon formed as a residue in the distillation of tar or occurring naturally (e.g. asphalt). The most famous source is Pitch Lake, Trinidad, where the oil sand has been exposed by erosion and the oil has evaporated, leaving natural pitch as residue.

pitchblende A form of uraninite, an ore of uranium. Pitchblende is the most important source of uranium. It is found chiefly in hydrothermal veins associated with small amounts of other elements such as lead, thorium, and cerium. The main areas in which it occurs include the Great Bear Lake, Canada, and Zaïre.

plagioclimax vegetation †A type of vegetation that is a more or less permanent by-product of man's activities over a long period. For example, repeated grazing following the burning of forest prevents the redevelopment of scrub and woodland, creating a plagioclimax grassland. Plagioclimax is analogous to subclimax vegetation, but is determined by protracted human interference rather than natural factors. The vegetational stages leading to plagioclimax are termed a *plagiosere*. Removal of the restraint allows the normal succession to resume. *Compare* subclimax vegetation. *See also* climax vegetation.

plain An extensive area of flat or gently undulating land, usually lowland. The wide grassland expanses of the North American Great Plains, the Canadian prairies, and the Russian steppes are typical plains.

plan A map of a comparatively small area drawn on a large scale and showing considerable detail. For example, large-scale Ordnance Survey (OS) maps such as the 1:2 500 and 1:1 250.

planation †The process of planing or creating a surface of low relief; such surfaces are known as *planation surfaces* and may be the end-product of

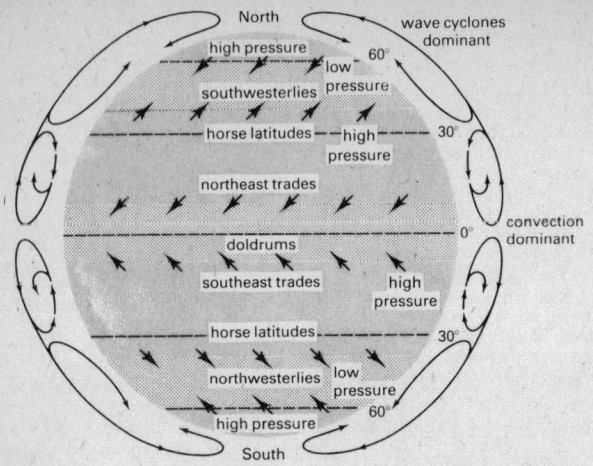

Planetary winds

marine or subaerial processes. During the last century many large planation surfaces were attributed to marine transgressions, but in the 20th century the emphasis has been on subaerial processes forming peneplains and pediplains. *See also* pediplain, peneplain.

plane table An instrument used in surveying that consists of a horizontal board fixed to a tripod. Paper is fixed to the board and after marking magnetic north sights are taken to distant objects using an alidade. Direction lines (rays) are then drawn on the paper. The plane table is then moved to the other end of a measured base line where the operation is repeated. The point at which rays intersect marks the position of an object. *See* alidade.

planetary winds The general circulation of surface winds throughout the world. The wind belts are basically controlled by the latitudinal pressure belts and by the forces produced by the rotation of the Earth. Moving N and S from the equator, the main wind belts are the trade winds, which

blow from the subtropical high pressure area (around latitude 30°) towards the equatorial low pressure area; the temperate westerlies, which blow from the subtropical high pressure area to the temperate low pressure area (around latitude 55°); and the polar easterlies, which blow from the polar high pressure area to the temperate low pressure area. This general pattern is disturbed by several factors, including the differential heating of land and sea, the development of waves on a front, and the relief of the land surface. *See also* general circulation of the atmosphere.

planimeter An instrument for measuring area on a map. A number of different models exist. The *wheel planimeter* uses a dial to record the distance travelled around the perimeter of an area; this figure multiplied by a known constant for the instrument gives the area. Another type, the *dot planimeter*, uses evenly spaced dots on a transparent surface. This is placed over the area to be measured and dots falling within the boundary are counted and multiplied by the area-factor each represents.

plankton Very small animals and plants (e.g. diatoms and algae) that form the drifting organic life of the surface layers of the ocean. Plankton is very important as a source of food for other marine animals such as fish and whales. The great fishing grounds of the world are found where water conditions encourage its growth and so attract the fish that feed on it; for example, the Grand Banks off the SE coast of Newfoundland. Similar organisms are found in fresh-water lakes.

planning The process of placing actions into a particular order to achieve an objective. Practically all areas of human activity have an element of planning within them. Perhaps the most widely understood use of the term refers to land-use planning and the production of a plan in the form of a map. *See also* economic planning.

planosol An intrazonal soil of poorly-drained areas, e.g. flat areas. Planosols are characterized by the eluviation of fine materials to form a claypan.

plantation An estate or farm in sub-tropical or tropical countries that is used for the specialist production of a crop such as bananas, coffee, rubber, tea, palm oil, or sugar cane, for export to countries of the developed world. Capital from multinational companies is frequently invested in the development of plantations, which depend on indigenous labour or workers brought in from elsewhere, e.g. the Tamil workers from S India in the rubber plantations of Malaysia.

plant succession †A progressive sequence of changes in the vegetation of an area. In all environments, succession begins with a pioneer stage in which plants colonize a bare surface. These plants modify the environment (e.g. by introducing humus) so that slightly more demanding species are able to replace them. At each seral stage in the succession the plant community improves the soil and other conditions so that the next, more demanding, plant community can succeed it. The composition and structure of the plant community, and associated animal life, become successively more complex until a stable stage or climax is reached. Rates of succession vary, but are generally slow, and decelerate towards climax. Natural succession may be interrupted, set back, or deflected by man's activities or natural events such as fire. *See also* climax vegetation, sere.

plastic deformation of ice †One of the processes of glacier flow in which

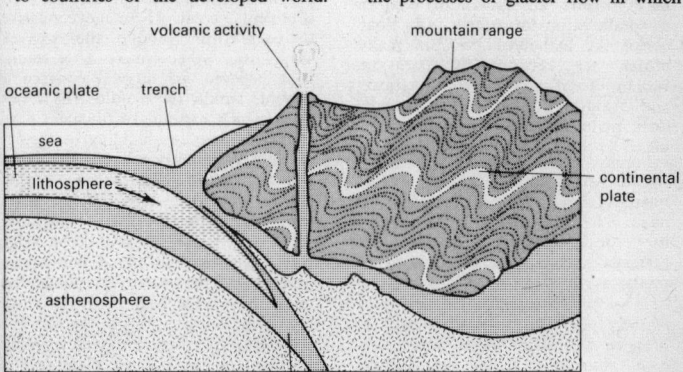

The development of a mountain range at a destructive plate margin

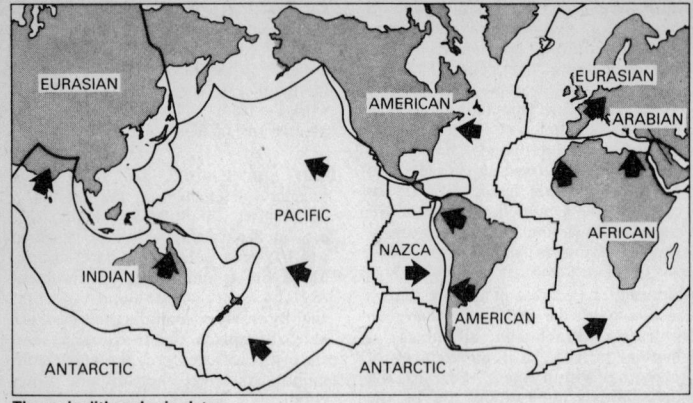

The major lithospheric plates

the ice crystals deform under pressure, breaking along one axis and then recrystallizing.

plastic relief map A three-dimensional map moulded in plastic to show the relief of an area.

plastic shading *See* hill shading.

plate One of the large rigid segments into which the Earth's crust is divided. The plates are composed of continental or oceanic lithospheric material, or a combination of both, and 'float' above the asthenosphere. Six major plates are recognized: Eurasian, Indian, Pacific, American, African, and Antarctic. *See also* continental drift, plate tectonics, sea-floor spreading.

plateau An extensive elevated area of relatively flat land. Widespread movements of the Earth's crust may result in vertical warping, which produces plateaus and rift valleys divided by faults, as in the Cordilleran plateaus of North America. *Intermontane plateaus* (e.g. the Tibetan plateau) lie between mountain ranges, while *piedmont plateaus* (e.g. the Appalachian Piedmont in the USA) are bounded by mountains on one side only. *Continental plateaus* (or tablelands) stand

above the surrounding areas, as in the Cape Province of South Africa, bounded by cliff faces. Rivers flowing across plateaus of near-horizontal strata in semiarid areas create canyons (e.g. the Grand Canyon in Colorado, USA). Widening of valley floors can carve plateaus into mesas and eventually into buttes. *See also* butte, mesa.

plate tectonics The theory that the outer shell of the Earth is made up of a number of relatively thin lithospheric plates composed of oceanic and continental crust. These move relative to each other above the weaker semiplastic asthenosphere as a result, it is believed, of thermal convection currents within the mantle. The zones along which earthquake (seismic), volcanic, or mountain-building (orogenic) activity is currently taking place coincide with the junctions between plates. Where two plates are moving apart from each other sea-floor spreading takes place along the mid-ocean ridge that separates them. In a continuous process magma wells up along the mid-ocean ridge and cools to form new oceanic crust. This is known as a *constructive plate margin*. Where a continental and an oceanic plate collide the oceanic plate is forced down into the mantle below the other. This is known as a *destructive plate margin* and is marked by a deep ocean trench. As the oceanic crust is forced downwards into the subduction zone it becomes molten and forced back to the surface of the Earth as a chain of volcanoes. The continental crust is contorted under the pressure into mountain chains, for example the Andes of South America. A third type of plate margin – a *conservative plate margin* – occurs when two plates slip past each other with no gain or loss of material. Collision between two continental plates results in the formation of fold mountain chains, for example the Alpine-Himalayan belt. *See also* continental drift, plate, sea-floor spreading.

platform 1. A piece of flat raised ground, such as a tableland.
2. *See* continental shelf.
3. *See* wave-cut platform.

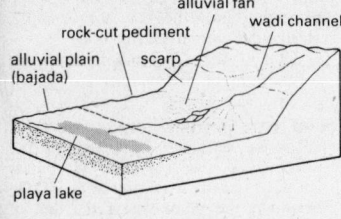

Playa

playa The low flat central area of a basin of inland drainage. Playas occur in areas of low rainfall, particularly where drainage to the sea has been prevented by crustal movements, as in Death Valley, California, and the Great Basin of Utah. During wet periods the centre may contain a *playa lake*, which is typically salty because of high evaporation. The lakes dry out to leave alkali flats or salinas. [Spanish: shore]

Pleistocene The lower geological epoch of the Quaternary period. It extended from the end of the Pliocene, about two million years ago, to the beginning of the Holocene. During the Pleistocene great fluctuations in temperature occurred with four cold periods (glacials) alternating with warmer periods (interglacials). In the British Isles the Pleistocene is represented by glacial deposits, which accumulated N of a line from the Thames estuary to the Bristol Channel, and by gravel and river terraces, which accumulated S of this line. During this time the geography of Britain assumed its present form. Fossils from this epoch include horses, elephants, and pigs. [Greek: most recent]

plimsoll line A series of lines on the hull of a ship indicating the legal limit to which the ship can be loaded. The heavier the load the lower the hull sinks into the water. Because water in different parts of the world has different buoyancies according to its salinity (for example, the Baltic Sea is less buoyant because it is nearly fresh water), a number of lines at different levels must be drawn.

Pliocene The last geological epoch of the Tertiary period. It followed the Miocene and extended for about two million years to the beginning of the Quaternary period. During this time the conditions that prevailed throughout the Miocene continued, in that the sea continued to withdraw from the area of the British Isles and the temperature continued to drop. Few rocks of Pliocene age were deposited in the British Isles; the epoch is represented mainly in East Anglia by shallow-water shell gravels known as crags. Most of the fossil forms found in Pliocene deposits still exist today. Elsewhere, the conditions were continental and the Pliocene is represented by erosion surfaces.

plucking A process of glacial erosion in which ice freezes to shattered rock on the bed of a glacier and carries it away. Ice and meltwater enter joints and fractures in the bedrock, loosening blocks, and glacier movement drags them away from the downstream (or lee) side of protruding rocks. *See also* roche moutonnée.

plug A body of igneous rock formed from magma, which has crystallized beneath the surface blocking a volcanic vent or duct. When not active, the central pipe of a volcano is usually plugged. Erosion may subsequently expose the plug, as at Edinburgh Castle, which stands on a volcanic plug of basalt.

plug volcano (plug dome) A cylindrical protrusion of very viscous lava, blocking the crater of a volcano or forcing rocks above it to dome up. The plug stands out because the lava was too viscous to spread, not because later erosion has exposed it as in the case of volcanic plugs. Several of the puys in the Auvergne region of France are plug volcanoes. *See also* puy.

plumbago *See* graphite.

plum rains *See* Bai-u season.

plunge pool A pool created at the foot of a waterfall by plunging water. It is mainly formed by abrasion from sediments, but in calcareous bedrock solution also plays an important role.

plutonic rock An igneous rock that cools in a large mass at great depth within the Earth's crust. Cooling takes place very gradually at great pressure resulting in coarse-textured rock in which the crystals are large and about the same size. Examples include granite, diorite, gabbro, and peridotite. *Compare* hypabyssal rock, extrusive rock.

pluvial (pluviose) Denoting rain or the action of rain. A *pluvial period* is a period of time, usually on a geological time scale, during which there is greater rainfall than in the periods preceding or following it.

pluviometric coefficient †An index calculated by expressing the mean monthly rainfall of a place as a ratio of the hypothetical monthly rainfall that would fall if the annual rainfall was evenly distributed throughout the year. This gives an indication of the raininess or otherwise of each month in comparison to other months.

podzol (podsol) A zonal soil that is formed in cool humid climates under heath and coniferous forests. It is characterized by an ashen coloured A horizon and an illuvial B horizon. Iron compounds and base salts are rapidly leached down from the A horizon into the B horizon where they accumulate, often forming a thin dark ironpan; this strongly contrasts with the pale sandy clayless zone immediately above. The processes by which podzol soils are formed are known as *podzolization*. Podzol soils extend in a broad belt across the USSR and North America. Elsewhere they occur as intrazonal soils, e.g. on sandstone and following forest clearance by man. †*See also* chelation. [Russian]

point 1. The end of a headland or cape projecting into the sea; for example, Hartland Point in Devon.
2. One of the 32 standard directions or divisions of the compass.

point bar (point-bar deposit) A ridge of coarse sediment deposited on the inside of a river meander. †It builds up downstream from the apex of the meander, the point where the line of fastest flow begins to shift towards the opposite undercut bank. A trough of slack water is left between the point bar and the inner bank, which gradually fills up with finer sediment.

polar air mass An air mass having its source region between latitudes 40° and 60°N and S (an air mass with its source area between latitude 60° and the pole being known as either an Arctic or an Antarctic air mass). There are two distinct types: polar maritime air (mP), which has its source region over oceanic areas (e.g. the N Atlantic), and polar continental air (cP), which originates over land (e.g. Siberia). The polar maritime air mass is cool and fairly moist in winter and warm and moist in summer, while the polar continental air mass is very cold and dry in winter and warm and dry in summer.

polar front †The frontal zone in the N Pacific and N Atlantic where cold polar air masses come into contact with warm tropical air masses. It is along this front that the depressions that dominate the weather of mid-latitude areas are produced. See Bjerknes polar front model.

polar outbreak A cold wave experienced when cold air from high latitudes moves into comparatively low latitudes, resulting in unusually low temperatures and, frequently, strong winds. This is always a danger to crops. See cold wave.

polar wind A wind that blows from the polar high-pressure belt towards the temperate low-pressure belt. It is a northeasterly wind in the N hemisphere and a southeasterly wind in the S hemisphere, and tends to extend to about latitude 55°, although this is very variable.

polder A piece of land reclaimed from the sea or a lake. The submerged land is surrounded by an embankment (dyke) and drained by pumping the water into canals raised above the land surface. Windmills were traditionally used for this purpose in the Netherlands and in the English Fens. The reclaimed land is normally highly fertile, but where sea water has been evacuated a gradual process of desalinization is necessary before normal crops and grasses can be planted. See also reclamation. [Dutch]

pole 1. Each of the two points – the North Pole and South Pole – that form the extremities of the Earth's axis of rotation. As the axis is perpendicular to the plane of the equator (0°), the latitude of the North Pole is therefore 90°N and the latitude of the South Pole is 90°S. The Pole Star is almost overhead at the North Pole. See also axis, Earth's.
2. See magnetic pole.

political geography The geography of political phenomena, including states and frontiers, their variations and inter-relationships, and their form and impact on the Earth's surface. [Coined by the geographer Norman Pounds]

pollen analysis †The identification and counting of the remains of pollen grains from samples of sediments as a means of studying past floras. Pollen is microscopic material produced by plants in great quantities. It becomes airborne and may subsequently be trapped in sedimentary deposits, such as lake muds and peat, from which core samples can be extracted for analysis. Pollen types from different plant genera have distinctive surface shapes and markings, which are preserved because of the highly resistant exterior of the pollen grain.

Pollen analysis enables a reconstruction to be made of past vegetation, especially that of the Quaternary period. It gives information on the species present and their relative abundances, changes through time, and correlations from place to place, from which climatic conditions can be inferred. The study of pollen analysis, together with the analysis of other microfossils, is known as *palynology*.

pollution The fouling of the environment by man, which makes it harmful for living organisms or reduces its amenity value. Pollution affects the environments of the atmosphere, e.g. exhaust fumes from cars and sulphur dioxide from industrial chimneys; the land, e.g. the use of dangerous chemicals as pesticides and fertilizers and the dumping of waste and litter; and water, e.g. oil spillage from tankers, untreated sewage in the sea, and industrial effluent poured into rivers. Noise may also be considered a form of pollution of the environment, e.g. the noise from jet aircraft and motorway traffic.

polyconic projection A modified conical map projection in which all parallels of latitude cut the central meridian at true-to-scale distances apart.

The distances between the meridians along each parallel are also true to scale. The projection is not equal area and there is much distortion at the edges making it unsuitable for maps of large areas. In a modified form it is used for the International Million Map. *See* conical projection.

polygonal karst (cockpit karst, kegelkarst) †A form of limestone scenery found in the humid tropics, consisting of deep depressions separated by steep pointed hills. The depressions are formed by surface erosion rather than collapse and form local basins of internal drainage often covered in alluvial deposits. Polygonal karst has been classified as conical, linear, pinnacle, or tower according to the exact form of the residual hills. *See* karst.

ponente A cool dry westerly wind experienced in S France and Corsica.

pools and riffles †The alternating bars of gravelly sediments (riffles) and deeper sections (pools) that occur along the course of many rivers. They tend to be regularly spaced at distances between bars of five–seven times the channel width. Experiments with laboratory models have shown that this pattern often comes before the development of meandering. In nature the distance between pools is about half the length of meanders.

population The total number of inhabitants of a city, country, or other unit of area. *Population geography* is the study of spatial variations in demographic processes and patterns (e.g. migration, distribution, growth, and composition) in relation to the environment.

population density A measure that relates the size of population to the area of land that it occupies. *Crude population density* is a measure of the average number of people per unit area (usually per square kilometre). The measure does not recognize the variations in density that may occur within the spatial units and so is of limited value. Improved density values may be obtained, for example, by relating numbers of people to inhabited areas or cultivated areas only.

population explosion A rapid growth in the numbers of a population. In the world as a whole accelerating growth has created anxiety concerning the future supply of food and other resources. It has been suggested that continued rapid unplanned growth will lead to uncontrollable problems on a world scale as resources become more limited.

porosity The ratio of the amount of space between the mineral grains to the total volume of a rock, soil, or sediment. It is usually expressed as a percentage of the volume. For example, the porosity of an evenly graded rock with perfectly packed grains is 27%. Sediments vary a great deal from less than 1% to over 50%. Sandstones range from 5–15%; clays, which are very porous, have a porosity reaching 50%.

port A place close to navigable waters where ships can shelter or load and unload. Smaller centres are often fishing ports that may also be used by pleasure craft, e.g. Brixham in Devon. Larger centres with quays and docks may be inland, e.g. Rouen on the River Seine, or seaports, e.g. Liverpool.

portage The moving of boats and goods overland from one navigable river to another, or from one part of a river to another. Portage routes were used by early explorers and settlers. In Canada and the USA a number of these routes mark the movement of traders and explorers, e.g. around the Great Lakes and along the St Lawrence valley.

positive movement of sea level A rise of sea level relative to the land that may be on a local or worldwide scale. A very marked worldwide rise of sea level follows every ice age as meltwater swells the oceans. This has been a major feature of the last 10 000 years during which the rising sea level has flooded the Irish Sea and English Channel, cutting Britain off from the continent and Ireland from Britain. It has also flooded the glaciated valleys of Norway creating the fiords. World sea level continues to rise.

Local rises in sea level can be caused by the sinking of the land in tectonic movements, e.g. downwarping in areas such as the Netherlands and E England where the crust has compensated for the isostatic rise of land to the N following the melting of the glaciers. Many delta areas are also sinking under the weight of sediments, as in the Mississippi Delta and Venice. *Compare* negative movement of sea level.

postglacial The interval of geological time extending from the end of the last ice age (the Pleistocene) to the present, approximately the last 10 000 years. It is also referred to as the Holocene or Recent epoch.

potash A salt of potassium or a mineral containing potassium. Salts such as potassium chloride or potassium magnesium chlorate are found in association with common salt at sites of salt lakes, for example the Dead Sea.

potential evapotranspiration index (PEI) A ratio expressing the relation between the amount of precipitation that falls to the amount of evapotranspiration (the combined total of evaporation and transpiration) and the amount of water otherwise available, e.g. ground water, at any particular place. It was introduced by C. W. Thornthwaite (1948) as the basis for his second classification of climate.

pothole 1. *See* sinkhole.
2. A round hole worn in rock stream beds by boulders, cobbles, or pebbles rotating in eddies.

power Any form of energy that can be harnessed to drive machinery or provide a force. The sources of power include electricity (hydroelectric, nuclear, thermal, etc.), running water, steam, wind, and sunlight. As conventional sources of power such as fossil fuels become more scarce new forms are being investigated, e.g. geothermal (heat from within the Earth's crust), tidal, and solar energy.

prairie The gently undulating grassy plains of the interior regions of North America. The main water supply to the prairies tends to be springtime snowmelt from the Rocky Mountains to the W. Limited summer rainfall and high temperatures generally restrict tree growth to along the water courses, but provide an ideal environment for cereal production. The prairies are one of the main 'granaries' of the world and the equivalent of the Russian steppe and South American pampas. [French: meadow]

prairie soil (brunizem) A zonal soil that develops under mid-latitude grasslands where rainfall is sufficient for leaching of soluble materials. Prairie soils are pedalfers with pH values slightly below neutral. They form under the taller grasses that the relatively moist conditions encourage. Humus production is as a result high, giving fertility and soil structure that is very suitable for agricultural use. *See also* chernozem.

Precambrian The earliest geological era, extending to the beginning of the Cambrian period, about 570 million years ago, and the system of rocks laid down during this era. The rocks have undergone much alteration and are usually metamorphosed and practically devoid of fossils, which makes correlation between areas difficult. The largest areas of exposed Precambrian rocks are the shield areas of Canada (the Laurentian Shield) and Scandinavia (the Baltic Shield).

precipitation In meteorology, the particles of water or ice that form within clouds and fall towards the Earth's surface. The main types are rain, drizzle, sleet, snow, and hail. Any of these may evaporate before reaching the ground surface, appearing as streamers below the cloud base, a phenomenon known as virga. Dew and hoar frost are frequently described as types of precipitation but are not strictly covered by the definition.

precipitation day A period of 24 hours during which at least 0.2 mm of precipitation falls. The term is synonymous with rain day but is a more accurate description, especially in countries where much of the precipitation falls as snow. See also rain day.

precipitation efficiency index †An index devised by C. W. Thornthwaite as the basis for his first classification of climate. It links mean monthly rainfall with mean monthly temperature to show how efficient the rainfall is in particular temperature regions. For example, low rainfall in high-temperature areas is less efficient than low rainfall in low-temperature areas as a result of the increased rate of evaporation. For each month the ratio is:

$$11.5(r/t - 10)10/9$$

where r is the mean monthly rainfall and t is the mean monthly temperature.

pressure See atmospheric pressure.

pressure gradient (barometric gradient) The rate of change of atmospheric pressure between two points on the Earth's surface. On a weather chart this is indicated by the spacing of the isobars, the gradient being 'steep' if they are close together, and 'gentle' if they are far apart. A steep gradient indicates strong winds and a gentle one slight winds.

pressure-plate anemometer An instrument for measuring the velocity of the wind, which consists of a metal plate suspended from a horizontal axis at right angles to the wind direction. The wind speed is indicated by the inclination of this plate to the vertical. It is not commonly used as it is not particularly accurate, especially when wind speed and direction are very variable.

pressure system An individual high- or low-pressure circulation system within the atmosphere, e.g. depression, anticyclone, col, ridge, secondary depression, wedge.

prevailing wind A wind having a higher directional frequency than any other at a particular place, for example, the southwesterly winds of the W British Isles. This can be identified easily on a wind rose, the most common wind direction having the longest radial. See wind rose.

primary industry An activity that obtains or makes available natural resources. Primary industries are not normally involved in processing these resources, e.g. farming, fishing, forestry, hunting, and mining. Only 3.1% of the total workforce of the UK is employed in primary industries. See also secondary industry, tertiary industry.

primary product A product that results from the activities of primary producers, e.g. farm produce, iron ore, timber, wool, and building stone are primary products.

primary wave See earthquake.

primate city †The largest city in a country or region. It is also the centre of political affairs, trade, and economic, social, and cultural activities. According to the rank-size rule the primate city should be twice the size of the second largest city. However, the ratio between the two cities is often considerably higher. The law of the primate city, which was put forward by the US geographer Mark Jefferson in 1939, recognized this primacy and attempted to explain how it might occur. He argued that 'once a city is larger than any other in its country, this fact gives it an impetus to grow that cannot affect any other city, and it draws away from them all in character as well as in size'. See also rank-size rule.

prime meridian *See* Greenwich meridian.

primeur A fruit or vegetable that is the first of a new season's crop. For example, in Brittany potatoes, cauliflowers, artichokes, and other crops are grown and reach the market ahead of similar crops from other parts of N France. [French]

principality A territory ruled by a prince or from which a prince obtains his title. Monaco and Liechtenstein are the only two principalities left in continental Europe. Wales became a principality in 1301 when Edward II was proclaimed the first English Prince of Wales.

prisere †*See* sere.

prismatic compass A magnetic compass used in surveying and for taking bearings. It consists of a circular card marked in degrees (0°–360°) attached to a concealed bar magnet, which swings on a central pivot. The card is read by means of a prism and sights. By holding the prism to the eye and lining up the sights with a distant object, the bearing appears as a number on the card immediately below the prism.

process–response system †A system that demonstrates the manner in which form is related to process. It involves the intersection of at least one morphological system and one cascading system. The systems are linked together and often have some components in common. For example, a lake in a desert region is a morphological system consisting of a certain area, depth, and volume of water. Rain falling onto the lake is a cascading system. The rain may increase the surface area of the lake, which in turn will result in increased evaporation. The volume of the lake and its surface area have thus increased in response to rainfall, i.e. the morphological system has changed in response to a change in the cascading input (rainfall). *See* cascading system, morphological system, system.

producer (*Biogeography*) An organism that creates foods from inorganic sources, chiefly from solar energy. Most plants and some bacteria are primary producers. †Producers thus form the first trophic level, and the organic material they produce is eaten by herbivores, primary consumers. *Compare* consumer, decomposer. †*See also* trophic level.

producer goods Goods that are needed for the production of other goods. For example, moulds are required to produce plastic shapes and pig iron is needed to make steel.

productivity 1. (*Biogeography*) The rate of production of organic matter in an ecosystem. Primary productivity occurs at the autotroph (producer) level and secondary productivity at the heterotroph level. Gross productivity is the total amount of organic matter produced at each level; net productivity is that remaining available to consumers and decomposers within the ecosystem after some has been used in respiration.
Gross primary productivity is controlled by factors such as the available light, climatic conditions (especially temperature and length of growing season), and the availability of nutrients. It is generally higher in terrestrial and tropical environments than in aquatic and polar environments. Secondary productivity is determined by the conversion of plant to animal matter.
2. (*Agriculture*) The yield that is obtained from a unit of land or the yield the land is capable of producing, i.e. the potential productivity. For example, barley yields in S England average 5800 kg per hectare.

profile of equilibrium †The long profile of a graded river. *See also* grade.

proglacial lake A lake ponded up in front of an ice sheet or glacier snout. Large lakes may be impounded where meltwater is held back by hills or ridges. When lake levels rise above the height of passes in the hills, overflow channels are formed. Small proglacial lakes can be found today ponded up by glacial moraines. Examples include Lake Agassiz in North America.

projection *See* map projection.

promontory A hilly headland that juts out into the sea.

protected state A territory that is internally self-governing with its own ruler but which is under the protection of a foreign country that manages its foreign relations. Tonga was a British protected state until its independence in 1970.

protectorate A territory that is subject, either through a treaty or a grant, to the control of a foreign country. Aden, Nigeria, British Somaliland, and Uganda were formerly protectorates of the UK.

province A major administrative division of a country, often with distinctive historical characteristics. For example, the province of Andalusia in Spain.

psychrometer A hygrometer for measuring atmospheric humidity in which a current of air is artificially induced in order to produce the maximum evaporation at all times at the wet-bulb thermometer, thus ensuring more reliable results. One method of producing the current of air is the use of a small fan (Assmann's psychrometer), while another requires the thermometers themselves to be whirled around in the air (the sling psychrometer). *See* hygrometer.

pteropod ooze An ooze on the ocean floor in which the chief constituent is the calcareous remains of pteropods, which are minute floating molluscs. Below a depth of about 2750 m the material is dissolved by the sea water; as a result the distribution of pteropod ooze is limited to the parts of the ocean floor that lie between the

depths of about 1000 m and 2750 m. It is found mainly around islands in the Atlantic Ocean, e.g. the Azores and Canary Islands. *See* ooze.

public participation The involvement of the public in the processes of political decision-making (usually beyond the election of representatives to a council). Public participation in the UK has developed particularly within the land-use planning process, since the findings of the Skeffington Report (1969). The process involves a vast area of public consultations on proposals at an informal level to more formal public inquiries.

pueblo 1. A town or village in Spain or Spanish America. **2.** A communal village or settlement of American Indians in the SW USA in which the typically flat-roofed dwellings are made of stone, rock, or adobe, or the inhabitant of such a village. [Spanish]

pumice A volcanic rock, usually derived from acid lava, which is full of rounded cavities produced by expanding gases trapped in the solidifying lava. It is sometimes light enough to float on water. *See also* scoria.

puna The highest parts of the plateaus of the Andes in Bolivia. It is a bleak region between 3000 m and 4000 m above sea level, mainly uninhabited and desolate. The summer sun may be intense, but shade temperatures are icy and the nights bitterly cold. Rivers are dry and lakes salty except during the three-month rainy season. The Indians wear thick clothing and ponchos or cloaks to protect themselves and husband sparse herds of llama and alpaca. Hardy barley and potato are almost the only crops that will survive the harsh climate. [From Quechuan]

push moraine A ridge of debris pushed up by an advancing ice sheet or glacier. Push moraines contain folds and faults as a result of the pushing action. They also tend to have a very mixed composition, perhaps made up with coarse sediments. These features make them clearly distinguishable from terminal moraines. *See also* moraine, terminal moraine.

push-pull factors †An explanatory concept relating to migration in which migratory movements are seen to be the result of the operation of two conflicting forces. The 'push' factors are those forces that encourage people to leave a particular area such as overcrowded living conditions, lack of employment, and low wages. The 'pull' factors are the economic and other attractions (real and imagined) offered by the new location. These may include, for example, improved employment opportunities, higher wages, better housing, and a more attractive environment. *See also* migration.

puszta An area of natural grassland that forms the westernmost part of the Eurasian steppe, mainly in Hungary. †More specifically, the term indicates an area of saline soils.
See steppe. [Hungarian: waste]

puy A distinctive steep-sided and dome-shaped mountain found in the Auvergne region of France. Puys are extinct volcanoes, some with craters and cones, others with just domes of lava remaining. Although probably dating from the Tertiary period, they have a remarkably fresh appearance. The largest example in the Auvergne region is the Puy de Dôme, which is 1465 m high. The term is also used to describe similar landforms in other parts of the world. [French]

P wave *See* earthquake.

Pygmy (Pigmy) A member of a tribe of dwarf peoples of equatorial Africa.

pyramidal peak (horn) A sharp isolated mountain peak resembling a pyramid, formed by the convergence of three or more cirques back-to-back.

For example, the Matterhorn in Switzerland. *See also* cirque.

pyroclast A fragment of igneous rock thrown out during a volcanic explosion. The various types of pyroclast include large bombs (over 64 mm in diameter), middle-sized lapilli (including Pele's tears), and fine ash (smaller than 2 mm in diameter).

pyrometer An instrument for measuring the temperature of molten lava.

Q

quadrant 1. An instrument formerly used in navigation and astronomy for measuring 90°, having the shape of a quarter-circle, graduated in degrees.
2. A quarter of the circumference of a circle; i.e. 90°.

quadrat A plot of ground (usually one square metre) that is taken at random as a representative sample of the organisms of the area. Within this the composition of organisms is noted, for example, the numbers of species (i.e. diversity), number of individuals per species (i.e. population density), and area occupied by each species (i.e. percentage cover). Sampling by quadrat is the most common method of quantifying vegetation. The term is also applied to the rigid frame used to delineate the area.

quagmire (quaking bog) A soft bog that gives way under the feet. It may consist entirely of bog mosses or cover a stagnant pond beneath, in which case it is a *floating bog*.

quaking bog *See* quagmire.

quantification The expression of values in precise terms, usually as sets of figures rather than as generalities or descriptions. For example, the rainfall of New Orleans can be described as fairly heavy with a maximum in summer; a quantification of this description would give the annual rainfall and monthly totals.

quarry An open excavation from which rock, clay, or sand is or has been obtained for building or other purposes.

quartz The crystalline form of silica. It is the most abundant of all minerals and is a common constituent of many igneous, metamorphic, and sedimentary rocks. Most sandstones are predominantly quartz (over 66%) and quartzite is almost wholly quartz (95%). Pure quartz is colourless and is also known as rock crystal. Trace amounts of other substances in quartz give rise to coloured varieties such as amethyst (purple), rose quartz (pink), and cairngorm (smoky-yellow or smoky-brown), which are all used in jewellery. In acid igneous rocks (e.g. granite) quartz forms over 66% of the rock. In some mineral veins it is found in association with economically important minerals such as gold. Quartz is used as an abrasive in sandpaper and toothpaste, and in many manufacturing processes including the making of pottery.

Quaternary The geological period extending from the end of the Tertiary period, about two million years ago, to the present day. It is composed of the Pleistocene and Holocene (Recent) epochs. During the Quaternary four advances of the ice sheets took place, separated by interglacial periods of warmer weather during which the ice sheets retreated, in North America, and N Europe and the Alps. Fossil mammals and plants were chiefly modern and man became dominant on land. The term Pleistocene is increasingly being used in place of Quaternary. *See also* Holocene, Pleistocene.

quebracho forest A type of evergreen vegetation of dense trees and scrub that is found in the dry Gran Chaco area of Paraguay and Argentina in South America. Quebracho forest merges into savanna with scattered trees.

quicksand Sand that is saturated with water and when walked on appears to liquefy; it may suck people and animals down. It is often found in river estuaries and along some coasts.

R

race 1. A strong rapid tidal current around a headland or in a channel where the tidal range is greater at one end than the other.
2. An artificial channel cut to or from a water wheel.
3. A large group of people who share a common ancestry and who have certain physical characteristics (e.g. skin colour, hair colour and type, and stature) in common. The chief races are Caucasoid, Mongoloid, and Negroid.

radar meteorology The collection of atmospheric data by the use of radar. Radar employs high-frequency radio waves, which are sent out as pulses and are reflected back to a receiver from objects they encounter. The distance between a target and the receiver can be determined. There are several ways in which radar is used in meteorology, the most important being the tracking of balloons to determine the speed and direction of upper winds, the tracking of clouds and areas of precipitation, and the detailed analysis of the movement of areas of precipitation by using the Doppler effect (the variation in signal caused by the movement of the target in relation to the receiver).

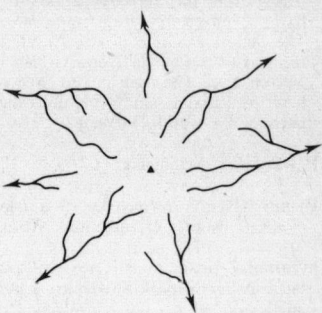

Radial drainage

radial drainage A drainage pattern in which streams radiate from a central peak or upland mass. Dome structures commonly develop radial drainage, as in the English Lake District.

radiation, electromagnetic The form in which radiant energy (e.g. heat) is transmitted between two bodies not in contact. It is emitted by all bodies, but in meteorology the two most important sources are the Sun and the surface of the Earth. The heat balance between these two is essential for the maintenance of life on Earth. †The Sun's electromagnetic radiation is mainly short wave between about 0.1 and 10.0 micrometres. It ranges from the short-wavelength and high-frequency gamma and x-rays, through the ultraviolet and visible spectrums, and into the infrared spectrum. Radiation from the Earth is weaker and is in the longer wavelengths (about 4.0–100.0 micrometres). *See also* insolation, solar constant.

radiation fog A type of fog normally seen as a shallow layer over low-lying ground at night or in the early morning. It is formed when the ground is cooled by radiation overnight and the layer of air immediately above the surface is chilled to below its dew point, when condensation takes place. It develops best when the air is moist, the night sky is clear allowing maximum heat loss through radiation, and there is a slight breeze (1–4 knots) to spread the cooling upwards. Such conditions are found most frequently in temperate latitudes when an anticyclone develops in autumn or spring. It generally clears quickly after dawn as the sunshine heats the air, or if the wind rises. It may be persistent if there is a temperature inversion, or in urban areas where large amounts of dust and smoke are trapped within the water droplets.

radioactive dating (radiometric dating) †A reliable method of dating rocks based on the rate of decay of a radio-active (parent) element into a stable (daughter) element. For any given radioactive element, this rate of decay is a constant and so the age of the rock can be calculated from the ratio present. Various radioactive elements are used: uranium, thorium, and lead, which all decay to lead; potassium, which decays to argon; and rubidium, which decays to strontium. The calculation of age is based upon the rate of decay, i.e. the fraction of the initial number of radioactive atoms that decay in one year. This is sometimes referred to as the half-life period, i.e. it is the period of time necessary for half of the number of radioactive atoms to decay. The most reliable dates are obtained from igneous rocks; with metamorphic rocks the date is that of the last metamorphism. *See also* radiocarbon dating.

radiocarbon dating (carbon dating) †A method of radioactive dating in which the amount of carbon-14 in material that was once living (e.g. wood, bone) is used to determine the age of the organism. The atmospheric carbon dioxide contains small quantities of the radioisotope carbon-14, which has a half-life of 5568 years. This is maintained at a constant rate throughout the world and as all living organisms absorb carbon dioxide they also absorb carbon-14. On the death of the organism the intake of carbon-14 ceases and the level begins to fall at a known rate as the result of beta decay. Measurements of the radioactivity of the material enable the concentration of carbon-14 and hence its age to be estimated. This method has been used to date organisms up to 70 000 years old. *See also* radioactive dating.

radiolarian ooze A type of ooze deposited on the ocean floor in which the chief constituent is the siliceous remains of minute shelled animals called radiolaria, which have a lattice-like structure. It is found mainly in deep tropical seas, especially the S Pacific between latitudes 5°S and 15°S, and in the Indian Ocean. *See* ooze.

radiometer *See* solarimeter.

radiometric dating †*See* radioactive dating.

radiosonde A hydrogen-filled balloon equipped with instruments for measuring and recording temperature, humidity, and atmospheric pressure in the upper atmosphere. It is fitted with a radio transmitter that telemeters the information collected to a ground station. This equipment is of great importance for gathering information about conditions at altitudes of 20–30 km. When the balloon bursts at high altitude the equipment is parachuted to Earth where it can be recovered. Radar tracking of the balloon gives additional information about upper winds. *See also* radar meteorology.

rain Precipitation consisting of drops of water formed by the coalescence of minute condensation droplets within clouds. In strict terms, these drops have a diameter ranging from about 0.5 mm to 5.0 mm, although smaller drops can be called rain if they are widely scattered. *See also* rainfall.

rainbow An arc of light, displaying the colours of the spectrum, that is observed when rays of sunlight pass through falling drops of rain. It is caused by the light being refracted and reflected as it enters the drops of water, and split into the colours of the spectrum. It is only visible when the Sun is at an elevation of 42° or less, and when the observer is standing with the Sun behind, and the rain in front of him. The primary bow is seen at an angular radius of about 42°, and the colour sequence is from red on the outside to violet on the inside. A fainter secondary bow, caused by double reflection in each raindrop, may sometimes be seen at an angular radius of about 54°, and in this the colour sequence is reversed. A similar optical phenomenon known as a *fogbow* may sometimes be seen when conditions are foggy; this appears white.

rain day In the UK, a period of 24 hours, starting at 0900 hours GMT, during which at least 0.25 mm of rain falls. In the USA it is defined as a day with measurable precipitation. *See also* wet day.

raindrop erosion *See* rainsplash.

rainfall The total amount of precipitation that falls over a given period, as measured in a rain gauge. It includes melted snow, hail, dew, hoar frost, and rime. Rainfall is classified according to the three modes of its formation: orographic, convectional, and cyclonic (or depressional). *See also* convectional rainfall, cyclonic rainfall, orographic rainfall.

rainforest *See* equatorial rainforest, monsoon forest.

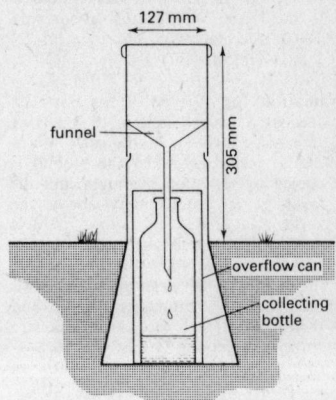

Rain gauge

rain gauge An instrument designed to measure rainfall. In its simplest form it consists of a funnel fitted into a collecting vessel. Any rain collected in the vessel over a set period of time is measured in a specially graduated measuring cylinder, an exercise that occurs twice daily at most meteorological stations. In order to obtain accurate readings the gauge must be carefully sited, away from shelter, where it cannot be knocked over, and at a

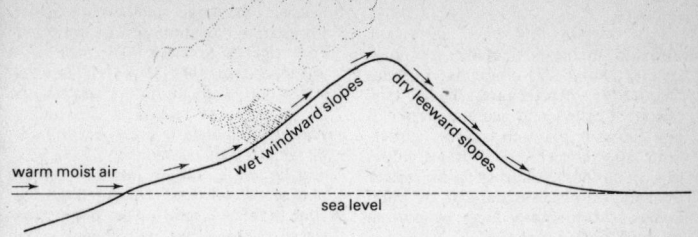

Rain shadow

height that prevents splashing from the ground. The protective outer casing must be close fitting in order to minimize evaporation losses. Most meteorological stations now use self-recording gauges such as the hyetograph, tipping bucket, and the float and syphon.

rain shadow An area of low rainfall in the lee of hills or mountain ranges. The prevailing rain-bearing winds deposit orographic (relief) rainfall on the windward side and crest of the upland; the descending air in the 'shadow' area produces less rainfall because it has less moisture and it is also warmed by its descent. In South America, the Andes produce a major rain shadow across the Argentinian pampas. *See also* orographic rainfall.

rain spell In the UK, a period of at least 15 consecutive days when at least 0.25 mm of rain falls on each day.

rainsplash (raindrop erosion) Soil erosion caused by falling raindrops. A series of impacts from the raindrops dislodges the soil particles and bounces or washes them downslope. If the process continues the impact of the raindrops compacts the soil and increases the surface flow of water; this reduces the water that actually percolates down into the soil. Rainsplash is only effective in areas with bare soils or poor vegetation and can be a great hazard in regions that experience heavy convective rainstorms. Farmers in the USA, Austra-

lia, New Zealand, and many parts of Africa have been particularly troubled by this form of erosion.

rain wash *See* sheet erosion.

raised beach A former beach that has been raised above the range of present-day tides by earth movements or by a fall in sea level. The old coastal cliffs often remain above the beach and a series of raised beaches can be formed if the earth movements have been repeated at intervals. In Scandinavia and N Canada isostatic movements following the melting of the ice sheets have created a number of raised beaches at different levels. Many raised beaches occur in Scotland, especially along the W coast, where they can be 800 m across. †*See also* isostasy.

ranching The type of farming in which livestock, particularly cattle, are reared on large-scale farms for sale. The term is used in North America, especially in the W states where the ranches are very large. [Spanish]

random sample †The selection of a representative subset of the total population being studied by a method that ensures that each member of the population has as much chance of being selected as part of the sample as any other. This makes it possible to infer statistically that the data used is valid for the total population. One method is to use random number tables to select the sample. *See also* sampling.

Randstad In the Netherlands, the crescent of towns and cities that includes Dordrecht, Rotterdam, Delft, the Hague, Leiden, and Harlem in the S and SE and curves to include Amsterdam and Utrecht. The Randstad is not a continual conurbation but rather a chain of places separated by rural areas. It contains a large proportion of the Dutch population.

range 1. A group of mountains or hills, especially when arranged in a chain or line.
2. The difference between the highest and the lowest of a series of values, for example, the *range of temperature* at a place. The *mean annual range* of temperature is the difference between the mean temperatures of the coldest and the warmest months; the *mean diurnal range* of temperature is the average difference between daily maximum and daily minimum temperatures over a certain period.

range of a good or service †The maximum distance over which people will travel to purchase a good or obtain a service offered by a central place. At some distance from the centre the increasing inconvenience (measured by time, cost, or trouble) will outweigh the value of obtaining the goods in that central place.

rank The position that a town or city occupies in the urban hierarchy. The concept of a hierarchy was first developed by W. Christaller in central place theory in which he recognized seven orders or ranks of central places ranging from market hamlet to regional capital city. A town may be ranked acording, for example, to its functions or population size but there is no general agreement on methods for identifying rank. *See* central place hierarchy. *See also* rank-size rule.

rank-size rule †A rule that is used to describe the size distribution of towns and cities in a country or region and especially the regular relationship between the large number of smaller towns, fewer medium-sized towns, and very few large cities. The rule was introduced in 1949 by G. K. Zipf. It states that if all the urban settlements in an area are ranked in descending order of population size (i.e. the largest city is ranked number 1, the second largest 2, and so on) the population of the nth town will be $1/n$th the size of the largest. The population sizes of the cities can therefore be arranged in the series 1, $\frac{1}{2}$, $\frac{1}{3}$, $\frac{1}{4}$ ... $1/n$. The rank-size rule can also be expressed in the formula:
$$Pn = P1/n$$
where Pn is the population of town rank n, $P1$ is the population of the largest city, and n is the rank order position. If the rank numbers are plotted against the population sizes for each respective town on a graph the curve takes the form of an inverted J. When a logarithmic scale is used for both axes on the graph the curve is converted to a straight line.

rape, oilseed A plant that grows to a height of about 1 m and has a yellow flower. The foliage can be used as a fodder crop or as green manure. The seed produces an edible oil used in industry. Oilseed rape is being grown on an increasing scale in East Anglia and S England.

rapids A stretch of swift-flowing water where a river bed suddenly becomes steeper. The water is turbulent and relatively shallow and the surface is often broken. Rapids may form where a stream crosses an outcrop of more resistant rock if the dip of the strata is steep or downstream (horizontal bedding favours waterfalls instead). †Waterfalls tend to degenerate into rapids as streams approach grade. Rapids may also result from rejuvenation downstream, especially if rejuvenation is relatively slight or slow. *See also* grade, rejuvenation.

rattan A type of liana (climbing plant) related to palms. It is common in

Indonesia and its long stout stems are used for ropes. *See* liana.

ravine A deep narrow gorge, often found in arid and semiarid areas with poor vegetation cover and occasional heavy convective rainstorms. A number of gullies join together to form ravines, which may then join larger river valleys.

raw material A material on which a manufacturing process is carried out, e.g. cotton lint is a raw material that is spun and woven to make cloth; bauxite is the raw material from which aluminium is made.

reach A length of river channel. The upper reaches of a drainage basin are the headwaters.

reafforestation (reforestation) The replanting of trees to form a forest in an area that was formerly forested. Large-scale reafforestation is usually of coniferous plantations to produce quick-growing softwood timber. *See also* deforestation.

Recent *See* Holocene.

recessional moraine A ridge of boulders and finer material deposited by a retreating glacier or ice sheet. Recessional moraines are deposited at right angles to the former direction of ice flow. They mark sites where ice retreat has halted for some time whilst ice flow has continued to pile up rock debris at the ice edge.

recharge area The area of land surface through which rainwater and snowmelt percolate to feed or recharge the ground-water body, e.g. the hillslopes feeding an artesian basin.

reclamation The making of flooded, derelict, or waste land suitable for agriculture and other uses. For example, the Dutch have reclaimed large areas once beneath the Zuider Zee for agriculture and settlement. Parts of the lower Swansea Valley, until recently an area of waste tips and

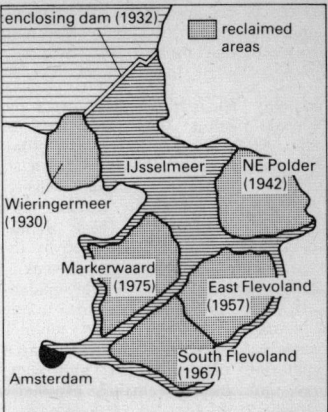

Reclamation of the Zuider Zee in the Netherlands

ruined buildings, have been made into a recreation park. Areas of sandy heathland in W Denmark have been reclaimed for farming by the heavy use of fertilizers.

recreation Activity that is done during leisure time for relaxation and enjoyment. Recreation and tourism have now become a major commercial activity throughout the world and many towns and cities depend upon such activities for their income. †Geographical studies of recreation are primarily concerned with recreational activities that take place away from home. These have focused on the demands for recreation, the patterns of recreational trips made, and the impact that recreational activities have on the environment.

rectangular drainage A pattern of drainage consisting of two main directions of flow at right angles to one another. The pattern is most commonly caused by streams following fault lines that have a rectangular pattern.

recycling The reusing of materials or products that have been discarded because they are no longer of value for their original use. Recycling is increasing in importance as some resources become scarce. Glass, metal, and paper are the most common materials that are recycled.

red clay The deposit that covers the abyssal plain in the deepest parts of the Atlantic, Pacific, and Indian oceans at depths of over 5000 m. It differs from ooze in that it consists mainly of inorganic rather than organic material; at these depths organic materials are dissolved by the sea water. The major constituent is aluminium silicate (a true clay material); the presence of iron oxide colours it red and it also contains some volcanic material, fine terrestrial material, meteoric dust, and some resistant animal remains, especially sharks' teeth.

redevelopment The development of an area that has previously experienced an earlier phase of building, which will involve the demolition and removal of these earlier buildings. Redevelopment is closely associated with the process of urban renewal where older parts of urban settlements are cleared of buildings and replaced by new. This process was particularly dominant in the postwar decades of the 1950s and 1960s but has been increasingly replaced by a policy of renovation and rehabilitation of existing buildings. *See also* urban renewal.

reef 1. A ridge of rocks in a sea or lake at or near the surface of the water. It may be permanently submerged, submerged at high tide, or normally just above the water. It may be of solid rock or pebbles, but the word is most commonly used for coral reefs built up in tropical seas by the hard calcareous material secreted by colonies of minute sea animals. Oysters and marine worms may also build reefs. *See also* coral reef.
2. A mineral vein, particularly a vein of quartz that contains gold

reflection coefficient *See* albedo.

refugee A person who is forced to seek safety in a foreign country, as a result chiefly of religious or political persecution and warfare. Examples of refugees include the 'Boat People', who were forced to leave Vietnam through persecution.

reg A desert plain covered with gravel and small pebbles; it is particularly well developed in parts of the Sahara. *See also* erg, hamada. [Arabic]

regelation †The refreezing of ice that has been melted under pressure in a glacier once the pressure is reduced. It is one of the processes that allow glaciers to move downslope. As ice pressure builds up in the glacier, the melting point of ice crystals is lowered and meltwater is formed. This is squeezed towards parts with less pressure, usually downslope, where it refreezes. Sometimes melting only affects the outer part of the crystals before regelation freezes the crystals together again.

regime The organization or pattern of a natural system. For example, a winter-rainfall regime refers to the season in which most of the rainfall occurs; a stream with a flashy regime is one that has very uneven levels of flow and is liable to flood suddenly.

region A unit of land that is considered to have a unique character or homogeneity, based on local features of geology, relief, soil, climate, vegetation, and human way of life. Examples of regions include the Hebrides, the Central Valley of California, and the lower Nile Valley. *See also* natural region, †formal region, functional region.

regional geography The study of the geography of regions. Traditionally, the main approach to this has been through chorography. Regional geography is concerned with the geographical inter-relationships between regions and the examination of the changes that take place in a region over time. *See also* chorography.

regionalism 1. A way of viewing the Earth's surface by dividing it up into regions. Regionalism can be based on geographically well-defined areas (e.g. the Lake District) or on areas with a strong unifying characteristic or identity (e.g. the West Country). It is a term often used by planners when identifying an area for future development. **2.** An identity with or feeling of group consciousness associated with a region.

regional plan A plan for the development of an area which in the UK would normally include a number of counties. Hence the South West region would comprise the counties of Cornwall, Devon, Somerset, and Avon, for example. Such plans may concern land-use issues (e.g. the main lines of communications and location of airports) or be more economically orientated (e.g. relating to employment).

regional science A field of study that links economics, geography, and planning. Regional scientists are interested primarily in theoretical and statistical approaches to human, especially economic, distributions.

regolith (waste mantle) A layer of unconsolidated weathered material lying above bedrock and below the ground surface. It includes the parent material from which soil is formed and the soil itself. It consists of weathered particles of the bedrock itself or mineral particles transported into the area by the action of water, wind, or ice.

regosol A thin azonal soil that is derived from a parent material of soft loose weathered material such as recent alluvium, loess, or dune sands. The soil is young and has no or little profile development; the A horizon directly overlies the C horizon and appears darker because of its organic content. *Compare* lithosol.

regur (black cotton soil) A soil found in the Deccan Plateau of India and in E Africa. Regurs are dark and alkaline as a result of a basaltic parent material. Their high clay content causes heaviness and swelling when wet and cracking when dry. Despite these disadvantages, cotton growing is widely practised on regurs. [Indo-Pakistan]

rejuvenation †A revival of erosive activity, especially of a river. *Dynamic rejuvenation* is an increase in the erosive capacity of rivers due to a fall in local or regional base level, e.g. a fall in lake level or sea level. The increased erosion begins at the downstream end of the river and works upstream, thus creating a new long profile. The point at which old and new long profiles meet is marked by a knickpoint and the old valley bottom or floodplain becomes a river terrace. Hillslope erosion processes are also rejuvenated by a fall in base level. *Static rejuvenation* is an increase in a river's erosive capacity resulting from an increase in the volume of its flow. This may be due to a change in local water balance, such as increased precipitation or deforestation reducing evaporative losses, which produces more runoff, or due to river capture, which introduces a new base level from an adjacent network.

rejuvenation head *See* knickpoint.

relative humidity The ratio, expressed as a percentage, between the amount of water vapour actually present in an air mass and the maximum amount that the air mass could hold at that temperature. Saturated air has a relative humidity of 100%. It varies inversely with temperature, given a fixed amount of water vapour, decreasing as the temperature increases and increasing as the temperature decreases. Relative humidity cannot be measured directly, but is calculated by applying the readings from a hygrometer to a set of pre-prepared tables.

relict landform A land surface feature created by processes no longer operating in that place, e.g. a glacial feature in Britain or a periglacial feature in Wisconsin, USA.

relief The differences in elevation or the physical outline of the land surface or ocean floor. *Relative relief* is the difference between highest and lowest elevations in an area. *Available relief* is the difference in height between an original upland surface and the bottom of adjacent graded valleys.

relief map A map showing the surface relief of an area. The height of the land can be shown by a variety of methods, which attempt to represent three dimensions on a plane surface. *See also* contour, hachures, hill shading, layer tinting, photo relief.

relief model A three-dimensional model showing the relief of an area. Most models are made with an exaggerated vertical scale compared with the horizontal. This highlights relief features such as hills and valleys.

relief rainfall *See* orographic rainfall.

relief road A road that has been built to relieve traffic congestion or excessive use of existing roads.

remote sensing †The use of aircraft or satellites to gather data about the surface of the Earth. The principal techniques include aerial photography, infrared imagery, multispectral imagery, and radar. Remote sensing is used in areas such as N Canada where detailed geological maps are not suitable.

rendzina An intrazonal soil that is developed on very calcareous parent material (i.e. limestones). It develops in humid to semiarid climates under grass and grass-and-tree vegetation. It is a shallow soil with a dark A horizon, which is rich in mull humus, directly overlying the C horizon of weathered bedrock. Calcium, which is well distributed by soil organisms, gives pH values of 7.0 to 8.5. *Compare* terra rossa. [Polish]

rent bid *See* bid rent theory.

resequent fault-line scarp A steep slope eroded along the line of a fault and facing the same direction as the downthrow of the initial fault.

reservoir A storage area for water, usually a river valley that has been dammed to retain water for one or more purposes, such as irrigation, industrial use, water supply, hydroelectric power, or recreation. For example, the Kielder Valley reservoir in Northumberland, which is N Europe's largest man-made lake, supplies large areas of N England with water and is also a recreational area.

residential town *See* dormitory town.

residual landform A mountain, hill, ridge, or isolated block of rock remaining after an extended period of erosion has lowered the surrounding land. For example, monadnocks are residual mountains rising above peneplains; inselbergs are typical residual hills found in tropical climates; and buttes, kopjes, and tors are small residual landforms. Sometimes the residuals are composed of more resistant rock, but more generally they are remnants of former interfluves.

resource A feature of the environment that is of value to man in one form or another. Natural resources include forests, rivers, minerals, solar energy, and the sea. Man-made resources include machinery, transport systems, and artificial fertilizers. Less tangible resources include attractive scenery, a healthy climate, and the skills and expertise of groups of people.

resource management The recognition of the finite resources of the Earth and the growing demand for them has led to an appreciation of the need for their careful management. This involves establishing the extent of existing resources, seeking new ones, and making more efficient use of existing resources.

retrogradation The steepening of a beach profile as breakers actively erode the beach.

reverse fault (reversed fault) A fault in which the fault plane is inclined at an angle between 45° and the vertical, and the upper beds are pushed up the fault plane relative to the beds below. This type of fault occurs as a result of very strong compressional forces. *Compare* thrust fault. *See* fault.

rhumb line (loxodrome) A line of constant bearing cutting all meridians at the same angle. A rhumb-line route can be longer than the great circle route if it does not follow the equator or stay close to a meridian. On the Mercator projection a rhumb line is shown as a straight line. *See also* great circle.

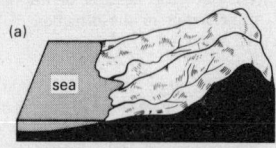

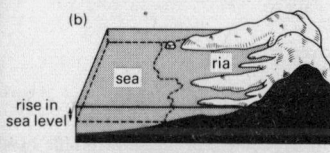

Development of a ria coastline

ria A drowned river valley forming a long funnel-shaped inlet of the sea. Rias often branch like river valleys and have V-shaped cross profiles. The valleys have been flooded by the sea as a result of the post-glacial rise in sea level or local subsidence of the land, or a combination of both. Rias are common in SW England, SW Ireland, and NW Spain. [Spanish]

ribbon development The outward residential extension of a town or city along main roads; it forms part of the process of urban sprawl. In the case of London, ribbon development was a particular feature of the 1930s when private-car ownership increased and public transport services improved but there were no strong planning controls on building. Attempts were made to control it by the Restriction of Ribbon Development Act (1935) and the Green Belt (London and Home Counties) Act of 1938 but the acts were largely ineffectual.

rice *See* paddy.

Richter scale A scale that indicates the quantity of energy released by an earthquake (i.e. the magnitude of the earthquake), devised by the US seismologist Charles F. Richter in 1935. The scale ranges from 0 to 9, but in practice there is no upper limit. The largest observed magnitude was 8.6.

ridge An elongated hill or narrow range of hills.

ridge and valley A form of relief in which approximately parallel ridges and valleys alternate. The chief example is the Ridge and Valley region of the Appalachian Mountains of the USA, which is distinguished by parallel ridges and valleys with a general NE–SW trend. The ridges are the remnants of folds formed during several periods of mountain building. Erosion has exposed weaker beds within some of the ridges and former lines of drainage are marked by wind gaps. This was a classic area for the development of geomorphological theory.

ridge of high pressure A narrow extension of an anticyclone, forming an area of high pressure, bounded on three sides by lower pressure areas. It is analogous to an upland ridge extending out from a highland area. It often gives a brief period of fine anticyclonic weather in the midst of generally wet depressional conditions. *Compare* trough of low pressure.

riding One of the three former administrative divisions of the geographical county of Yorkshire: West Riding, East Riding, and North Riding.

rift valley A flat-bottomed valley formed by the sinking of the ground between two nearly parallel faults or two parallel series of step faults. Valley sides are steep and follow the fault lines or sets of fault lines, which also form the edge of the mountain masses. Evidence of volcanic activity is often found along the fault lines. There are conflicting views as to the origins of rift valleys, including tension in the Earth's crust, compression, or the cracking of a crustal dome along the crest. Currently it is believed that rift valleys result from tectonic plates moving apart at constructive plate margins and that mid-ocean ridges are rift valleys. Major rift valleys include the great rift extending from Syria 4000 km through the Sea of Galilee and the Dead Sea to the Gulf of Aqaba, the Red Sea, and the East African rift to the Zambezi. Smaller rift valleys include those of the Rhine between Basle and Mainz and those of the Central Lowlands of Scotland.

rill erosion Erosion of the soil surface by shallow short-lived channels known as rills. Rills form during storms when water flowing over the surface of the soil is deep enough to become turbulent and erosive or when water flowing underground re-emerges on lower hillslopes and washes away the soil. If the rainfall continues the rills may be further eroded and enlarged to form gullies. *See also* gully, overland flow.

rime A deposit of white opaque ice crystals, of low density and containing many air spaces. It usually forms when supercooled droplets of water in fog or fine drizzle are driven by a light wind against a cold surface onto which they freeze. The frozen droplets may form patterns similar to those displayed by snowflakes, and these are often called frost feathers.

ring road A road that passes around or through the outskirts of a city in order to relieve traffic congestion by providing an alternative route for through traffic or traffic wishing to cross the built-up area. For example, the North Circular road around London.

ripple marks Small roughly parallel surface wave forms commonly found on a beach, a sea bed, or a river bed. They develop in sand or fine gravel at right angles to the direction of flow of the water or wind.

river A large stream of fresh water flowing downhill within a channel to enter another river or a lake or sea. Rivers join to form a river system or river network. *See also* channel, tributary.

river basin *See* basin.

river capture *See* capture, river.

river profile The shape or profile of a river channel, especially the long profile. *See* cross profile, long profile.

river regime *See* regime.

river terrace *See* terrace.

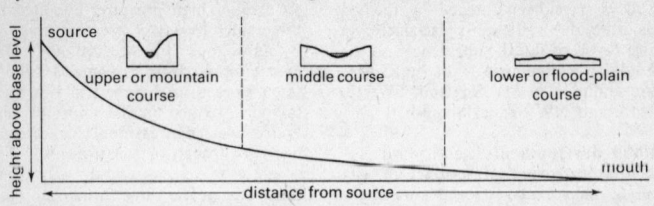

The ideal long profile and cross profiles of a river

riviera A strip of the Mediterranean coast extending from Marseilles (France) to La Spezia (Italy). It is a famous tourist area containing resorts such as Nice and Cannes. The word has now been applied to other scenic holiday coasts; for example, the Cornish riviera.

road metal Stone that is broken into small pieces and bound together with tar to make a hard surface. The broken stone is known as macadam after John Loudon McAdam who discovered the process in the 19th century. The stone must be tough and hard enough to withstand the weights of heavy vehicles, e.g. granites from the Penmaenmawr district of N Wales.

Roaring Forties The area lying between latitudes 40°S and 50°S where the prevailing westerly winds (the brave west winds) blow with strength and regularity across uninterrupted stretches of ocean. It is an area of storms and heavy rainfall, which are caused by the steady westerly procession of depressions. Sometimes the name is applied to the winds themselves.

robber economy The exploitation of nonrenewable resources or of renewable resources that are in danger of being used up unless adequate conservation measures are taken. Nonrenewable resources include fossil fuels and minerals. Renewable resources that can become scarce include forests and their products and some species of fish such as the pilchard.

roche moutonnée A glacially eroded mound of rock, which resembles a lawyer's wig. Stones and boulders embedded in the glacier ice have worn down the upstream side of the rock leaving a gently sloping smooth surface with gouge marks or striations in it. The downstream side is steep and rough owing to meltwater – which is often formed under increased pressure on the upstream side – freezing under less pressure, shattering the rock, and encouraging glacial plucking. Roches moutonnées are common features of glaciated areas, e.g. Llanberis Pass in N Wales. *See also* plucking, striations. [French, literally: fleecy (woolly) rock]

rock Any aggregate of mineral particles that forms part of the Earth's crust. It may be consolidated (e.g. granite, limestone) or unconsolidated (e.g. clay, mud, sand). Rocks are classified into three major groups: igneous, metamorphic, and sedimentary. Each of these groups is subdivided into many different types, each with its own composition, character, and age. *See* igneous rock, metamorphic rock, sedimentary rock.

rocketsonde A device that has the same functions as a radiosonde, i.e. the collection and transmission by radio to surface receivers of data concerning temperature, humidity, and pressure in the upper atmosphere, but which is projected into position by rocket rather than carried by balloon. *See* radiosonde.

rockfall A landslide of rocks free falling from a bare rock cliff. Most rockfalls occur as a result of the thawing of ice and in Norway they are most common in spring and autumn. *See also* mass movement.

rock flour Finely ground rock material created by abrasion on the bed of a glacier due to stones and boulders embedded in the ice. Meltwater issuing from a glacier is milky white because of the rock flour it carries.

rocking stone (logan stone, logging stone) A large boulder that stands apparently precariously balanced and may be rocked by hand. It may be a form of tor formed by weathering and erosion of the surrounding rock, or it may be an erratic carried there by an ice sheet. The Logan Stone, a granite tor near Land's End in Cornwall, is an example of a rocking stone. *See also* erratic, tor.

rock pavement *See* ruware.

rock salt (halite) The crystalline form of common salt (sodium chloride, NaCl). It occurs as an evaporite deposit as the result of evaporation of enclosed or partially enclosed bodies of sea water or salt lakes and is often found in association with other mineral salts. Rock-salt beds were formed at several times throughout the geological record; for example, during the Silurian and Carboniferous (New York State, USA), the Triassic (Cheshire), and the Tertiary (Poland). Rock salt is extracted either by mining or by pumping brine to the surface where the salt is recovered by evaporation. Common salt is used in the cooking and preservation of food, and in several manufacturing processes such as the making of glass and soap. *See* evaporite.

rockslide A landslide of weathered rock falling or sliding down a bedding joint or fault surface on a hillside. Examples include the rockslide in the Madison River valley of Montana, USA, on 17 August, 1959, which was triggered off by a strong earthquake. A buttress of dolomite supporting schists and gneiss sheared away from the mountain and 80 000 000 tons of debris slid downslope damming the Madison River.

root crop A plant that is cultivated mainly for its root, which may be used as food for humans or animals. For example, turnips, carrots, parsnips, and swedes are eaten by both humans and animals, mangolds are fed to animals, while sugar beet is processed to extract the sugar content, the residue providing cattle feed.

Rossby wave (long wave) A large-scale wave motion that occurs at an altitude of over 10 km in the upper air westerlies, around latitudes 30° to 40°N and S. The formation of these waves was studied by the Swedish-born American meteorologist C. G. Rossby in the late 1930s and early 1940s. He demonstrated their connections with mid-latitude weather patterns and also showed how they help to carry warm air

poleward and cold air equatorward, thus aiding the global heat transfer.

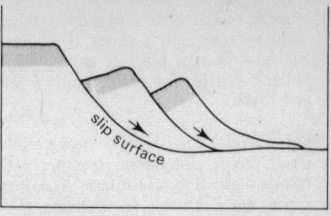

Rotational slip

rotational slip (rotational slide) A form of landslide in which the sliding mass rotates with the bottom slipping further forward than the top.

rotation of crops A systematic succession of different crops on a piece of land in order to avoid soil exhaustion. Crops are rotated in a cycle that usually varies between two and eight years. Rotation also enables weeds, diseases, and pests to be controlled more effectively. Nitrogen-fixing plants, such as clover or beans, normally follow wheat or barley, which use up large amounts of nitrogen. *See also* Norfolk rotation.

rotation of Earth The rotation of the Earth from W to E around its axis, so that the stars, the Sun, and the Moon appear to rise in the E and set in the W.

†The speed of rotation (despite extremely small variations) remains remarkably constant. From the moment a star crosses the observer's meridian (i.e. at its highest point in the sky) to its next crossing is 23 hours 56 minutes 4.09 seconds (the sidereal day used by astronomers is based on this). As the Earth moves along its orbit the Sun appears to move eastwards compared with the 'fixed' stars. Thus, after rotating once relative to the stars, the Earth must rotate eastwards for a further 3 minutes 56 seconds of time to complete one rotation relative to the Sun.

rough grazing Natural moorland, scrubland, salt marsh, or hill pasture that can be used for grazing but has not been improved in any way. Such land is not normally fenced in and may be grazed during the summer months by sheep or cattle from nearby farms. On Dartmoor, for example, wild ponies and sheep share the rough grazing on the higher ground.

rubber, natural An elastic material that is obtained from the milky latex of rubber trees, especially from the species *Hevea brasiliensis*. The liquid is coagulated by adding ammonia and then rolled into sheets for export. The rubber tree grows in equatorial regions where the rainfall is 2500 mm annually and temperatures average 26°C. Over 90% of the world's rubber comes from plantations or smallholdings in SE Asia, particularly Malaysia and Indonesia. Some is grown in the Amazon Basin of South America, the original home of the rubber tree.

rubber, synthetic Rubber that has been made from chemicals, which are mainly derived from crude oil. Germany and the USA developed the industry during World War II when supplies of natural rubber from SE Asia were cut off. *Styrene-butadiene rubber* (SBR) is the most common synthetic rubber and is used in tyres. Others include the softer *butyl rubber*, which is used for inner tubes, and *neoprene*, which is used for protective clothing. The largest producers today are the USA, where the main producing region is in Texas close to the oil refineries, and the USSR. About 92% of the rubber consumed in the USA is synthetic.

rudaceous Denoting the coarsest of the clastic sedimentary deposits, i.e. those in which the size of the constituent grains is over 2 mm (on the Wentworth scale). These sediments include the gravels, conglomerates, and breccias. *See* clastic.

runoff The portion of rainwater or meltwater from snow that drains into rivers, as opposed to evaporating or being lost through transpiration by plants. *Direct runoff* enters the rivers during storms having passed over the surface as *surface runoff* or through the soil as throughflow (interflow); *delayed runoff* passes through the soil and regolith by slower routes; and baseflow (low flow) is maintained by slow drainage into rivers from the normal water table.

rural-urban continuum †A concept that recognizes that there is no specific point at which rurality disappears and urbanity begins and that the largest urban agglomerations, small nucleated settlements, and isolated dispersed dwellings are located at points along a continuum. The division into urban and rural is therefore artificial and arbitrary.

rural-urban fringe (rurban fringe) The interface between a town and the surrounding countryside. It is a zone of transition with distinctive characteristics resulting from its partial assimilation into the growing urban area. The fringe contains a mixture of urban and rural land uses including agriculture, out-of-town shopping centres, derelict land, mineral workings, sports grounds, market gardens, old villages, and new housing estates. Socially, the residents tend to be segregated into distinctive groups – the older rural groups, council-house tenants, and owner-occupiers of new private housing. Many newcomers work in and identify with the city.

ruware (rock pavement) A low rounded whaleback-shaped rock standing a few metres above a plain in the tropics. It is believed to be an exposed part of the basal surface of weathering and, as erosion of the loose weathered material around exposes it more, it becomes an inselberg. *See also* basal surface of weathering, inselberg.

rye A hardy cereal grass grown on the poorer soils of N and E Europe and the N central states of the USA. The grain is used to make a form of whisky and is also milled to produce a saddle

dark brown flour that is used for making bread. The grain can also be fed to livestock.

S

saddle A broad shallow gap in a mountain ridge forming a pass or col.

saddle reef In mining, a lens-shaped body of ore-bearing igneous rock intruded into other rocks.

saeter (seter) In Scandinavia, a mountain pasture to which cattle are sent during the summer months. The pastureland is used from June to September after the snow has melted and usually includes a simple hut where the herdsman lives. In Switzerland, these upland pastures are called alps. [Norwegian]

sagebrush A scrub vegetation of semi-desert areas in the W USA, particularly Utah, Arizona, and Nevada (the 'Sagebrush State'), and Mexico. Sagebrush comprises low small-leaved drought-resistant herbs, especially *Artemisia tridentata.*

St Elmo's fire (corposant) A form of lightning consisting of a more or less continuous electrical discharge from objects such as aircraft wings, lightning conductors, ships' rigging, etc., which is visible as a luminous glow around the object. *See* lightning.

sakiyeh A primitive irrigation device for lifting water from a channel or river onto land at a higher level. It consists of a wooden water wheel with buckets instead of paddles attached to a spindle and cog wheels. The wheel is turned by a draught animal such as a donkey or water buffalo, which follows a circular path round the wheel. Water raised in the buckets pours out as the wheel descends. [Arabic]

salina A very salty or dried up playa lake at the centre of a basin of inland drainage in an arid region, chiefly in

SW North America. *See also* playa. [Spanish]

saline soils †A group of intrazonal soils that contain high concentrations of salts such as common salt (NaCl). They often occur in semiarid and arid areas where there is strong evaporation and where the parent material or ground water contains salt. Near salt lakes saline soils merge with pure salt crusts. Saline soils pose problems for irrigation. *See also* solonchak, solonetz.

salinity The saltiness of water, usually expressed in parts per thousand ($^o/_{oo}$). This varies greatly throughout the seas of the world, especially in the surface layers. The main determining factors are the rate of evaporation and the rate of influx of fresh water (rivers and precipitation). In the open seas there is a rough latitudinal pattern. The most saline areas are in the trade wind belts, which are fairly dry and have a high rate of evaporation. The salinity reduces towards the equator, where there is a large influx of fresh water (rainfall), and towards the poles, where low temperatures result in low evaporation. The average salinity of the world's oceans is about $35^o/_{oo}$. The greatest variations occur in the partially enclosed seas, ranging from about $8^o/_{oo}$ in the Baltic Sea to over $40^o/_{oo}$ in the Red Sea. The low salinity of the Baltic is a result of low evaporation and the inflow of numerous rivers. The high salinity of the Red Sea is due to very high evaporation rates and virtually no influx of fresh water because of the surrounding desert. Salinity is greatest in totally enclosed seas (salt lakes); for example, in the Dead Sea it is about $250^o/_{oo}$. The most common salt in sea water is sodium chloride (common salt); others include magnesium chloride, magnesium sulphate, and calcium sulphate.

salinization †The process by which soils are enriched with salt. Salinization may occur as a result of evaporation from the soil surface, which draws up salts in solution by capillary movement. It may also be induced by man, for example, irrigation practised where evaporation is intense produces salinization unless counteracted by regular flushing, deep ploughing, or chemical treatment. *See also* saline soils.

Salpausselkä A major ridge of sand and gravel running across Finland from E to W. It is a set of terminal moraines formed by Quaternary ice sheets. [Finnish]

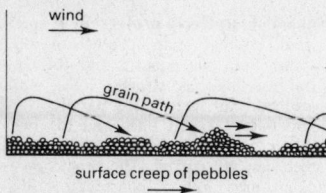

The saltation of sand particles

saltation A process of sediment movement in which loose grains bounce and jump across the surface. It occurs in sand moved by the wind in the desert and in sand and slightly coarser particles lifted by rivers.

salt, common *See* rock salt.

salt dome A mass of rock salt, roughly circular in shape, that has been intruded into sedimentary rock strata. Salt domes are formed by salt slowly flowing into the dome from thick evaporite deposits. The salt is relatively plastic and moves upwards through the overlying rocks rather like an intrusive igneous rock. The overlying rocks are arched upwards over the salt dome and the surrounding rocks are highly deformed. These structures are economically important because they are associated with deposits of oil; salt domes are found in many parts of the world, for example in the Gulf States of the USA and under the North Sea.

salt flat A salt-covered surface that was formed when a lake in an area of internal drainage in an arid area dried

up. The salt flats of Lake Bonneville in Utah, USA, famous for attempts at the land speed record, are the remnants of a Quaternary meltwater lake.

salt lake A lake in which high rates of evaporation cause the dissolved salts brought in by the streams feeding the lake to become highly concentrated. Such lakes usually lack an outlet to the sea and are situated in inland drainage basins, especially in arid areas such as the deserts of the W USA and central Asia. Examples include the Great Salt Lake, Utah, and the Dead Sea on the border between Israel and Jordan.

salt lick A naturally occurring deposit of salt, which is visited by wild animals to lick at the salt. Salt licks occur in climates where there are high evaporation rates, such as the Serengeti Plains of N Tanzania.

salt marsh A marsh along a low-lying coastline, which is occasionally covered at high tide. It is formed by an accumulation of mud and fine sand, which has been colonized by salt-tolerant plants. Silt is trapped by the plant growth and the surface of the marsh is slowly built up so that it is flooded less and less frequently by the tide. The older marshes are characterized by deep meandering channels and salt pans. Many old marshes, such as the Dutch polders and Romney Marsh in S England, have been reclaimed.

salt pan A pool on the surface of a coastal salt marsh that may be occasionally replenished by the tide. Salt pans tend to dry up more frequently on the older higher marshes and may eventually be invaded by vegetation, especially if a creek or soil pipe cuts back and drains the pan. *See also* salt marsh.

sampling †The technique that is used to obtain a subset of data that is representative of the total population or total area that is under study. The selection process needs to ensure that each member of the population has as much chance of being chosen in the sample as any other to avoid bias. A number of methods are used to avoid the sample being biased, the most common method being random sampling. *See also* random sample.

samun (samoon) A hot dry wind of the föhn type that blows in Iran down from the mountains of Kurdistan and across the plains. *See* föhn. [Persian]

sand Particles of rock of a size intermediate between silt and gravel; internationally defined as particles with diameters ranging from 0.06 to 2.00 mm. Fine sand ranges up to 0.2 mm, medium sand up to 0.6 mm, and coarse sand up to 2.00 mm. Most sands are mainly composed of the hard-wearing mineral quartz.

sandbank A bank of sand and other sediment found in the sea near the coast and in estuaries. Sandbanks are formed where the supply of sediment is fairly abundant and the currents are not strong enough to shift it once it has been deposited. Sandbanks are often exposed at low tide.

sandstone A clastic sedimentary rock consisting primarily of grains of quartz, together with grains of mica, feldspar, and other minerals in small quantities. The grain size is between 1/16 mm and 2 mm in diameter. The particles may be bound together by a natural cement (e.g. calcite) or by compaction. The colour of sandstone varies from dark brown through red to white according to the presence of other minerals. The sand accumulates through the action of wind or water and is laid down in shallow seas, estuaries, deltas, low-lying coasts, and deserts. Sandstones may be classified calcareous, siliceous, or ferruginous according to the nature of the material cementing the grains together. Examples of sandstones in the geological record include the Devonian Old Red Sandstone and the Permo-Triassic New Red Sandstone.

sandstorm A storm in a desert or semiarid area when strong winds lift and transport relatively large grains of sand. The particles are rarely raised more than about 25 m above the ground surface as a result of their size and are not transported far outside the dry areas. The most important effects are the erosive power of the wind-blown particles and the construction of sand dunes.

sandur *See* outwash plain. [Icelandic]

Sanson-Flamstead projection A modified conical map projection that shows both the equator and the central meridian as straight lines. Both are true to scale. The parallels are horizontal straight lines and the meridians on each side of the central meridian are curved. The projection is equal area but is considerably distorted at the edges. It is sometimes used in atlases for countries astride the equator such as Africa or South America.

Santa Ana A föhn-type wind, which is hot, dry, and frequently dust-laden, that blows from a northerly or northeasterly direction from the ranges of the Sierra Nevada across the deserts of California. It occurs mainly in winter and spring, during which it can cause considerable damage to crops and to the blossom and buds on fruit trees.

saprolite A deeply weathered rock that retains the outward structural form of the original rock, such as jointing, although many of the original minerals have broken down. It is a feature of tropical weathering and has been regarded as indicating an intermediate stage in the lowering of tropical planation surfaces by subsequent erosion of the saprolite.

saprophyte †An organism that absorbs its food from dead or decaying matter on which it is biologically dependent (e.g. fungus growing on a dead tree). Many bacteria and some higher plants are saprophytic. Saprophytes are important decomposers in food chains, breaking down materials created by primary and secondary producers, and thus recycling essential nutrients.

sarsen (greywether) An irregular hard sandstone boulder found chiefly on the chalk downs of S England. Sarsens remain as relicts of the Tertiary sandstone rocks that formerly covered the downs, the softer sands between them having been eroded away. The sarsens are still found within the Tertiary Reading and Bagshot beds. The boulders are also commonly known as greywethers from their resemblance to sheep when seen from a distance.

satellites, meteorological See meteorological satellites.

satellite town A town that is located close to a major city, with which it is closely associated. In the UK, the term was originally applied to a town, similar in character to a garden city, which was built near to an existing city.

saturated adiabatic lapse rate (SALR) †The rate at which a rising parcel of saturated air cools as a result of adiabatic expansion. This rate varies with the actual temperature of the air, but averages about 0.5°C per 100 m. The saturated air cools at a slower rate than unsaturated air as a result of the release of latent heat into the air as vapour is condensed into water droplets. The saturated adiabatic lapse rate tends to be less than the environmental lapse rate, i.e. the rising air cools more slowly than the surrounding air is cooling, therefore conditions within the atmosphere are said to be unstable. In general, when a parcel of air starts to rise it is unsaturated and cools at the dry adiabatic lapse rate, but eventually it cools sufficiently to reach its dew point, condensation begins, and the air becomes saturated. Beyond this point it cools at the saturated adiabatic lapse rate. See lapse rate. See also dry adiabatic lapse rate.

saturation In meteorology, the state of a mass of air, at a particular temperature, when it contains as much water vapour as it can possibly hold (i.e. as many molecules of water enter the air as leave it). The amount of water vapour required to saturate the air depends on temperature and pressure: the amount increases as temperature increases. Condensation in the atmosphere occurs when saturated air is cooled, thus losing the capacity to retain all the water vapour within it.

savanna (savannah, savana) Open tropical and subtropical grassland with scattered bushes and trees, which covers extensive parts of Africa, N Australia, and South America (where it is known as the llanos in Venezuela and Colombia and the campos in Brazil). In Africa and Australia it is bordered by rainforest towards the equator and by hot desert on its opposite margin. Within these limits savanna varies considerably from tree-less plains dominated by grasses up to 3 m in height (e.g. elephant grass, especially *Pennisetum purpureum*) to relatively well-wooded areas. Savanna tree species (e.g. baobab and acacia) conserve moisture by shedding leaves at the outset of drought (the cooler season), when grasses also wither. †Fire, grazing by large herbivores, and poor soils, as well as climatic factors, are responsible for the formation and continuation of the savanna vegetation. See campo, llano. See also grassland. [Carib]

scale The measure that is used on a map to represent corresponding distances upon the Earth's surface. It is usually shown as a line marked at regular intervals, the graduations representing proportionately larger distances on the ground. Sometimes it is also written as a ratio, for example 1:50 000, which signifies that 1 unit of measure on the map represents 50 000 on the Earth's surface.

scar 1. A cliff or crag, essentially bare of soil.

2. The steep slope left behind by a landslip. A *meander scar* is a steep undercut slope created by a present or past meander of a river cutting into a steep terrace scarp or hillslope.

scarp (escarpment, scarp face, scarp slope) The steep slope of a cuesta, forming a long barrier rising above the surrounding land. The *scarp slope* and the *scarp face* refer specifically to the steep slope, whereas the escarpment may loosely refer to the overall ridge. *See also* cuesta, fault-line scarp.

scarp-foot spring A ground-water spring that emerges from the bottom of a steep slope or scarp, usually at the junction of different rock strata. Commonly the scarp is formed by more porous but resistant rock, such as the oolitic limestone of the English Cotswolds, and the rock at the base of the scarp is more impermeable, such as the Cotswold Liassic clay strata; this causes water draining through the cuesta to emerge laterally at the foot.

scatter diagram †A diagram using dots plotted on a graph to show the amount of correlation between two sets of statistical data. Each value is plotted against scales drawn on two axes and represented by a dot. It can be used to show what sort of relationship exists between two variables such as the percentage change in population and the percentage change in the number of dwellings in a group of parishes. If the dots show a random scatter no correlation exists between the two variables. If there is a definite grouping, or location within a narrow linear zone, then a relationship between the variables exists. The more closely the points on a scatter diagram conform to a straight line the stronger is the correlation between the two variables.

Schattenseite The shady side of a mountain valley. It is synonymous with the French term ubac. [German]

schist A coarse-grained metamorphic rock in which recrystallization has taken place resulting in a parallel layered arrangement of the minerals, notably mica, and a foliated or wavy texture. The major minerals of which the schist is composed have a flaky or fibrous appearance. Schists generally result from regional metamorphism.

scirocco *See* sirocco.

sclerophyll †A plant that has hardened leathery foliage, either in the form of broad leaves (e.g. holly) or needles (e.g. pine), as an adaptation against excessive transpiration (water loss). Sclerophyllous scrub or woodland inhabits areas where drought is recurrent, such as in the Mediterranean region.

scoria (*plural* scoriae) A volcanic rock, derived from basic lava, which is full of cavities produced by gases trapped in the solidifying lava. It has a higher density than pumice and will not float on water. Scoriae are often found in large flat masses up to 3 m across; these were soft enough to spread out like pancakes when they hit the ground.

scour The process of localized erosional deepening by moving water (e.g. rivers, tides) or ice. †*Scour-and-fill* is a sequence of over-deepening of a river channel followed by deposition of debris that is commonly found in turbulent short-lived streams in arid and semiarid regions.

scree (talus) A collection of angular rock fragments that builds up at the foot of a steep mountain slope, commonly in the form of a cone or fan. A *scree slope* (talus slope) is the slope formed by these deposits. It is one of the few landforms that can be produced by weathering and mass movement alone, often largely through a combination of frost-shattering and gravity.

screen, meteorological *See* Stevenson screen.

screen temperature *See* shade temperature.

scrub 1. A type of vegetation that consists chiefly of low-growing aromatic herbs, frequently evergreen. It is characteristic of areas in which water shortages occur, including tropical semideserts (e.g. spinifex vegetation in Australia) and temperate areas with seasonal drought (e.g. maquis in S France).
2. The tangled shrubby growth of such plants as hawthorn, gorse, and young birches, where these invade open ground (e.g. heathland) in the British Isles.

scud (fractostratus, St fra) A type of low cloud that appears as ragged wind-driven fragments below the base of rain-bearing nimbostratus (Ns) clouds.

sea 1. The expanse of water that covers much of the Earth's surface.
2. A division of an ocean. This may be partly enclosed by land masses (e.g. the Caribbean Sea and the Mediterranean Sea).
3. A large completely landlocked area of salt water, e.g. the Dead Sea and Caspian Sea.

sea breeze A local wind that blows from the sea to the land and is the daytime equivalent of the land breeze, which blows the opposite way at night. It is caused by the greater heating of the land relative to the sea during the day, producing low pressure over the land into which cooler air moves from over the sea. It reaches its maximum strength in mid-afternoon when the temperature difference is greatest, and dies out as the Sun's heat diminishes in the evening. As with a land breeze, it is best developed in the tropics when conditions are calm and the sky is clear, but can be observed under similar conditions in temperate areas. It is very variable and reaches only a few kilometres inland, but can be very useful in bringing cooler conditions to uncomfortably hot areas during the day. *See illustration at* land breeze.

sea-floor spreading The theory developed during the early 1960s that explains how lithospheric plates move apart from one another. This movement takes place along the mid-oceanic ridges (constructive plate margins). Magma rising from the Earth's mantle flows out along the rift of a mid-ocean ridge and solidifies to form new oceanic crust. This process continues with new material constantly welling up as the sides of the lithospheric plates move apart. This explains the relatively young age of the oceanic crust and the fact that it becomes progressively older with increasing distance from the mid-ocean ridge. The theory was substantiated by palaeomagnetic evidence. Throughout geological history periodic reversals of the Earth's magnetic field have taken place. The magnetic polarity at a rock's formation becomes 'frozen' into its structure. From studies of the ocean floor it was found that symmetrical stripes of reversed and normal magnetism lie parallel to the mid-ocean ridges. These could be dated confirming that the oceanic crust becomes progressively older with increasing distance from the mid-ocean ridges. The rate of sea-floor spreading has been measured as being in the order of 2.5 to 5 cm per year. *See* plate tectonics.

sea level *See* mean sea level.

season One of the periods into which the year can be divided as a result of climatic conditions, which are largely due to changes in the duration and intensity of solar radiation. In the mid-latitudes the seasons are spring (new growth springing up), summer (the hottest season), autumn or fall (ripening of crops and fall of leaf), and winter (the cold season in which plants are dormant). The tropical pattern is commonly a wet season (in which the Sun is overhead) and a dry season. In S Asia these become the summer monsoon and winter monsoon.
The cause of seasonal change is that the Earth's axis is inclined at an angle of $66\frac{1}{2}°$ to the plane of its orbit, and points constantly towards the Pole

Star. Thus in June the North Pole of the axis is tilted towards the Sun so that solar radiation is concentrated in the N hemisphere (summer) and in December the position is reversed.

second 1. Symbol: s The SI base unit of time: 1/60 of a minute.
2. Symbol: ″ A unit of angle: 1/3600 of a degree; 1/60 of a minute.

secondary depression A relatively small low-pressure area that develops near a large depression and travels with or around it (in an anticlockwise direction in the N hemisphere). It frequently develops on the trailing cold front, and is then called a cold front wave. It may become more intense than the large depression, resulting in heavy rainfall and stormy conditions, and eventually absorb it.

secondary growth †Plant life that develops after the removal (e.g. by clearing or fire) of the previous vegetation. It is, therefore, often an initial stage of plant succession towards a renewed climatic climax vegetation. It is characteristically quick growing. Examples include bamboo, scrub, and heath.

secondary industry An industry that manufactures goods by processing raw materials, assembling components, or producing commodities of value to man. The range of industries covered is considerable and includes steel production, car manufacture, printing, and construction. About 37% of the UK workforce is employed in secondary industries. *See also* primary industry, tertiary industry.

secondary sere †*See* sere.

secondary wave *See* earthquake.

section An imaginary vertical cut through the surface of the Earth to expose the underlying geological structure. Profiles of landscape features such as the cross profile of a river valley are sometimes called sections. Strictly speaking the term section is

correctly used only when the geological structure is shown.

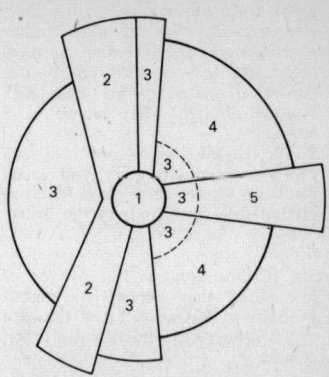

1 CBD
2 wholesale and light manufacturing
3 low-class residential
4 medium-class residential
5 high-class residential

Sector model of urban land use

sector model of urban land use †A theory on the internal structure of an urban area that was developed by the US geographer H. Hoyt in 1939. The model is an extension of the concentric model of urban land use but adds another dimension – direction – to that of distance. Hoyt believed that the structure of an urban area is determined by the location of routes radiating out from the city centre. Variations in the degrees of accessibility to the town or city centre between these routes would create corresponding variations in land and rental values, which would influence the land-use pattern. Functional zones are shown as sectors with wedges of residential areas radiating outwards from the central business district. High-class residential areas are located in the most convenient situations. Middle-class residential areas would, according to Hoyt, be located on either side of the wedge of high-class residences, with the low-class residential areas pushed furthest away. *See also* concentric model of urban land use, multiple nuclei model of urban land use.

sedentary agriculture The farming of land at a fixed location instead of migrating from one site to another. Nearly all the world's agricultural systems are sedentary, the two main exceptions being shifting cultivation and nomadic herding.

sediment Particles deposited by a fluid. Sediments are most commonly composed of rock particles, but organic sediments or mixed organic and mineral sediments also occur. *Aeolian sediments* are those deposited by the wind; *fluvial, marine,* and *fluvioglacial sediments* result from deposition by water. Sediments range in size from clays (less than 0.002 mm in diameter), through silt, sand, gravel, cobbles, and boulders.

sedimentary rock A rock composed of sediments and generally having a layered (stratified) appearance. Sedimentary rocks form one of the three major groups of rocks that together make up the Earth's crust (the other two groups being igneous and metamorphic rocks). The sediments are largely derived from pre-existing rocks that have been broken down and then transported by water, wind, or ice. Rocks formed from these sediments are known as clastic sedimentary rocks. They are divided into three groups according to the size of the constituent particles: rudaceous or pebbly (e.g. conglomerate), arenaceous or sandy (e.g. sandstone), and argillaceous or clayey (e.g. shale). Organic sedimentary rocks are those derived from the remains of plants and animals (e.g. the skeletons of fish or microorganisms) and include limestone and coal. Sedimentary rocks may also be of chemical origin, for example evaporites, which are formed from sediments precipitated from a saturated solution and dried out by evaporation.
Shale, sandstone, and limestone are the most abundant of the sedimentary rocks. *Compare* igneous rock, metamorphic rock.

sediment yield The amount of sediment removed by rivers from their drainage basin over a period of time, usually in tonnes per year.

seeding, cloud *See* cloud seeding.

seepage 1. The movement of water through the pores of the soil, sediments, or porous rocks. Seepage is diffuse and relatively slow compared with flow through fissures, springs, or natural pipes.
2. The slow oozing out of oil onto the ground surface.

seiche A regular fluctuation in the level of water in a lake, harbour, bay, or other enclosed or partly enclosed bodies of water. It may be caused by abrupt changes in barometric pressure and winds. The period of oscillation, which generally ranges from a few minutes to several hours, is usually related to the dimensions of the water body. [Swiss French]

seif dune (longitudinal dune) A desert sand dune in the form of a long ridge, parallel to the prevailing wind. Seif dunes are approximately six times as wide as they are high and tend to run in parallel chains with knife-edge crests. Whilst the prevailing wind builds up their length, cross winds build up the height and width. In Iran they can reach 210 m in height. Some seif dunes may form from barchans modified by strong cross winds transverse to the prevailing wind. [Arabic]

seismic focus The place of origin of an earthquake in the Earth's crust from which the earthquake waves move outwards. The focus can be a point or an area such as a fault.

seismic sea wave *See* tsunami.

seismic wave *See* earthquake.

seismology The scientific study of earthquakes, their causes, effects, and distribution. The instruments used are the *seismograph*, which records the magnitude of the oscillations of earthquake waves, and the *seismometer*, which records the movement of the rocks in a particular direction. [From Greek *seismos*: earthquake]

selective logging The periodic cutting down of mature trees or those growing too close together in a forest to allow optimum use to be made of the forest resources. This is most commonly practised where a forest consists of a variety of trees of different ages.

selva The equatorial rainforest of the Amazon Basin in South America. The term is often applied to the area itself, and sometimes extended to rainforest vegetation in general. *See* equatorial rainforest. [From Latin *silva*: forest]

semiarid climate (semidesert climate) The climate of those areas of the world in the transitional zone between true desert and tropical grassland (in the tropics) and between true desert and either Mediterranean or temperate grassland (outside the tropics). Characteristically the mean annual rainfall is between 100 mm and 300 mm, but it is extremely irregular both in distribution and intensity. The natural vegetation consists of coarse grasses and shrubs and small trees able to withstand long periods of drought.

seminatural vegetation A vegetation type that owes some characteristics to the natural conditions of soil and climate, but that has been modified by human activity, particularly farming. †Seminatural vegetation usually comprises some indigenous plant species, but in proportions or at a stage of succession determined by recent biotic influence, such as grazing. Savanna grasslands and W European heaths are examples. *See also* natural vegetation.

sequent occupance A concept in geography in which changes in human settlement are seen as a succession of stages. It is likened to plant succession

in botany. [Coined by Derwent Whittlesey in 1929]

sérac A pillar or ridge of ice between crevasses formed on the surface of a glacier as it passes over a steeper section of valley floor. *See also* crevasse. [French]

seral stage †*See* sere.

sere |The series of changes that takes place in a plant community in the succession from pioneer plants to a climax community. Each step in the succession is called a *seral stage*. Seres are categorized according to the environment in which they develop. A *prisere* (primary sere) is a plant succession that begins with pioneer vegetation colonizing a newly created surface, such as a lava flow, and progresses through succession to a climax state. The plant communities involved vary according to the nature of the original environment. A *secondary sere* develops on a surface that has already been vegetated, such as abandoned cultivated land. *See also* hydrosere, xerosere.

sericulture The rearing of silkworms to produce raw silk. This is a labour intensive form of farming using cheap skilled workers. Fresh mulberry leaves are fed to the silkworms and the thread unravelled from the cocoons on small spinning machines. The main silk-producing countries are China, Korea, Japan, and Italy.

series In geology, the rock layers deposited during one epoch.

serir A desert surface covered with gravel and cobbles in Libya and Egypt. It is synonymous with the reg of Algeria. [Arabic]

serozem *See* sierozem.

service centre A place that provides goods and services for the surrounding area. The services provided may include shops, banks, commercial offices, places of entertainment, and professional services, such as solicitors, accountants, and estate agents. Service centres vary in size and in the facilities they offer, e.g. Croydon is a large service centre for S London whereas Orpington is a much smaller one.

service industry An industry that provides a facility or service instead of manufactured goods, e.g. telecommunications, banking, tourism, and nursing.

seter *See* saeter.

settlement Any permanently occupied human dwelling place. Although an isolated occupied hut may be described as a settlement, the word more usually indicates a community of dwellings and associated buildings, ranging from a hamlet to a conurbation. Settlements tend to form where there is easy access to resources or services such as water supply, fertile soil, mineral wealth, shops, industries, and transport facilities. *See also* central place.

settlement hierarchy The categorization of settlements according to such factors as size, status, and the range of facilities provided. †An important concept that qualifies the hierarchical concept is the *settlement continuum*, which proposes that each category of the hierarchy merges gradually into the next. For example, hamlet into village, village into town, and town into city. *See also* central place hierarchy.

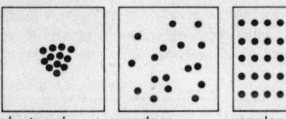

clustered   random   regular

Settlement patterns

settlement pattern The spatial distribution and organization of settlements. At one time geographers were concerned essentially with the study of patterns in terms of the relationship of the settlement to its physical environment and its historical evolution. Settlement geographers are now more concerned with a search for order – in the spacing of settlements and in

their internal organization – and for greater understanding of the processes that help to shape settlements, including cultural influences and factors of perception. *See also* dispersed settlement, nucleated settlement.

sextant An instrument used in navigation for measuring angles, chiefly the altitudes of celestial bodies. It consists of a telescope to which a protractor is attached. The observer measures the elevation above the horizon of the celestial body. By consulting a nautical almanac and a chronometer the observer's position in degrees of latitude and longitude can be calculated.

shade temperature (screen temperature) The temperature of the air as measured in a Stevenson screen where the thermometer is protected from direct sunlight, radiation from the ground and buildings, and rainfall, but is in freely circulating air. The shade temperature is used in meteorological records and weather forecasting except when otherwise stated.

shaft mining The extraction of minerals from under the ground by sinking a vertical shaft into the ground and then cutting horizontal tunnels to reach the seams or veins. More than one shaft is required for each mine to provide adequate ventilation. Most coal in the UK comes from deep mines with shafts over 800 m deep.

shale An argillaceous sedimentary rock composed of layers of clay and silt particles with a grain size of 1/16 mm to 1/1024 mm in diameter. It is one of the most abundant sedimentary rocks. Shale is a fissile rock, i.e. it splits easily along bedding planes into thin flakes of rock. This results from the alignment of the clay minerals parallel with the bedding planes. Examples are found throughout the geological record, the Ludlow Shale and the Wenlock Shale coming from the Silurian.

shamal A steady hot dry northwesterly wind that blows mainly in summer across the plains of Iraq. It is caused by air being drawn into the intense summer low-pressure area over NW India. It is often dust laden, causing hazy skies over wide areas. [Arabic]

shanty town (squatter settlement) An unplanned settlement composed of dwellings built on land that is not owned or rented by the inhabitants; it may be either within a city or, more usually, on the edge of a city. Shanty towns are found in particular in Third-World countries where rapid population growth together with migration from rural areas to the cities has created tremendous demands for living space. Examples include the bustees of Calcutta, the favellas of Brazil, and the barriadas of Peru.

sharecropping A type of agricultural tenure in which the tenant farmer hands a proportion of his output over to the landlord as rent. In return the landlord often provides the seed, equipment, accommodation, and livestock. The proportion paid as rent varies from place to place. This practice is found in such areas as the Midwest of the USA, Canada, and France (where it is known as *métayage*).

shear A force tending to deform layers of rock by moving the layers laterally over each other.

sheet erosion (sheet wash, rain wash) Erosion by water flowing over a relatively smooth slope. It is important in pediment formation. It is a slow process and may be caused by water flowing as overland flow in a sheet or in shallow rills, which form and reform at different locations from storm to storm. *See also* sheetflow, overland flow.

sheetflow (sheetflood) The flow of water over the surface of the land in the form of thin continuous sheets. Very shallow sheetflow may not result in erosion, but deeper flows are turbulent and erosive causing sheet erosion, particularly where vegetation cover is

poor. *See also* overland flow, sheet erosion.

sheet lightning The diffuse glow that is seen when a discharge of forked lightning is obscured from the observer by clouds or precipitation.

shield An extensive, rigid, and very permanent part of the Earth's crust that is composed of Precambrian rocks (more than 590 million years old), which are the oldest rocks on the Earth's surface. The relief of shields tends to be relatively subdued because more recent mountain-building movements have had little effect on them and sometimes, as in the case of the Canadian (Laurentian) and Baltic (Fennoscandia) Shields, partly due to erosion by continental ice sheets.

shield volcano A volcano formed by the extrusion of fluid basaltic lava, which spreads out over a large area so as not to create a true cone. Shield volcanoes may, however, form on top of one another to create high mountains. The Hawaiian islands were formed in this way and Mauna Loa is now 4170 m above sea level.

shieling A summer pasture on the hills of the Scottish Highlands and Islands to which cattle and sheep were driven to graze on the sedge and grass. Small stone huts were used as homes by the people accompanying the animals and one of the activities was butter making. This type of transhumance is no longer practised. *See also* saeter. [Scottish]

shifting cultivation An agricultural system in which a patch of forest is cleared of trees and most vegetation to be cultivated for a few years until the fertility of the soil is seriously reduced. The site is then abandoned and another is cleared elsewhere. Cleared vegetation is usually burned (slash and burn) and crops planted in the fertile ashes. True shifting cultivation is practised by nomadic tribes but the term can also be applied to the cultivation in which people living in a permanent village periodically clear, cultivate, then leave fallow the plots of land surrounding their village. Certain cash crops are also grown by shifting cultivation methods; when yields fall the land is abandoned and a fresh site is cleared. Shifting cultivation is practised in equatorial rainforest regions in Central America, Brazil, Zäire, India, Sri Lanka, Indonesia, and the Philippines.

shingle A deposit of coarse gravel or pebbles, especially on a beach. Coarse shingle ridges known as storm beaches are often formed at the upper limit of the beach by storm waves.

ship canal A canal that is wide and deep enough to be used by oceangoing vessels. Most ship canals have been cut to shorten sea journeys, e.g. the Suez and Panama canals. In some cases the canal links an inland city with the sea, e.g. the North Sea Canal to Amsterdam and the Manchester Ship Canal. *See also* waterway.

shire A county in the UK. Historically, a shire was a unit of local government. Many of the counties of England that do not carry the suffix -shire were separate kingdoms at one time; for example, Sussex and Kent.

shoal A ridge of sediments, such as sand, mud, or pebbles, causing shallow water in a river, lake, or sea.

shopper goods and services The range of goods and services that are provided by the shops of a service centre. The main categories of goods are foodstuffs, clothing, and household and luxury goods, while services include hairdressing, shoe repairs, and travel agents. †The range of these goods and services at a number of centres is used as one of the criteria by which a hierarchy of the service centres within a region can be established.
See also service centre.

shore 1. Land bordering an expanse of water, particularly a lake or sea.

2. The area between the high- and low-water marks.
The *shoreline* is often regarded as being synonymous with the coastline. *See also* coastline, foreshore.

shott (chott) A salt lake or salina found in N Africa. *See also* salina, salt lake. [French]

shoulder A broad ridge, bulge, or prominence.

sial The relatively light rocks that form the continental crust. These are composed predominantly of silica and aluminium, i.e. acid rocks such as granite. *See also* sima. [*si*lica–*al*uminium]

sierozem (serozem, grey desert soil) A zonal soil formed in cool or temperate arid areas under scrub vegetation, the litter of which is sparse; the overall grey colour of the soil reflects this paucity of organic matter. Calcium carbonate accumulates at or just below the surface, sometimes as a concentrated pan, and the whole soil profile is alkaline. Sierozem soils occur in Colorado, New Mexico, and Utah (USA), and in Turkestan (USSR). [Russian: grey]

sierra A long mountain range with a jagged profile, typical of arid and semiarid environments where vegetation is thin. The term is used in Spain, the SW USA, Mexico, and South America, e.g. the Sierra Nevada, USA, and Sierra Madre, Mexico. [Spanish: saw, mountain ridge]

silage Green fodder crops that are stored in a pit or a tall container (silo) to be used as animal fodder. Grass, clover, lucerne, corn foliage, and other green crops are chopped up, sprayed with molasses, and allowed to ferment under pressure from the layers above. Silage keeps well and is a valuable milk-producing feedstuff.

silica The mineral silicon dioxide (SiO_2), the most abundant of all minerals in the Earth's crust. It occurs naturally in at least nine different forms (polymorphs). The three main forms are quartz, trydymite, and cristobalite. Other forms include coesite and stishovite, which are found near meteorite impact craters; chalcedony (e.g. agate, flint), the cryptocrystalline form of silica; and opal, the hydrous amorphous form of silica. *See* quartz. *See also* silicate minerals.

silicate minerals A group of minerals that constitutes about one third of all minerals and consists of silicates of calcium, magnesium, aluminium, and other metals. All are based on a structural unit known as the SiO_4 tetrahedron. The silicate minerals include the feldspars, mica, clays, talc, chlorite, amphiboles, pyroxenes, olivine, garnet, and beryl. Many are economically important.

silicon The second most abundant element in the Earth's crust, after oxygen. It forms a major constituent of nearly all rock-forming minerals. Pure silicon is of major use in the electronics industry as a semiconductor.

sill A near-horizontal intrusion of igneous rock along a bedding plane between two rock layers. Sometimes the magma breaks across the layers of rock and continues spreading along another bedding plane. The cooled rock forms a tabular sheet more or less parallel to the surrounding layers of rock. Thicknesses vary from a few metres to over 60 metres, the average being about 30 metres. The best-known example in the British Isles is the Great Whin Sill in N England, which is composed of dolerite and extends over 3900 km² from the Northumberland coast to the W Pennines. *See also* dyke.

silt Fine grains of mineral material between clay and sand in size, with diameters between 0.06 and 0.002 mm. Silts are deposited with clays as mud in quiet water. The particles are formed both by weathering and abrasion of rocks. Silts are classified with clay as argillaceous sediments. *See also* sediment.

Silurian The geological period extending from the end of the Ordovician period, about 440 million years ago, to the beginning of the Devonian, about 395 million years ago. Conditions at this time were mainly marine with shallow seas and during the Silurian the first true fishes appeared. The first evidence of land plants also appeared at this time. The Silurian is generally divided into three series: the Llandovery, Wenlock, and Ludlow. In the British Isles Silurian rocks occur in Shropshire, Wales, the Lake District, the S Uplands of Scotland, and NE Ireland. [Named by the British geologist R. I. Murchison in 1835 after the ancient tribe of the Silures who inhabited the Welsh Borderland where he conducted his early fieldwork].

silver A greyish-white metallic element. It occurs in nature as the metal, frequently in association with other metals such as copper and gold, as argentite (Ag_2S), and in association with lead, zinc, and copper ores. Silver is readily workable and is used in coinage, silver plate, and jewellery. Pure silver has the highest thermal and electrical conductivity of all the metals and is used in some printed electrical circuits. Silver salts are used in medicine and photography. Mexico, the USA, and Canada are the major producers.

silviculture The cultivation of trees, including the theoretical and practical aspects of forestry. In the UK the Forestry Commission is responsible for the establishment and maintenance of forest land that is not privately owned. Great emphasis has been put on quick-growing conifers to produce large quantities of softwood.

sima The relatively dense rocks that underly the lighter continental sial and the floors of the oceans to form the oceanic crust. These are composed predominantly of silica and magnesium, i.e. basic rocks such as basalt. [*si*lica—*ma*gnesium]

simoom (simoon) A hot dry wind experienced in N Africa, especially in the N Sahara, in summer. It is a convectional wind caused by the intense heating of the ground and therefore gusts and swirls about, often resulting in localized duststorms and sandstorms. [Arabic]

simulation models †The construction of a generalized system based on situations existing in reality in order to test hypotheses and formulate theories. This may be done with physically constructed models, diagrams, mathematical formulae, computer programs, or verbal reasoning. For example, input-output models for assessing what changes the growth of agricultural output will have on the economy of a country have been designed to help plan economic development in countries of the Third World. The models are mathematical simulations using a set of mathematical assertions from which consequences can be derived by logical mathematical argument.

sinkhole (sink) A saucer-shaped depression into which water drains, typical of limestone areas. Rainwater and stream water containing carbon dioxide from the air form dilute carbonic acid; this attacks the limestone causing solution, weakening, and eventual collapse so that the sinkholes grow larger and deeper. *See also* swallow hole.

sinter A crust-like mineral deposit found around springs. Where cold water springs issue from limestone rocks release of pressure can cause water supersaturated with calcium salts to deposit *calcareous sinter* (travertine) crusts. *Siliceous sinter* (geyserite) is formed only around hot springs or geysers where the water has been sufficiently hot to dissolve large quantities of silica (quartz). *See also* geyser.

sirocco (scirocco) A southerly wind that blows from N Africa across the Mediterranean Sea to Sicily and S Italy, preceding a depression moving E through the Mediterranean basin. It is hot and very dry on the N African coast as a result of having blown off the desert, but after crossing the sea and picking up moisture it is often very humid and enervating when it reaches Sicily and Italy. [Italian]

site The area of land upon which a building or settlement is built. The initial choice of a site for a settlement depended on its meeting certain everyday needs. Important factors included a water supply and the availability of potential farm land, building materials, and fuel, the need for defence, and communications. *See also* defensive site, dry-point site, wet-point site.

site of special scientific interest A site that because of its fauna, flora, geological, or physiographical features needs special protection. Proposals for building or other developments that affect these areas have to be submitted to the local authority concerned for comment, although powers to prevent any damage to the sites are limited. They have been designated by the Nature Conservancy Council since 1949.

situation The location or position of a place in relation to its surroundings.

skeletal soil *See* azonal soil.

skerry A small rocky island or reef of rock that may at times, such as high tide or stormy weather, be submerged below the sea. Examples include those off W Norway and SE Sweden. [From Old Norse *sker*]

skew (skewness) †The extent to which a frequency curve is asymmetrical. The curve is said to have a *positive skew* when the modal class (i.e. the class containing the largest number of values) lies off-centre to the left and a *negative skew* when it lies off-centre to the right.

sky cover In meteorological observations, the proportion of the sky that is obscured by cloud. In the UK the unit of measurement is the okta (ranging from 0, a cloudless sky, to 8, a totally obscured sky). Internationally, the sky cover is measured in tenths.

slash Swamp-forest vegetation of coastal regions of the SE USA. Much of it has been planted with water-tolerant pines and cypresses.

slate A fine-grained metamorphic rock in which the arrangement of flaky minerals such as mica into parallel layers gives the rock its characteristic structure known as flaky cleavage. Slates are formed by the metamorphism of mudstones, siltstones, and other fine-grained deposits. For example, the silvery green slates of the Lake District are formed from beds of fine volcanic ashes. The slate is easily split along the flaky cleavage planes, making it valuable as a roofing material, for example the Llanberis slates of the Snowdonia area, North Wales.

sleet 1. In the UK, a form of precipitation consisting of either partly melted snowflakes or rain and snow falling together.
2. In the USA, a form of precipitation consisting of frozen raindrops that have subsequently partially remelted.

slickenside A clean-cut rock surface that has been either polished or striated by friction along the fault plane between moving fault blocks. The linear grooves or ridges are formed when two surfaces under pressure move over one another in close contact and they develop parallel to the direction of movement.

slip 1. *See* landslide.
2. In a geological fault, the amount of oblique movement between the upthrow and downthrow sides. *See also* fault.

slip face The front advancing side of a moving dune, either in a sandy desert or in the bed of a river. Sand slides over the top of the dune and down the face to accumulate at the foot of the dune. The profile of the slip face is slightly concave due to the eddying effect of wind. *See also* dune.

slip-off slope The slope on the inside of a meander bend on a river. The slope is very gentle compared with the undercut slope, which is being actively eroded, on the outside of the meander. The term 'slip off' arises from the fact that the main current slips away from this bank leaving relatively quiet water in which sediments may be deposited to form point bars. *See also* point bar.

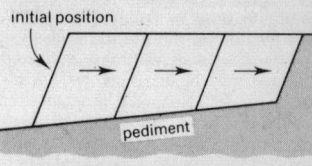

initial position

pediment

parallel retreat (backwearing)

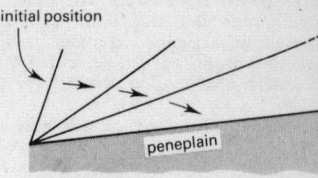

initial position

peneplain

slope decline (downwearing)

Slope retreat

slope retreat The backwearing and/or downwearing of a slope through erosion. Hillslopes evolve in a variety of ways as a result of erosion. A classic theoretical argument in geomorphology centred on the question of whether slopes retreat without changing their steepness, known as *parallel retreat* (backwearing), or whether they become less steep, i.e. the top of the slope retreating more than the bottom, known as *slope decline* (downwearing). The slope decline theory was proposed by W. M. Davis but was later challenged by the parallel-retreat theories of W. Penck and L. C. King. In fact, both can occur in the same climate depending on lithology and on the relative rates of erosion on the slope and removal of material from the foot of the slope.

slum A deteriorating urban area, usually a part of an inner city area, that is characterized by dilapidated housing, overcrowding, poverty, social disadvantage, and other forms of physical and social deprivation. It is also an area in which immigrant ethnic minorities may congregate. *See also* inner city.

slump A mass movement of rock debris or soil, often lubricated by water, that rotates backwards as it slips on a curved concave surface. It leaves a scar on the hillslope from which it has sheared. *See also* rotational slip.

small circle Any imaginary circle on the Earth's surface, the plane of which does not pass through the centre of the Earth. All parallels of latitude except the equator are small circles decreasing in size towards the poles. *Compare* great circle.

smallholding A farm with an area of less than approximately 20 hectares that is cultivated as a unit by the owner or a tenant. In the UK, the term 'farm' is used for large units of agricultural land and gives a degree of status that the word smallholding does not.

smelting A process for extracting a metal from its ore by using very high temperatures. Pig iron is obtained by heating iron ore with coke and limestone in a blast furnace. In the smelting of iron, copper, zinc, and other metals large amounts of power are used, which often involves transporting the ore long distances to the power source. *See also* blast furnace.

smog (smoke fog) A form of fog that occurs in areas where the air contains a large amount of smoke. The smoke particles provide a high concentration of nuclei around which condensation occurs. Condensation can occur around these nuclei even when the air is not saturated and therefore it forms earlier, becomes denser, and lasts longer than fog that develops in unpolluted air. The smoke gives the fog an acrid taste because of the chemicals, especially sulphur dioxide, contained within it. Large industrial and residential areas (e.g. London and Pittsburgh) have suffered worst in the past, but smog is now rare in these cities as a result of the widespread introduction of smokeless fuels.

snout The downslope end of a glacier. It frequently contains a cave from which issues a meltwater stream.

snow A form of precipitation consisting of crystals of ice. It is produced when condensation takes place at a temperature below freezing point, so that the minute crystals (spicules) of ice form directly from the water vapour. These may fall as they are, but more commonly they combine together to form *snowflakes*, which display an infinite variety of patterns, all basically hexagonal in shape. If the temperature is sufficiently low the snowflakes will reach the ground, but often they melt and turn into rain as they fall through warmer air. When temperatures are near or at 0°C the snowflakes tend to aggregate forming large wet flakes, but if the temperature is very low (e.g. in Antarctica) the flakes tend to be small and dry (powder snow). With the former, a depth of about 6 mm is equivalent to 1 mm of rainfall, but with the latter the ratio is about 30 to 1. In compiling meteorological records a sample depth of snow is collected and melted to determine the equivalent rainfall, and this figure is used.

snow avalanche *See* avalanche.

snow bridge A bridge of compacted snow crossing a crevasse or a bergschrund on a glacier.

snowdrift Snow that has been blown by the wind and accumulated as a bank, either against obstacles such as hedges or buildings, or in sheltered spots such as road or railway cuttings. Drifting occurs most readily when the snow is dry and powdery as wet snow tends to stick when it hits the surface of an object. Road and railway cuttings are often protected by snow fences, barriers, or snow sheds to prevent snowdrifts.

snowfield An area of permanent snow cover found mainly in sheltered places in regions of present-day glaciation. This perennial snow may feed glaciers, but more slowly than the main accumulation area.

snow line The lower limit of snow cover in a landscape. Most commonly it refers to the *annual snow line* below which snow melts away in summer. It is often difficult to define because it is rarely in exactly the same place from year to year. The annual snow line is therefore an average line. It lies at about 5000–7000 m above sea level in tropical areas and decreases polewards to reach sea level in the Arctic Seas, Greenland, and Antarctica. †On glaciers, it is the line between net accumulation above, and net loss (ablation) below. However, on some glaciers in the high Arctic meltwaters may freeze on the surface of the glacier below the visible snow line. In this case the snow line is not the true line between net gain and net loss, which is called the *equilibrium line*.

snow patch erosion †*See* nivation.

social geography A branch of geography that is concerned with the study of the spatial arrangement of social phenomena in relation to the total environment. Main focuses of study within social geography are population studies, the geographical study of towns and cities, rural settlement studies, and the distribution of social groups and their relationships to different environments.

soffoni (soffione, suffione) A volcanic vent in the Earth's crust from which steam and sulphurous vapours rise. The name was originally used in the old volcanic areas in Italy. *See also* fumarole. [Italian]

softwood Wood that is easily cut and relatively lightweight. Trees producing softwood grow quickly, e.g. sitka spruce can reach a height of 30 m in 30 years. Softwood is in demand for use as building timber, paper pulp, and chipboard. The main softwood trees in the UK are Scots pine, spruce, fir, and larch. *Compare* hardwood.

soil The surface material covering much of the Earth, composed of mineral particles and humus, water, and air, in which plants grow. Soil formation results from weathering processes acting on (1) parent material that supplies minerals; and (2) organic remains (humus) derived from vegetation. Usually, soil also contains water and air, which occupy pore spaces. Invertebrates living in the soil, especially earthworms, aid soil formation by mixing it. The time over which processes have acted is an important factor. The character of a particular soil depends on such factors as its texture, structure, and pH value. Most soils vary with depth, displaying a series of horizons that together constitute the soil profile; this forms the basis of soil classifications. Major categories of soils often correspond to climatic zones and are associated with certain vegetation. Local variations in soil frequently reflect changes in geological or topographical characteristics. *See also* azonal soil, intrazonal soil, zonal soil, soil erosion, soil horizon, soil profile, soil texture.

soil creep *See* creep.

soil erosion The removal of soil, either mechanically or chemically. Wind and water are the main agents of soil erosion. Wind may remove the finer soil particles by deflation. Water may cause surface erosion by overland flow, in the form of sheet erosion, rilling, and gullying. Subsurface erosion of soil may be caused by movements of water through pipeflow or throughflow. Soil may also be eroded by mass movements, which vary from the relatively slow soil creep to the rapid earthflows and landslides.

Soil erosion is frequently caused or accelerated by man, for example by the removal of the protective vegetation cover. Preventative measures include contour ploughing and the planting of cover crops. *See also* deflation, gully, mass movement, overland flow, rill erosion, sheet erosion, throughflow.

soil horizon A well-defined layer within the soil profile, parallel to the local ground surface. Three main horizons – A, B, and C – are characteristic of many free-draining soils. †Such horizons are visually distinctive, reflecting their different physical and chemical properties, which result from soil-forming processes including weathering, the introduction of humus, and movement of minerals.
See also A horizon, B horizon, C horizon, soil profile.

soil profile A vertical section of soil, from the ground surface down to the parent material, which includes all the depth subject to soil-forming processes. The profile of a soil can be

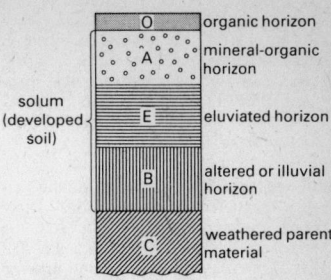

Soil profile

determined from a specially-dug soil pit. They usually show different layers (horizons), from which the soil is classified. *See also* soil horizon. †*Compare* pedon.

soil series A collection of soils that have very similar profiles and are developed over the same parent material. Soil series are subdivisions of soil groups determined by local geological

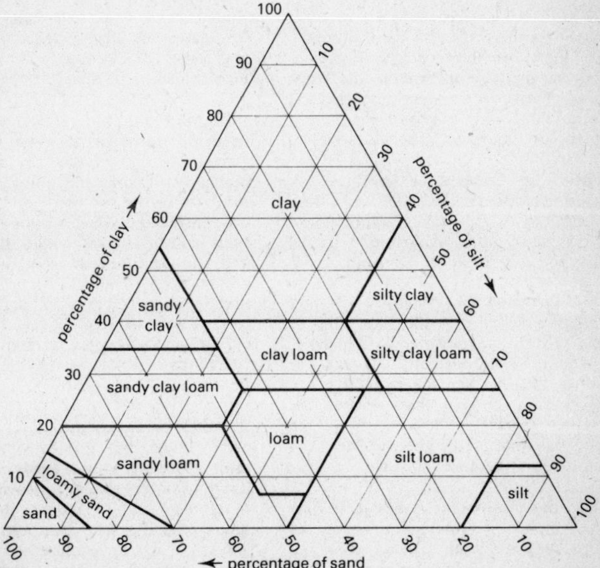

Soil texture

← percentage of sand

conditions. Members of a soil series differ only in the texture of their A horizons (surface horizon). The soil series are used as a basis for soil mapping. They are differentiated by being given local names, e.g. Windsor Series. *Compare* soil type.

soil texture The proportion of the various particle sizes in a soil. Three chief grades of soil particle size are recognized: sand (2–0.2 mm), silt (0.2 –0.002 mm), and very fine-grained clays (less than 0.002 mm). Most soils contain some of each size category. A well-balanced mixture (i.e. a loam soil) gives the optimum conditions of water holding, temperature, and supply of plant nutrients. It is often possible to identify the soil texture by its 'feel'; for example, sands feel gritty to touch. †More detailed laboratory analysis enables soils to be precisely plotted within a triangular diagram of textures. Texture results principally from the mineral composition of the parent material.

soil type A subdivision of a soil series based on the texture of the A horizon. Soil types are the most specific classification used in description and mapping. Their titles incorporate the soil series label plus a textural term, e.g. 'loam' or 'sand'. *See also* soil series.

solano An easterly or southeasterly wind experienced in SE Spain and Gibraltar, which brings hot humid conditions and occasional rain in the summer. *See* levanter. [Spanish]

solar constant The intensity of solar radiation received at the outer limit of the Earth's atmosphere, which is thought to be virtually constant at about 1.35 kW per square metre. The proportion of this that reaches the Earth's surface is determined by the physical state of the atmosphere at the time. *See* insolation.

solar day (apparent solar day) The interval between the Sun crossing the observer's meridian (i.e. at noon, its highest point in the sky) and the next

crossing. Because the Sun's motion is not uniform this varies widely and is impracticable for accurate timekeeping. *See also* local time.

solarimeter (radiometer) An instrument that measures the intensity of solar radiation. There are several different types but they all depend on the principle that the temperature of a body rises when it absorbs radiation. The most simple instrument consists of a thermometer with a black bulb housed in a vacuum, which is exposed to the direct sunlight. The radiation passes through the vacuum and most is absorbed by the black bulb.

solar system The Sun and the nine planets and other heavenly bodies that revolve around it in elliptical orbits. All the orbits, except Pluto's, are nearly in the same plane as the Earth's orbit, so that their paths among the 'fixed' stars appear to follow the ecliptic (the Sun's apparent path through the heavens).

solfatara A volcano that emits sulphurous gases and steam, rather than lava and ashes. Such volcanoes may be near extinction. [Italian]

solifluction (solifluxion) A form of mass movement characteristic of periglacial areas. It occurs in areas of permafrost. During the spring melt the top layer of the soil becomes saturated with meltwater, which originates from snow above and ice within the layer and cannot drain away because the ground below remains frozen. The saturated soil becomes weak and begins to flow downslope, often creating typical *solifluction terraces* in river valleys. Freeze-thaw processes may assist the movement.

solonchak (white alkali soil) †An intrazonal saline soil formed where hot arid conditions cause evaporation of soil moisture, bringing dissolved salts to the surface by capillarity from the saline ground water. This results in a greyish crusty appearance of the soil profile. Solonchaks occur in desert or

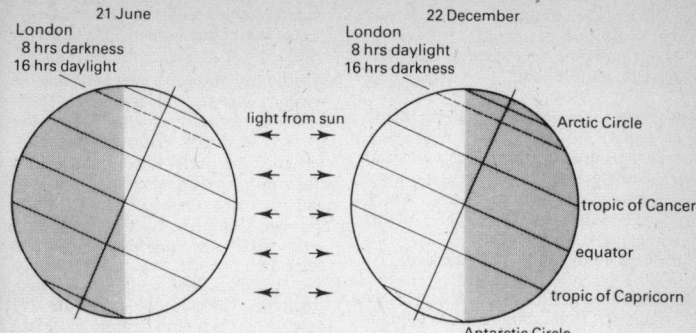

21 June
London
8 hrs darkness
16 hrs daylight

light from sun

22 December
London
8 hrs daylight
16 hrs darkness

Arctic Circle

tropic of Cancer

equator

tropic of Capricorn

Antarctic Circle

Summer and winter solstice in the northern hemisphere

semiarid climates where vegetation is sparse, for example, in the Great Basin of Utah, USA, and around the Caspian Sea. *Compare* solonetz. *See also* saline soils. [Russian]

solonetz (black alkali soil) †An intrazonal saline soil formed where relatively heavy or periodic rainfall leaches salts out of the surface layer into the lower illuvial B horizon. The surface horizons become structureless with dispersed humus and clay, which gives the soil a dark appearance. *Compare* solonchak. *See also* saline soils.

solstice One of the two dates in the year on which the Sun reaches its greatest altitude N or S of the equator and is directly overhead along one of the lines of the tropics. On about 21 June the Sun is overhead on the tropic of Cancer (23½°N), the summer solstice and longest day in the N hemisphere and winter solstice and shortest day in the S hemisphere. On about 22 December the Sun is overhead on the tropic of Capricorn (23½°S), the summer solstice and longest day in the S hemisphere and winter solstice and shortest day in the N hemisphere.

solution A form of chemical weathering in which mineral materials are dissolved by water. Weak acids in the water, such as carbonic acid from rainfall containing carbon dioxide

from the air or humic acid from decayed organic material, increase the amount of solution, as do greater amounts of water flowing or seeping past the minerals. Salt and gypsum are very soluble; calcium, sodium, magnesium, and potassium are moderately soluble; silica (quartz) is fairly insoluble except at high temperatures in the tropics. *See also* carbonation.

Sonnenseite The sunny side of a valley. It is synonymous with the French term adret. *See also* adret. [German]

sorghum A tropical grass, of which there are a number of varieties, that yields a cereal crop. Some varieties are grown for the sugar in the stem. In the USA it is grown in Texas, Oklahoma, and Kansas. It is also extensively cultivated in the dry season in S Africa (where it is known as Kaffir corn), India (durra), and China. *See also* millet.

sough In N England, especially Derbyshire, a horizontal passage or drainage level in a mine. Lead mining was once carried out in Derbyshire and the workings with their soughs can still be seen.

sound 1. A passage of water, wider than a strait, between an island and the mainland or linking two bodies of water. For example, the Sound of

Mull, between the island of Mull and the W Scottish mainland.
2. An inlet or long arm of the sea, e.g. Plymouth Sound.

sounding A measurement of the depth of water. Formerly, this was taken with a lead weight on a line known as a sounding lead. It is now done more accurately by *echo sounders*, which transmit vibrations to the floor of the sea or lake and the times taken for the echoes to return are recorded.

source The point at which a river originates. It is commonly a groundwater spring but may also be a lake, glacier, resurgence from a cave, or marshy area.

southeast trades The trade winds that are experienced in the S hemisphere. *See* trade winds.

southerly burster A cold wave or polar outbreak that occurs usually in spring or summer and brings unseasonal low temperatures, strong winds, and thunderstorms to SE Australia, especially to the coast of New South Wales, where the mountains to the W intensify the effect. The cold air is drawn in from the S behind an easterly moving depression. A rapid fall in temperature of 10°C or more is common and is very damaging to crops. The strong winds are usually dry and carry dust, but their approach may be marked by a line squall and thunderstorms.

southern lights *See* aurora.

South Pole *See* pole.

sovkhoz A large collective farm in the USSR that is owned and operated by government agencies. These state farms are the principal means of cultivating marginal land and promoting advanced methods of agriculture. Their average size is 10 000 hectares and they have a workforce of about 800. They form a large proportion of the grain farms in W Siberia and N Kazakhstan. In other areas the emphasis of these farms is on dairying, fruit

and vegetables, sugar beet, cotton, etc. *See also* kibbutz, kolkhoz. [Russian]

soya bean (soybean) A plant that is widely cultivated for its seeds. These seeds are planted annually and the plant is used as a green manure or as forage. The seeds have widespread uses; they contain about 35% protein and are eaten whole, as flour, and as soy sauce. Soya-bean oil is used in a wide variety of products, including margarine, paints, and cooking oils. The residue is used as fodder for livestock and as a meat substitute for man. As a leguminous crop the plant adds nitrogen to the soil. It is grown in the S states of the USA, China, and Japan.

spa A place that is noted for the medicinal properties of its mineral springs. [Named after Spa, a watering place near Liège in Belgium]

spatial interaction †The movement, contact, and linkages between geographic areas. It can be applied to geographical phenomena such as the movement of commodities, migration, and the journey to work; traffic and telecommunications between cities; and the transfer of capital between areas as in development programmes. [Coined by E. L. Ullman]

spatter cone A low mound or flow of lava around a fissure on the side of a Hawaiian type of volcano. The spattering of lava is caused by a high gas content. *See also* Hawaiian eruption.

Spearman's rank correlation test †A statistical technique used in geography for assessing the degree of association between sets of data. It is not based upon actual values but on their rank order and therefore gives only a crude index of correlation. The formula is:
$$r_s = 1 - 6\Sigma d^2/n^3 - n$$
where r_s is Spearman's rank correlation coefficient, d is the difference in rank and value of two sets of data, and n is the number of pairs being compared.

The test can be used, for example, to assess the degree of relationship between crop yields and altitude in a particular region.

specific humidity The humidity of the atmosphere expressed as the ratio of the weight of water vapour (in grams) to the total weight (in kilograms) of a given volume of air. This varies from about 0.2 gm/kg in very dry cold arctic air to over 18.0 gm/kg in hot humid tropical air.

speleology (spelaeology) The scientific study of caves, including their geology, flora, and fauna, and the practice of caving.

sphere of influence 1. (*Urban Geography*) The area surrounding a town or city within which that urban centre has major cultural, social, and economic influence. The inhabitants of the sphere of influence depend upon the urban centre for a range of services. The term is more or less synonymous with hinterland, umland, and urban field.
2. (*Political Geography*) A region or territory in which a political power claims to have a special interest for political or economic purpose. For example, Poland lies within the sphere of influence of the USSR.

spheroidal weathering A form of exfoliation of rocks in which concentric layers are removed from the surface creating rounded blocks. †The process is thought to begin underground when ground water percolating along joints in the rock causes chemical changes to take place. The water attacks, in particular, the rough corners of blocks and causes a layering in the outer shell of the blocks parallel to the surface. When the overlying material is removed the shell-like layers fall away.
See exfoliation.

spinifex (porcupine grass) A tufted very spiny dense-growing plant that grows in Australian deserts. Spinifex is drought resistant and often develops on sand dunes, which it stabilizes.

spit A long deposit of sand or shingle or both extending out into the sea. Spits are built up by longshore drift of material and are commonly formed where strong opposing currents coincide with plentiful supplies of sediments. They may form at headlands between bays, but some of the largest form across river mouths, where they sometimes bend inland at the tip following the turbulent currents at the junction of river and sea. *See also* longshore drift.

splash erosion *See* rainsplash.

spoil The waste material from mining or quarrying. Spoil heaps are found in many parts of the UK, e.g. the china-clay workings near St Austell in Cornwall. The mounds of waste material from coalmines are usually called tips.

spot height The height above a datum (usually mean sea level) of a point on the ground surface that has been precisely measured, represented on Ordnance Survey (OS) maps as a dot with the height in figures beside it. Spot heights, unlike bench marks, are not marked on the ground. *See also* bench mark.

spread effects †*See* cumulative causation model.

spring 1. (*Geomorphology*) A flow of water from the ground. It occurs where the water table intersects the ground surface. For example, at the base of a scarp slope where a permeable rock, such as chalk, overlies an impermeable rock, such as clay (a scarp-foot spring). *Perennial springs* tap ground water and frequently occur in bedrock. Shallow springs or springs in highly permeable rock like limestone may only flow in very wet conditions, e.g. seasonally or in heavy storms, and are described as intermittent. In general, the deeper the source of the spring the less its flow fluctuates.

2. (*Season*) The season of the year between winter and summer. In the N hemisphere it is defined as extending from about 21 March, the spring or vernal equinox, to about 21 June, the summer solstice; in the S hemisphere it extends from about 22 September to about 22 December. More commonly spring is taken to include the months of March, April, and May in the N hemisphere. It is the season of crop sowing when vegetation springs into growth.

spring line A chain of springs issuing at similar sites in the landscape. Spring lines commonly mark the junction of rock strata with contrasting lithologies, especially where ground water is forced to drain laterally above an impermeable rock. They are commonly found at the foot of scarp slopes where they frequently have attracted settlement in the past, resulting in lines of villages (*spring-line villages*).

spring tide A tide with the greatest range of rise and fall either side of the mean tide level. Spring tides occur twice a month around full moon and new moon when the Sun's gravitational pull acts in the same direction as the Moon's. Because they bring waves higher up the beach, greater erosion and damage can be caused if storms coincide with spring tides. *Compare* neap tide. *See also* tide.

sprinkler irrigation The watering of crops by using plastic or metal pipes to which fine nozzles are attached. The system is very flexible and takes a number of forms, e.g. water under pressure is sprayed from a rotating jet, or pipes fixed above the crops create artificial showers when required. Sprinkler irrigation is used in SE England during dry spells in summer for a variety of field crops.

spur A projecting part of a hill, especially a convex portion of a hillside between two hollows. *See also* interlocking spur, truncated spur.

squall A short-lived strong wind that rises suddenly and lasts for a few minutes before dying as quickly as it appeared. It is most commonly associated with a front or an area of convection and thus is often accompanied by heavy rain or hail. There is frequently a rapid shift in wind direction. In the USA it is more narrowly defined as a wind speed of at least 16 knots, lasting for at least 2 minutes. *See also* line squall.

squatter settlement *See* shanty town.

stability, atmospheric (stable equilibrium) †The state of the atmosphere where a parcel of air rises, becomes cooler and denser than its surroundings, and sinks back down to its original level. The stability of the atmosphere depends on the relationship between the lapse rates of the moving parcel of air and of the environment through which it moves. If the parcel of air has a lapse rate greater than the environmental lapse rate of the air through which it moves, stable air conditions will exist. *Compare* instability, atmospheric. *See* lapse rate.

stack A pillar of rock or small steep islet on the foreshore or near to the coastline, which has been cut off from the mainland by wave erosion. The Needles off the Isle of Wight are a classic example of stacks.

stage 1. The relative position reached in a sequence of development, such as youth, maturity, and old age, the three stages in W. M. Davis's cycle of erosion. Davis saw all landforms as a product of the trilogy: structure, process, and stage. Stage cannot be directly translated into time as development commonly proceeds quicker in the early stages.
2. The level of water in a river channel.
3. In geology, a stratigraphic division of rocks formed during the same age, frequently defined by having the same fossil assemblage.

stages of growth †A model put forward by the American economic historian W. W. Rostow in which the process of economic development of an area is seen as a series of five well-defined stages that the society goes through on the way from a 'traditional society' to a 'society of high mass consumption'. The five stages are:
(1) *Stage one* A traditional society with limited science and technology and a hierarchical social structure.
(2) *Stage two* The preconditions for the next stage of development (i.e. take-off) are established; production begins to increase, investments are made, and entrepreneurs are able to take up opportunities.
(3) *Stage three* The 'take-off' stage in which sustained growth takes place over a period of 10–30 years.
(4) *Stage four* The 'drive to maturity' in which further growth takes place and is transmitted to all parts of the economy.
(5) *Stage five* The age of 'high mass consumption' is established.
The theory has been criticized largely on the grounds that it ignores social factors.

staith (staithe) In N England, a coastal jetty that is used for loading coal into colliers. The term is used along the coast of Northumberland and Durham where local coal is loaded onto colliers, which supply the power stations along the River Thames.

stalactite A candle-like growth of mineral matter hanging from the roof of a limestone cave. Stalactites are formed as water draining through the rock above and containing dissolved minerals, especially calcium cabonate, evaporates and deposits the minerals. They may be thousands of years old and many were largely formed under conditions wetter than the present, especially during the melting of the ice sheets at the end of the last ice age. They are commonly accompanied by stalagmites. *See also* stalagmite.

stalagmite A mineral column on the floor of a limestone cave, generally shorter and fatter than the stalactites on the cave's ceiling. Stalagmites are formed mainly of calcium carbonate, which is deposited as drips from the stalactites evaporate on the floor. *See also* stalactite.

stand In biogeography, part of a plant community that typifies the vegetation and is chosen as an actual and representative sample for fieldwork. Several such stands comprise the local floral association. A stand implies some uniformity of composition and structure, and often refers to a single-species vegetation as in commercial plantations and crops. *See also* association.

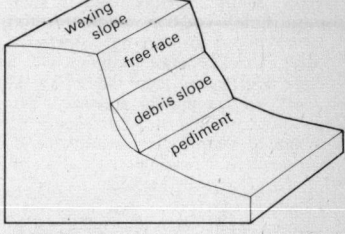

Hillslope profile

standard hillslope †An ideal or model hillslope consisting of four components: an upper convex portion steepening downslope, termed the 'waxing' slope; a 'free face', a cliff of bare bedrock; a straight 'debris slope' covered in weathered rock materials derived from the free face; and a lower concave pediment or 'waning' slope. L. C. King proposed that this scheme occurs irrespective of climate. However, not all components are necessarily present. The two middle components are most often absent, especially in less resistant rocks.

standard meridian *See* Greenwich meridian.

standard parallel A particular parallel of latitude that is selected for calculating and drawing a map projection. It is always true to scale. On a conical map projection it is the parallel where the globe is touched by a cone fitted over it; for example, the simple conical with one standard parallel. Conical projections with two standard parallels (e.g. 30°N and 60°N) are based on a cone intersecting the globe at two different latitudes. The polyconic projection in which every parallel is a standard parallel is derived from the concept of an infinite number of cones touching the globe.

standard time The mean time of a meridian, which is adopted as the time system for the whole of an area or country. This is necessary because of the confusion that would arise if every place used its own local time. Each country therefore bases its standard time in either whole or half hours ahead or behind Greenwich Mean Time. In some countries with a large W–E extent more than one *standard time zone* may be necessary. This is a longitudinal belt of about 15° longitude centred on a central meridian which is a multiple of 15° E or W of the Greenwich Meridian (0°). (15° longitude corresponds to 1 hour difference in mean solar time.) For example, the USA has seven time zones (Eastern, Central, Mountain, Pacific, Yukon, Alaska–Hawaii, and Bering) and Australia has three. In practice the zones are often altered in shape for the convenience of the inhabitants. Most countries adopted a standard time following the international Meridian Conference of 1884, which was held in Washington. *See also* local time.

staple 1. A principal item that is produced, consumed, or traded in an area.
2. The length and quality of a textile fibre. Short-staple cotton, which is rather coarse, is grown in India whereas finer cotton, with a long staple of 50 mm or more, is grown in Egypt and the USA.

star dune A star-shaped dune with a pyramid-like peak at the centre, which is often quite high and stable.

state An independent political community that occupies an area of land with well-defined political boundaries; for example, the UK. The term is also applied in some countries to the units of local government in a federation; for example, the USA is composed of 50 states.

steam fog A type of fog that is formed when cold air passes over warmer water and the vapour escaping from the water surface is quickly recondensed in the cold air, making the water surface appear to steam. It is usually wispy and rarely deeper than 10 m, as it quickly evaporates again in the drier air above, although it may accumulate below a night-time temperature inversion. It is often seen over a hot road surface after a sudden shower, but is best developed on cold winter mornings over lakes, canals, etc. It can be widespread and dense over Arctic seas, where it is called Arctic sea smoke or frost smoke. Occasionally, when the air is very cold, the vapour is converted directly into minute ice crystals, forming ice fog.

steel An alloy of iron with small amounts of carbon (usually less than 1%). It is made in a furnace by removing impurities from pig iron by passing oxygen over or through the molten metal. The type of furnace used depends on the amount of phosphorus in the pig iron and the quality of the steel required. One method is the basic oxygen process in which oxygen is blown onto molten pig iron in a pear-shaped furnace. The impurities combine with lime, which is poured into the furnace, and make a floating slag on top of the steel. Other methods of steel production include the Bessemer process, the open-hearth process, and the electric-arc furnace. *See also* Bessemer process.

step fault A series of parallel faults with repeated downthrow in the same direction, producing a stepped slope, for example the W slopes of the Vosges and the E slopes of the Black Forest, bordering the Rhine rift valley.

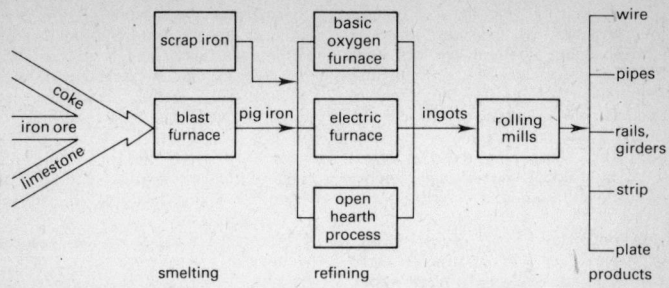

Iron and steel production

steppe Temperate natural grassland that occupies vast areas of E Europe and central Asia, including much of the USSR. Its total E–W extent is about 5000 km. Steppe grassland is generally drier than the North American prairies and, especially in its E parts, becomes semiarid as it approaches the deserts of central Asia. *See also* grassland, puszta. [Russian]

stereographic projection A zenithal map projection that has the centre of projection on the circumference of the globe diametrically opposite the point of contact. This reduces the rate of increase in the diameter of parallels compared with the gnomonic projection and therefore preserves shape, except towards the extremities. *See also* gnomonic projection.

stereoscope An instrument that enables the viewer to obtain a three-dimensional image from two adjacent aerial photographs. In its simplest form it consists of a pair of lenses, one for each eye, carried on a frame with folding legs. The lenses separate the line of sight of the two eyes and slightly magnify the picture. A stereo pair of photographs is placed under the stereoscope and positioned so that a three-dimensional picture is seen of part of them. A more elaborate stereoscope uses binoculars and mirrors so that the whole photograph can be seen three-dimensionally.

Stevenson screen The standard shelter for meteorological instruments at a weather station. It consists of a white painted wooden box with louvred sides and a double roof, the base of the box being 1 m above the ground. It allows the true shade temperature to be measured by allowing free circulation of the air, while protecting the instruments from such outside influences as direct sunlight, radiation from the ground and buildings, rainfall, and strong winds. At all weather stations it contains a standard thermometer, wet- and dry-bulb thermometers, and maximum and minimum thermometers; at larger stations it also contains a hygrograph and a thermograph.

stochastic process †A form of predictive model that depends upon some element of chance probability for its working. This is often achieved through the use of random numbers. Simulation models are stochastic models. [The term was introduced in 1957 by J. Neyman]

stock An igneous intrusion similar to a batholith but smaller, having a surface exposure of less than 100 km². It is probably formed by molten rock moving away from the main body of magma, and some stocks have been shown to be upward extensions of a batholith. *Compare* boss.

stock-grazing index †A measure for comparing the number of animals grazing in one locality with the number grazing elsewhere. To make this possible the livestock must be converted into standard livestock units for that region according to the feed requirements of the different types of animals. For example, in one classification 1 unit equals 1 dairy cow, or 5 pigs, or 7 sheep. The index is calculated by finding the average number of stock units per unit of area.

stone polygons (stone circles) †Rings of stones found in periglacial areas. The rings are polygonal (many sided), commonly with 4–6 sides, rather than circular. It is believed that alternate frost heaving and melting causes overturning in the soil similar to the convection that takes place in a warm fluid. The heaving and flow of the soil pushes the larger particles outwards, like froth in boiling milk. The stone rings extend vertically downwards in a wedge shape around each cell. Active polygons have little or no vegetation in the middle. *See also* patterned ground, periglacial, stone stripes.

stone stripes †A form of patterned ground formed in periglacial regions consisting of parallel lines of stones with fine soil between. They are of similar origin to stone polygons and are formed by sorting of the soil particles as the soil heaves and falls during freezing and thawing. They occur on steeper slopes than stone polygons where the effects of gravity add an element of soil creep, which stretches the heaving overturning cells down the hillside. *See also* patterned ground, periglacial, stone polygons.

stop-and-go determinism †A theory that was introduced by T. Griffith Taylor who maintained that man is able to accelerate, slow down, or stop the progress of a country but that the best economic programme for any country has been largely determined by nature. Stop-and-go determinists believe that man should not depart from the direction of economic development indicated by the natural environment if he wishes to achieve progress and it is the geographer's concern to interpret the nature of the programme that has been environmentally determined. Since man is able to alter the rate but not the direction of progress he is like a traffic controller with the power to signal 'stop' and 'go'.

storm 1. A severe atmospheric disturbance with very strong winds. For example, a thunderstorm, rainstorm, or duststorm.
2. A wind of force 11 on the Beaufort wind scale.

storm beach A collection of coarse material, pebbles, and boulders at the upper edge of the beach profile, above the level of spring tides and often some metres above the land behind. The storm beach, which is built up by material thrown up by the strong forward movement of storm waters, remains there as a natural sea defence only breached in very exceptional storms.

storm surge Exceptionally high sea water, which may develop rapidly during a storm. It may occur when strong onshore winds push waves and water higher up the shore, exaggerating tidal levels, during already high spring tides. Water levels may be further increased by low atmospheric pressure. Serious flooding can result as on the East Anglian coast in 1953 when a 2.75 m surge was added to already high tides. The Thames Flood Barrier at Woolwich has been constructed to prevent the flooding of London during storm surges. *See also* spring tide.

stoss and lee topography A landscape found in formerly glaciated areas in which glacial till is deposited behind a resistant intrusive bedrock obstruction – the stoss. The local country rock may also be less eroded in the lee of the stoss. The feature is similar to crag and tail. *See also* crag and tail, till.

strait (straits) A narrow sea passage that links two larger areas of sea, for example the Strait of Gibraltar. *See also* sound.

strath A broad valley. The term is typically applied to flat-bottomed valleys in Scotland. [Scottish]

stratification 1. (*Biogeography*) †The layered arrangement of vegetation within a plant community. For example, mature woodland may show the following layers:
(1) scattered emergent trees
(2) general canopy
(3) understorey of shrubs and saplings
(4) herb or field layer
(5) ground flora (e.g. mosses)
This vertical structure exists because plants compete for, and are adapted to, the different light intensities available at different levels.
2. (*Geology*) The accumulation and division of sedimentary rocks into horizontal layers or beds known as strata. This is the primary structure of sedimentary rocks and is due to the deposition of layer upon layer of differing or alternating types of sediments. The term is sometimes applied to similar structures in igneous rocks.

stratiform Denoting layered or sheet clouds that are usually formed by widespread cooling in a stable atmosphere, e.g. stratus, nimbostratus, and stratocumulus.

stratigraphy (historical geology) The branch of geology concerned with the division of the rocks of the Earth's crust into systems, and the study and description of those divisions. It involves the study of the sequence of rocks in time and the subdivision of systems into series and groups, the character of the rocks, the fossil assemblages, and the correlation of rocks from different areas, both local and worldwide.

stratocumulus (Sc) A sheet or layer cloud, grey or whitish in colour, with rounded masses or rolls visible within the general layer.

stratopause A discontinuity in the atmosphere, at an altitude of about 50 km, where the temperature is about 0°C. Below this level, in the strato-

sphere, the temperature tends to increase with altitude, while above, in the mesosphere, it tends to decrease with altitude.

stratosphere That layer of the Earth's atmosphere that lies between the tropopause, at an average altitude of about 8 km, and the stratopause, at about 50 km. Within the stratosphere the temperature increases with altitude from about −60°C at the tropopause to about 0°C at the stratopause. Little weather is generated within the stratosphere as there is very little water vapour and virtually no dust present.

stratovolcano *See* composite cone.

stratum (*plural* strata) A layer or bed of sedimentary rock.

stratus (St) A low type of layer cloud with a fairly uniform base that forms at heights below 2400 m. It is generally grey coloured and is often thin enough for the disc of the Sun to be seen through it. The cloud sometimes produces a light drizzle.

stream A body of flowing water, ranging in scale from a rill to a river. †Streams typically run in stream channels, whose geometry closely reflects the type of *stream flow*, e.g. the average discharge, whether perennial (continuous) or ephemeral (intermittent), or of violent (flashy) or relatively even regime, and the size and frequency of flood flows. *See also* hydraulic geometry, regime.

stream number †The number of streams in each order for a given drainage basin. If the number of streams in each order is counted it is found that the stream number decreases with increasing stream order in a regular pattern; this is one of the *laws of stream numbers*. If stream order is plotted against stream number on semilogarithmic paper the points roughly form a straight line (i.e. an exponential curve). *See also* bifurcation ratio, stream order.

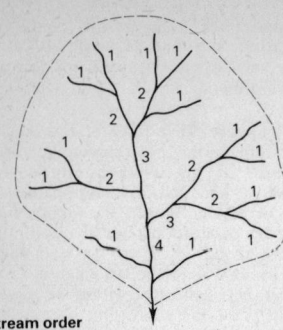

Stream order

stream order †A classification of streams in which a stream is given a number according to its position in a drainage network. The idea of stream ordering was originally proposed by R. E. Horton but the system most commonly used is that modified by A. N. Strahler. At the sources of the stream network, the first finger tip tributaries are termed *first order streams*; two first order streams join to make a *second order stream*; two second order streams join to make a *third order stream* and so on. Orders only increase when two streams of the same order meet so, for example, once a second order stream has been created no amount of extra first order streams joining it will affect its order. *See also* stream number, bifurcation ratio.

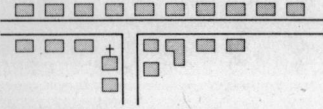

Street village

street village A linear form of village that has grown up along a line of communication. In England many of these developed along main roads in medieval times. A linear village may not necessarily have developed along a routeway; for example, the linear villages in the English Fens and Dutch polders grew along embankments and dykes avoiding the poorly drained areas on either side. *Compare* green village.

striations (striae) Grooves on exposed rock surfaces caused by the passage of rock fragments embedded in the base of ice sheets and glaciers. They are best preserved on fine-grained rocks that have been resistant to subsequent weathering and are often used to study the direction of ice movement during the ice ages.

strike The compass direction of a horizontal line along a rock stratum at right angles to the true dip. *See also* dip.

strike fault A fault in which the direction of movement is parallel to the strike of the layers of rock that it affects.

strike-slip fault *See* tear fault.

string bog †A poorly drained area found chiefly in tundra environments that consists of parallel chain-like ridges alternating with sedgy or pool-filled hollows. String bogs usually occupy gentle regular slopes, but are aligned at right angles to drainage. Solifluction or differential frost heave may be responsible for the formation of the feature.

strip cultivation (strip cropping) The growing of different crops on parallel narrow strips of ground, usually following the contour pattern of the land. This type of cultivation reduces the risk of soil erosion particularly if, on sloping ground, cover crops such as clover are alternated with crops that leave some of the soil exposed, e.g. maize. Strip cultivation is practised in hilly areas of the USA, e.g. in Oklahoma.

strip mining *See* opencast mining.

structure The arrangement and positioning of the rocks of the Earth's crust, including the hardness and texture of the rocks and their form before erosion.

structure plan In the UK, a plan that shows the framework for the development of an area (usually a county) for

10–15 years in the future. It deals with broad land-use issues, with regard to national and regional policies which affect it and the resources available for carrying it out. The plan is primarily a written statement of policy, with illustrations rather than a map-based plan. *See also* local plan.

subclimax vegetation †A plant community that is prevented from reaching a true climax vegetation. This is caused by localized natural features of the environment of a more or less permanent nature, such as persistently waterlogged soils or high ground exposed to wind, which prevent the growth of climax woodland. The subclimax community so produced differs markedly from the general climax vegetation of the area. Subclimax is thus a deflection from, rather than representative of, normal seral development. *Compare* plagioclimax vegetation. *See also* climax vegetation, sere.

subduction zone The area where one lithospheric plate (oceanic) is being overridden by another plate (either oceanic or continental) and pushed down into the mantle. This is usually marked by a deep oceanic trench and accompanied by volcanic and seismic activity. *See also* plate tectonics.

subhumid Denoting those transitional climatic regions lying between the semiarid and the well-watered areas of the world. Most of these areas also have seasonal rainfall, and this results in a grassland vegetation with very few trees; for example, the central savanna and the Mediterranean.

sublimation The change of state of a substance directly from solid to gas (or vice versa) without an intermediate liquid stage. In meteorology, it is most important for the formation of ice crystals directly from water vapour within the atmosphere, without an intervening liquid state, when the air temperature is below freezing.

sublittoral Denoting that area of the oceans lying between the low-water mark and the pelagic zone, i.e. that area that is affected by the land. It corresponds roughly with the continental shelf and continental slope. *Compare* pelagic.

submarine canyon A narrow deep depression in the sea floor with steep slopes, similar to a terrestrial canyon. Submarine canyons are normally found fairly near the coast, on the continental shelf and slope, and many seem to be extensions of present-day river courses. This has suggested that they may have been formed as river valleys, which have subsequently been drowned by a relative rise in sea level. Others, however, have no obvious connection with any land feature and their formation has not yet been adequately explained. Turbidity currents have been suggested as an alternative cause.

submarine ridge Any ridge on the sea floor that rises above the general level of the abyssal plain. The mid-ocean ridges are the largest, but other smaller ones may result from the drowning of a former land surface. *See* mid-ocean ridge.

submerged forest The remains of a former forest (e.g. peat, tree stumps) that now lies below the sea and may be uncovered at low tide along a sea shore. Submerged forests indicate that the sea level has risen in relation to the land surface. Examples occur off E England.

submergence The drowning of land by the sea, either as a result of a rise in sea level or a subsidence of the land. Submergent (drowned) features include rias, fiords, and Dalmatian coastlines. *Compare* emergence.

subpolar (cold temperate) Denoting those transitional areas that lie between the tundra towards the poles and the mid-latitude climates towards the equator. In the N hemisphere this includes much of N Asia and N North America, while in the S hemisphere it is almost entirely oceanic.

Summers are short and cool, winters are long and very cold. Precipitation is slight, but as evaporation rates are low as a result of the low temperatures there is sufficient moisture to support the great coniferous forests of the world.

subsequent stream A stream that follows a course roughly at right angles to the consequent streams that flowed down the original land slope. Subsequent streams develop during the erosion of gently tilting sedimentary rock strata where a more resistant capping is destroyed and rivers begin to excavate valleys along more erodible rocks, e.g. clay vales. *See also* consequent stream.

subsidence 1. (*Geology*) The sinking of a portion of the Earth's surface. It is caused naturally by subsurface erosion, either in caves or fissures in bedrock or through pipeflow and throughflow in the soil. Subsidence in peat lands is caused by the drainage, drying out, and oxidation of the organic material. Widespread subsidence may be caused by intervention by man; for example, by pumping out ground water for drinking, as around Tokyo, Japan, or by mining, as on Cannock Chase, Staffordshire.
2. (*Meteorology*) The downward movement of the air. Subsidence is widespread in anticyclones and the subtropical and polar high pressure belts in which air converges above and diverges near the ground. Subsidence dries the air (i.e. lowers the relative humidity) as it sinks into the higher pressure environment nearer the ground and warms up (by adiabatic warming). Persistent subsidence causes desert conditions. *See also* anticyclone, adiabatic warming.

subsistence agriculture (subsistence farming) A type of farming in which the output is consumed almost entirely by the farmer and his family leaving only a small proportion for sale. When there is a small surplus it is sold at a local market so that the family can buy basic provisions such as cooking pots and clothing. Subsistence farming is found in many countries of the Third World, such as India, Sri Lanka, and Nigeria. The crops produced are known as *subsistence crops.*

subsoil A layer at the base of soil, which underlies the topsoil. In general it corresponds with the B horizon in the soil profile. The term is commonly used in agriculture to describe the material not disturbed by normal ploughing. †Subsoil is less exposed to soil-forming processes. As a result it has a higher mineral content. It also has correspondingly less organic matter (humus), fewer soil organisms, and less penetration by plant roots. *Compare* topsoil. *See also* soil.

subtropical Denoting those areas lying between the tropics and approximately latitudes 40°N and S. They have distinct warm and cool seasons, unlike the tropical climates, and they are generally warmer than the temperate lands, having no month with a mean temperature below about 6°C. The Mediterranean areas of the world and the so-called 'cotton-belt climates', such as the SE USA, fall within this zone. The term is also applied to one of the important high-pressure belts of the world that lies within these latitudes.

suburb The outer part of an urban area that consists chiefly of residential housing. Suburbs are characterized by low housing densities and few other types of land use other than residential. They are dependent on the city for employment (hence a high level of commuting), shopping facilities, services, and other facilities.

succession, plant †*See* plant succession.

sudd A floating vegetation characteristic of the White Nile in N Africa. It consists of plants, especially papyrus, that have broken loose from surrounding swamps. It may extend for up to 25 km and accumulate to a depth of several metres considerably hindering navigation. [Arabic]

suffione *See* soffoni.

sugar beet A plant with a green foliage and large carrot-shaped white root, which grows best in moist mild climates. It is cultivated for the sugar content of its roots, which can be over 20% by weight. The juice is extracted and treated to produce sugar crystals; the green tops provide fodder and the dried pulp is also fed to livestock. The main growing areas are E and NW Europe and the central and W states of the USA.

sugar cane A tall bamboo-like grass from which sugar is obtained. The grass is mainly grown on plantations in tropical and subtropical regions where there is a long wet season followed by a cool dry season, during which the cane ripens and develops about 10–15% sucrose content. The canes, 3–8 m high, are cut and crushed between rollers to obtain the sugar solution. This is refined to produce sugar crystals; a sideproduct, molasses, is used as an animal feedstuff and is fermented to produce rum. The main producing countries include Cuba, Brazil, Peru, South Africa, Indonesia, India, Australia, and China.

sugar loaf A steep-sided mountain with a rounded crest, for example Sugar Loaf Mountain in Rio de Janeiro, Brazil. It is a type of inselberg, i.e. a residual hill formed by deep tropical weathering. *See also* inselberg.

sumatra A sudden strong squall that occurs mainly at night during the SW monsoon in the Strait of Malacca, Malaysia. It is usually accompanied by violent thunderstorms, a sharp drop in temperature, and a shift in the wind direction from southerly to westerly.

summer The warmest season of the year. In the N hemisphere it is defined as extending from about 21 June, the summer solstice, to about 22

September, the autumn equinox; in the S hemisphere it extends from about 21 December to about 21 March. It is popularly assumed to include the months of June, July, and August in the N hemisphere. *See also* season.

summit The highest point on a hill, mountain, or line of communication, such as a railway.

sump An area that collects water, especially a small pit at the bottom of a mine or excavation. In caving, a flooded area of stationary water often filling or nearly filling a section of cave.

Sun The central body in the solar system and the Earth's nearest star. The Sun is a sphere composed of at least 60 elements in a gaseous state, of which hydrogen and helium predominate. It has a temperature of about 6000°C at the visible surface and about 20 000 000°C at the centre, where thermonuclear reactions are converting hydrogen into helium and radiating the energy that supplies the light and heat essential for life on Earth. †The Sun has a diameter of 1 392 000 km, and a mass of 1.990×10^{30} kg, and is a mean distance of 149 600 000 km from the Earth. It is likely that variations in solar radiation cause climatic changes on Earth, especially the 11-year cycle of sunspot maxima, which have been correlated with rainfall maxima.

sunshine That part of the Sun's electromagnetic radiation lying within the visible spectrum that reaches the Earth's surface. It is essential for life and therefore its strength and its duration are of importance. The duration varies greatly from place to place and depends on several factors, the most important being latitude and the cloudiness of the sky. It is measured by a sunshine recorder – the Campbell–Stokes recorder – and can be measured either in hours per day (or any other convenient period of time), or as a percentage of the total possible sunshine at the particular place. The duration can be as high as 80% in cloudless tropical deserts, or as low as 20% in cloudy west coast temperate areas.

supercooling The process whereby water can be cooled well below its freezing point (0°C) and yet remain in the liquid state. This is common in the atmosphere as a result of the rarity of freezing nuclei and is important in producing certain weather phenomena. Supercooling is fundamental to the Bergeron–Findeisen theory of precipitation formation. When supercooled droplets come into contact with a cold surface they instantly freeze; this process is responsible for the formation of 'icing' on aircraft wings and of rime.

superficial deposit A deposit of material laid down above the solid bedrock as a result of having been transported to its present position and deposited by geomorphological processes. Superficial deposits include glacial and fluvioglacial drift, alluvium (e.g. river terraces), estuarine and deltaic deposits, and wind-blown sand and loess. Sometimes deposits such as clay-with-flints and other deposits that result from weathering of the bedrock *in situ* are called superficial deposits although this is not strictly correct.

superimposed drainage (superimposition) A river network that runs illogically across resistant and nonresistant rocks alike. The river originated on rocks covering those presently exposed and has maintained its original route whilst downcutting into the different underlying rocks and structures. Only superimposed drainage that retains this discordance with structure and lithology can be readily recognized as such; other cases of superimposition may not be obvious because they appear accordant. *See also* discordant drainage.

supersaturation The condition when the air is cooled to below its dew point without condensation taking place, i.e. the air contains more water vapour than is required to saturate it. This is rare in nature as the air nearly always contains nuclei (dust, smoke, salt, etc.) that encourage condensation, but it can occur in very pure air, especially at very high altitudes.

surf A mass of foaming water made by waves breaking over rocks or in shallow water.

surface flow *See* overland flow.

surface wave *See* earthquake.

survey The measuring of features on the Earth's surface to obtain data from which a map, chart, or plan can be made. The major surveying methods include triangulation, traversing, chain surveying, plane tabling, and levelling. In the UK, the government department that carries out this work is known as the Ordnance Survey.

suspension A method of transport of solid particles in which they are carried in a fluid, such as moving air or water. *See also* suspension load.

suspension load (suspended load) That part of the material transported by moving water, such as a river, that is carried by the fluid. The size of particles that can be carried in suspension increases with the velocity of flow and its turbulence and tractive force. Some particles may only remain in suspension for short distances. Generally suspension load accounts for 70–90% of a river's total load of solids.

swallow hole (swallet) A vertical shaft in limestone rock down which a surface river disappears, frequently as a waterfall, creating an underground stream. For example, Gaping Gill at Ingleborough, North Yorkshire. *See also* sinkhole.

swamp An area of persistently muddy and waterlogged ground. Swamps occur, for example, along the margins of streams and ponds or where drainage collects. In the British Isles swamps usually have a rich growth of herbs, grasses – especially reeds (*Phragmites*) – sedges, and rushes. Trees are generally absent. *See also* swamp forest.

swamp forest Vegetation of persistently waterlogged ground that develops a tree cover. Swamp forest occurs in North America, and includes such species as the swamp cypress.

swash The forward movement of sea water up the beach after the breaking of a wave. Some of the water percolates into the beach and the rest flows back down the beach as backwash. The turbulent water washes sand and gravel up the beach. The most effective transport of sand occurs when waves are relatively flat and the circulating motion within them is elliptically flattened with more force directed up the beach. *Swash channels* are created by water returning down the beach and scouring a channel. When these are formed on the ebb tide they remain after the tide has gone out. *See also* backwash.

S wave *See* earthquake.

swell The regular undulating movement of the surface of the open sea. It results from waves that have moved out of the area in which they were generated and have ceased increasing in size. Their wavelengths become longer and their previously sharp crests become rounded and flattened.

symbiosis A close association between two or more different organisms, plant or animal, which are dependent upon one another. For example, the algal/fungal relationship comprising lichen, in which the association is constant and indispensable – neither organism can exist separately. Symbiotic relationships usually bring benefits to both parties. Together with predation and competition, symbiosis is a fundamental way in which organisms interact within a community.

syncline A basin-shaped structure in sedimentary rocks in which the strata dip towards each other. Synclines alternate with anticlines and result from compressional forces acting on the rocks. In strongly folded rocks the limbs of the syncline are steep but some synclines form broad shallow basins, for example the London Basin. *Compare* anticline. *See* fold.

synoptic chart *See* weather chart.

system †A set of objects or attributes that are linked together in some discernible relationship. The concept was introduced into geography during the 1960s from other sciences, where it had already been widely used. Virtually all systems in geography are *open systems*; i.e. energy and matter are able to cross the boundaries of the system. The open system requires an energy supply to maintain it. It is self-regulating and tends to progress towards a condition of stability known as *steady state*, regulating itself through homeostatic (equilibrating) adjustments (i.e. feedback mechanisms). Over periods of time the system maintains optimum magnitudes. The open system also behaves equifinally, i.e. different initial conditions eventually result in similar end results. Examples of open systems in physical geography include the plant-soil system, a drainage basin, and the hydrological cycle. In human geography an example is a city region with such features as land-use zones, settlements, and transport links that involve the movement of people, goods, services, and wealth into and out of the region. A *closed system* is a system in which there is no exchange of energy across the boundaries. *See also* cascading system, morphological system, process–response system.

system of cities †A set of cities and towns that are closely linked together with social and economic changes in one having major repercussions for other places in the system. Implicit in the concept of a system of cities is recognition of a level of interdependence between its component parts.

systems analysis †The identification and interpretation of a set of objects, their functions, the way they are arranged, and the structured relationships and organizations linking them to form a system. In systems analysis the conventional subject boundaries are crossed as interrelationships within a system are explored. *See also* system.

T

tableland A raised plateau bounded by steep cliff-like slopes. For example, the South African tableland.

tacheometer An instrument used in surveying for the rapid measurement of distances. It consists of a theodolite adapted for the measurement of distances as well as horizontal and vertical angles. *Tacheometry* is the measurement of distances and angles using a tacheometer.

taiga The coniferous forest that covers large tracts of Siberia (N Asia). The region lies between tundra to the N and steppe to the S. The term is often also applied to the coniferous forests of North America. *See* coniferous forest. [Russian]

talus Weathered rock fragments accumulated at the base of a slope below a cliff face. The term is often regarded as being synonymous with scree and may be applied to all sizes of debris, including silt and clay. *See* colluvium, scree.

talweg †*See* thalweg.

talwind A wind that blows up a valley, i.e. an anabatic wind. It is the reverse of the berg wind. [German: valley wind]

tank A cistern-like feature, water-storage pool, or reservoir, particularly one used for small-scale irrigation such as those in India and Sri Lanka. †Recent models of river systems often consider that stream flow is generated by a series of reservoirs or tanks that store a certain amount of water before they overflow. For example, the soil cover may be considered as one tank, which can store a certain amount of rainwater. Such schemes are called *tank models*.

tanker A ship or motor vehicle designed to carry liquid in bulk, especially crude oil. Large ocean-going tankers are more economical than small ones and the size has been increasing rapidly in recent years with six tankers in the world's fleet each able to carry more than half a million tonnes of crude oil. Some 30 more can carry over 400 000 tonnes.

tariff A tax imposed by a country on commodities that are being imported or, less frequently, exported. For example, in the UK there is a tariff on imported cars. One of the aims of the EEC is the elimination of tariffs between member nations.

tarn A small lake formed in a steep-sided hollow or cirque at the head of a former valley glacier. The hollow is enlarged and deepened by rotational slipping of the ice with shattered rock fragments embodied in its base. This typically leaves a rock threshold, behind which water collects once the ice has thawed. Some tarns are also ponded up by ridges of moraine deposited at the final stage of melting of the glacier. *See* cirque, moraine.

tar sand *See* oil sand.

tea The dried leaves and shoots from an evergreen shrub (*Camellia sinensis*), from which the beverage is prepared. The shrub grows best on hillsides where rainfall is about 1300 mm and temperatures are above 22°C during the growing season. A number of pickings by hand takes place during the growing season so a large labour force is required. In Assam (India) and Sri Lanka, the chief exporters of tea, it is mainly grown on large plantations whereas in China and Japan it is grown on smallholdings. Two main types of tea are produced: Indian tea, which is allowed to ferment, and China (or 'green') tea, which is only partially fermented. It is also grown in other parts of SE Asia, S Africa, and parts of South America.

tear fault (strike-slip fault, transcurrent fault, wrench fault) A fault in which the displacement of the rocks is chiefly horizontal rather than vertical. The San Andreas fault of California is an example of a tear fault.

tectonic Denoting the forces responsible for the deformation of the Earth's crust, including the growth of mountain ranges and volcanism, and the landforms, structures, etc., resulting from these forces, e.g. tectonic earthquakes. *See also* plate tectonics.

temperate One of the original Greek divisions of the world based on temperature, lying between the torrid and the frigid zones, and meaning an area where there are no extremes of temperature. The temperate zone is now known more accurately as the mid-latitude area, because only parts of it are truly temperate, the interiors of the continents having very extreme temperatures. The parts of the world that are now described as being temperate are the mid-latitude areas of the S hemisphere, where there are no great land masses, and the mid-latitude oceans and coastal areas of the N hemisphere. (In the Köppen climatic classification, the temperate areas are prefixed by the letter C.)

temperature An index of the degree of heat of a body, substance, or medium such as the atmosphere. It is a fundamental component of meteorology and climatology and every weather station takes various temperature readings every day. The actual temperature of a particular place at a particular time depends on a number of factors, including latitude, time of year, time of day, sky cover, wind, altitude, proximity to the sea, etc. It is measured by a thermometer. Several *temperature scales* have been devised; the Celsius

(centigrade) scale is the one generally used internationally in meteorology. Other scales include Fahrenheit, kelvin (used in scientific work), and absolute. The global range of recorded temperature is from a low of about $-53°C$ in Antarctica to a high of about $58°C$ in the Sahara. *See also* absolute temperature, Celsius scale, Fahrenheit scale, kelvin.

temperature inversion *See* inversion of temperature.

temporary base level †A limiting level down to which a river will erode, but which it cannot exceed until the cause of that limitation is removed. For example, a lake will act as a local base level for tributary streams and slope development until the outlet of the lake is lowered and a new lower base level takes control. Sea level is the only absolute base level, although in practice this changes over geological time. *See* base level.

tension A stress produced in a body as a result of two forces pulling against each other. Such forces are set up in igneous rocks during cooling, giving rise to tension joints. Tension forces in the Earth's crust produce normal faults and rift valleys. *Compare* compression.

terai In E India and Bangladesh, a belt of marshy ground and vegetation on the lower parts of alluvial fans. Much terai has been cleared for human settlement. [Urdu/Hindi]

terminal moraine (end moraine) A ridge of material (chiefly till) deposited against the snout of a glacier and left behind as the glacier retreats. It marks the maximum extent of the glacier or ice sheet. Its composition varies from fine clays to coarse boulders. *See* moraine, till.

termite mound The nest, made from mud or plant debris, that houses a colony of termites (a tropical type of ant). Millions of individuals in a highly organized social system inhabit such mounds, which rise several metres from the ground surface. They appear to improve the structure of the soil but also cause some soil creep.

terms of trade The relative level or ratio between the price obtained for exports compared with the price obtained for imports. A fall in the relative price of imported raw materials will tend to improve the terms of trade for a manufacturing country like the UK. By contrast, countries of the Third World exporting raw materials are likely to experience a decline in the terms of trade unless an immediate fall takes place in the price of the manufactured goods they import.

terrace A relatively flat strip of land that forms a shelf or bench. *River terraces* usually occur paired, one on either side of the channel, and are the remains of a former alluvial floodplain that has been dissected by renewed downcutting by a river. This may result from rejuvenation of the river or from a climatic change. Subsequent terraces may form steps on either side of a river, for example, the River Thames has three recognized pairs of terraces. †*Solifluction terraces* are formed under periglacial conditions by the slippage of loose material down a hillslope. These can be confused with river terraces.
See also wave-built terrace.

Terraced cultivation

terrace cultivation The growing of crops on level steps or terraces that have been constructed on hillsides. Banks or walls are built to hold the soil in place and, where paddy is grown, to enable water to stand on the fields. Terrace cultivation takes place where land is scarce, e.g. China, Indonesia, Japan, or where rainfall is limited, e.g. S Italy, Greece.

terrain An area of land considered in terms of its physical characteristics, hence rough terrain or smooth terrain. †*Terrain analysis* is based on the scientific classification of landscape units considering the characteristics of the soils, slope angles, geomorphic context, and processes of development.

terra rossa A red intrazonal soil that occurs on limestones in Mediterranean climates. It generally has a clay-loam texture. Terra rossa soils are the insoluble residues from the chemical breakdown of limestone. †Where these residues accumulate, in crevices and depressions, the dissociation of clays produces ferric oxides, which are responsible for the red coloration, and leaching causes acidity. The development of terra rossa thus requires a wet season of moderate rainfall.
Compare rendzina. [Italian]

terra roxa A deep dark reddish-purple coloured soil that is formed over a parent material of basic igneous rocks in Brazil, especially in São Paulo state. It is rich in humus and is used for coffee plantations. [Portuguese: purple soil]

terrigenous deposit A deposit on the ocean floor that has been derived from the land masses bordering the sea. With the exception of very fine wind-blown material, these deposits are confined to the sublittoral zone. There tends to be a grading of material, the coarsest being deposited first, near the shoreline, and the finest being carried further from the land, towards the edge of the pelagic zone.
Compare pelagic deposit.

territorial seas (territorial waters) The strip of water surrounding the coast of a country and over which full political control can be exercised. A 3-nautical mile (5-kilometre) strip was originally designated by international law but during the last 30 years more and more states have extended their territorial seas, most to 19.5 km (12 miles). This results from the desire of many countries to secure offshore resources, such as fisheries and oil deposits. For example, the North Sea was divided up following the discovery of extensive oil deposits.

territory 1. an area of land.
2. An area of land subject to the administration of the country in which it lies.
3. A region of a country that enjoys less autonomy than most of the country's regions, especially in a federal state, e.g. Yukon Territory in Canada and Northern Territory in Australia.

Tertiary The first geological period of the Cenozoic era, extending from the end of the Cretaceous period, about 65 million years ago, to the beginning of the Quaternary period, about 2 million years ago. This system was formerly regarded as comprising one of the four eras, divided into the four periods of Eocene, Oligocene, Miocene, and Pliocene, but it is now more commonly regarded as being a period. It is divided into the Palaeocene, Eocene, Miocene, and Pliocene epochs. During the Tertiary period modern invertebrates and mammals evolved. *See also* Eocene, Miocene, Oligocene, Palaeocene, Pliocene.

tertiary industry Those types of employment that are concerned with transport, trade, communications, the professions, administration, and other services, and not with farming, extractive, or manufacturing industries. In the UK 59% of the workforce is employed in tertiary industries, many more than in manufacturing or primary industries. It is less extensive in Third World countries, e.g. in Thailand only 17% of the workforce is employed in tertiary industries. *See also* primary industry, secondary industry.

texture, soil *See* soil texture.

thalweg (talweg) †The line of a valley bottom. In a river channel the thalweg is the line of the deepest part of the channel bed and the term is used to denote the long profile of the river. In

geomorphological mapping it is the meeting point between two approaching and opposite slopes. [German]

thaw 1. The change of physical state from ice to liquid water that occurs when the temperature rises above freezing point (0°C).
2. The period during which snow, ice, and frozen ground melt. This may refer to a relatively small event in temperate areas after a cold spell or, in high latitudes, to the end of the long winter freeze.

thaw lake A lake that occupies a depression caused by subsidence when a mass of buried ice or frozen ground in permafrost areas melts.

theodolite An instrument used in surveying for measuring angular distances in both the vertical and horizontal planes. It consists of a telescope that can be rotated in both the horizontal and the vertical planes. It is normally mounted on a tripod and must be levelled before use. It is extremely accurate and all readings are made by means of verniers or micrometers. It is used for triangulation surveys.

thermal In meteorology, a rising current of air (convection current) in the atmosphere. It is caused by the intense heating of the ground and the subsequent heating of the air above it. The hot air rises, and as it does so, it cools and may produce clouds, rain, and thunderstorms.

thermal electricity Electricity that is obtained from burning fossil fuels to make steam, which in turn drives generators. The fuels used in the UK are coal, oil, and natural gas. In some countries peat and lignite are also available. Of the thermal electricity used in the UK, 73% comes from coal, 15% from oil, and 0.5% from natural gas.

thermal equator An imaginary line around the world that joins the places on each meridian having the highest mean surface air temperature for any particular period. As land areas tend to absorb more heat than the oceanic areas the thermal equator normally lies to the N of the geographic equator, because of the imbalance of land and sea areas between the N and S hemispheres. It shifts continuously, but its major movement is seasonal, following the apparent N–S movement of the overhead Sun. It frequently crosses the geographic equator, looping northwards over the land masses and southwards over the oceans, especially during the S summer.

thermal spring *See* hot spring.

thermograph A self-recording thermometer. It consists of a heat-sensitive mechanism to which is connected a pen resting on a rotating drum; this produces a continuous trace of the changing temperature (a *thermogram*). The most common type of thermometer used is the bimetallic coil type, which expands and contracts with rising and falling temperature, the movement being transmitted through a lever to the pen. *See also* thermometer.

thermokarst †A landscape in periglacial areas that resembles the true karst landscape of limestone regions, but which is formed by the melting of buried ice masses left by retreating glaciers and ice sheets. The thermokarst consists of an uneven surface formed in glacial till, and includes kettle holes, thaw lakes, and tunnels formed by the meltwaters. *See also* kettle hole, thaw lake.

thermometer An instrument for measuring temperature. There are several types, the most common consisting of a sealed glass capillary tube containing a liquid (usually mercury or alcohol), which is fed by a reservoir at the base. As the temperature rises the liquid expands and rises up the tube; as it falls the liquid contracts and moves down the tube. The tube is graduated in the required temperature scale, and the temperature is read directly from the meniscus of the liquid. †Other

types of thermometer include the bimetallic thermometer, which depends on the expansion and contraction of thin strips of metal with temperature changes, and the resistance thermometer, in which changes in the resistance of conductors (e.g. thin strips of metal such as platinum or copper) to the passage of an electrical current as a result of changes in temperature are measured.

thermosphere The uppermost layer of the atmosphere, extending from the mesopause, at an altitude of about 85 km, to the outer limits of the atmosphere. Within it the temperature increases with altitude from about −100°C at the mesopause to over 150°C. The thermosphere forms part of the ionosphere. *See also* ionosphere.

Third World The underdeveloped countries that make up most of the continents of Africa, Asia, and central and South America. These countries contrast economically with the developed rich advanced industrial countries of Western Europe, North America, and Japan, and with the countries of the Sino-Soviet bloc with their managed economies.

thorn forest A type of tropical and subtropical vegetation of semiarid areas that is dominated by dense but scrubby trees, the leaves of which are modified into narrow thorns. Some thorn forest is deciduous. These xerophytic features enable the vegetation to survive low and unreliable rainfall (about 250–500 mm per year). Thorn forest may be part of a mainly grassland vegetation (e.g. caatinga in Brazil) or transitional between savanna and true forest.

Thornthwaite climatic classification †A division of the world into climatic areas devised by C. W. Thornthwaite in 1931. Unlike the Köppen classification and most other climatic classifications, it is based on the efficiency of the precipitation rather than on temperature considerations. It is based on two climatic factors, a precipitation efficiency index and a thermal efficiency index. The system is complex and is rarely used in practice.

threshold In urban geography, a minimum level of demand that is required to maintain a particular activity. It is often expressed in terms of population numbers. It is largely a theoretical concept since it is not easy to evaluate the minimum level required for the establishment of a function. The *threshold population* (demand threshold) is the minimum number of people required to support a central place function or service. 'Convenience' goods such as basic food items have low threshold populations and are consequently commonly found in small local centres as well as in central shopping areas. 'Shopping' goods such as furniture, large items of clothing, and major electrical goods require a higher threshold population and compete for sites in the higher order shopping centres in the town.

throughflow The lateral movement of water through the soil above the level of the water table. †*Unsaturated throughflow* occurs when the soil pores are not full of water and the movement takes place through the thin films of water covering the soil crumbs. *Saturated throughflow* occurs when the soil matrix is saturated with water and is usually the only form of throughflow capable of transmitting water quickly enough to contribute stormflow to the streams.

throw The vertical change of level of rock strata as a result of faulting. The *downthrow* is the block of rock that has been displaced downwards following faulting; the *upthrow* is the block that has been displaced upwards.

thrust fault A type of reverse fault in which the fault plane is inclined at a low angle (i.e. between 45° and the horizontal). The upper rock strata move along the fault plane over the lower rock strata, sometimes for many kilometres. This type of fault results from strong compressional forces and

is commonly developed in the overthrust recumbent folds that are characteristic of the Alps. *See also* nappe.

thrust plane The fault plane running down the middle of a recumbent fold or nappe structure and along which the upper limb of the fold has been pushed forward over the lower limb. *See also* nappe.

thunder The sound that follows a lightning flash. It is produced when the sudden intense heat generated by a flash of lightning causes the air to expand explosively. The sound is always heard following the lightning flash as the speed of sound waves is considerably slower than that of light. The distance between an observer and a thunderstorm can be measured by counting the seconds between the flash and the thunder; a time difference of 3 seconds indicates that the source is about 1 km distant. Thunder is rarely heard at a distance of 30 km.

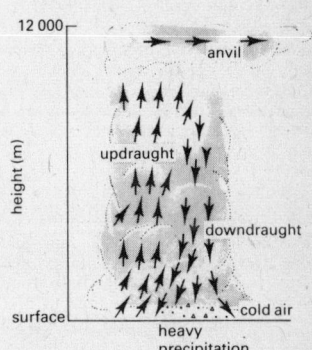

Thunderstorm

thunderstorm A storm accompanied by lightning and thunder that develops in large cumulonimbus clouds, which result from the rapid ascent of air under very unstable conditions. Such conditions normally occur either at the cold front of a depression or when the ground is intensely heated, and there must always be sufficient moisture in the air for cloud formation. The equatorial areas produce the high-

251

est frequency of thunderstorms as the air is usually very humid and the hot sun produces the necessary upcurrents of air. Very heavy rainfall or hail occur, accompanied by the discharge of electrical energy, producing lightning and thunder. The processes leading to the generation of such huge electrical charges within the clouds are yet to be fully understood, although there have been many attempts to explain it since Benjamin Franklin first demonstrated in 1752 that lightning is an electrical phenomenon. *See also* lightning.

tidal current The horizontal movement of the waters of the sea caused by the tide. It is most noticeable when the movement of the water is restricted; for example, in an estuary, bay, or strait. It can reach a considerable speed when conditions are right (over 22 km per hour in some Norwegian fjords). Such currents are reversible, moving one way with the flood tide and the other with the ebb tide. While some tidal currents are useful for 'scouring' silt out of shipping channels, others can be strong enough to hinder the movement of shipping. In some river estuaries the meeting of a strong tidal current and the river flow will produce a tidal bore. *See also* bore.

tidal range The measured difference between the water levels at high and low tide at a particular place and time. It is very variable, both from place to place and from time to time at a particular place, and depends on such factors as the phase of the Moon, the latitude, the shape of the coastline, etc. Enclosed seas (e.g. the Baltic Sea) have a very small range (less than 2 m), while a narrowing estuary or bay that opens into the ocean may have a very large range. For example, the Bay of Fundy in E Canada has a tidal range of over 19 m.

tide The slight oscillations of sea level that occur approximately twice a day

and attain exaggerated proportions in marginal seas, straits, and estuaries. Tides are extremely complex but their major cause is the gravitational pull of the Moon and, to a lesser extent, of the Sun. As the Moon travels in its orbit in the same direction as the Earth's rotation a period of 24 hours 50 minutes elapses between two successive occasions when the Moon is vertically above a point. High and low tides each occur twice during this period, giving an interval of almost $12\frac{1}{2}$ hours between successive high (or low) tides. The highest tidal range in open oceans occurs when the Sun, Earth, and Moon are aligned (at new and full Moon) and the gravitational pulls of the Sun and Moon are combined; the lowest range occurs when the Sun, Earth, and Moon are at right angles and the gravitational pulls are in opposition. Near coastlines this relatively simple pattern is complicated by the effects of the morphology of the coastline. Each part of the coastline has features that produce different effects. For example, small tidal ranges occur in enclosed seas, high tidal ranges in estuaries leading to the open sea, and double tides around offshore islands (e.g. at Southampton and around the Isle of Wight). *See* neap tide, spring tide.

tierra In Spanish-speaking countries, an area of land. In tropical South and Central America, the landscape is often divided into three climatic regions according to altitude:

(1) The *tierra caliente* occupies the hot humid lowlands below about 900 m and has a natural vegetation of tropical rainforest.

(2) The *tierra templada* extends between about 900 m and 2100 m and has considerably less rainfall and less luxuriant vegetation, although temperatures are relatively warm and uniform.

(3) The *tierra fria* lies above 2100 m and has a vegetation rising from coniferous forests to grasslands on the highest ground.
[Spanish: land]

till (boulder clay) Mineral material deposited by glaciers and ice sheets. Till is derived from glacial erosion of loose deposits and bedrock and varies in composition from fine clays (rock flour) to coarse boulders. It is typically unstratified. Vast amounts of till were deposited in Europe and the British Isles during the Pleistocene ice ages and form the parent material for many soils developed since. *See also* moraine.

tillage Land used for growing crops, as distinct from pastureland, or the process of cultivating land so that it can be used for raising crops.

till fabric analysis †A method of determining the direction of former ice flows based on analysis of the size, shape, and orientation of coarse rock fragments in a glacial till deposit. *See* till.

tilt block A section of rocks that has been lifted and tilted relative to neighbouring strata. The blocks are typically large features forming mountain ranges bounded by faults, i.e. tilted block mountains or horsts. Examples of tilt blocks include the North Pennines, which is tilted towards the North Sea, and the Central Massif of France. The Great Basin ranges in the USA contain many tilt-block mountain ranges. *See* horst.

timber industry The occupation of obtaining timber from the world's forests. This is carried out in a variety of ways depending on the character of the forest, the relief, and other environmental factors. It is a form of 'robber economy' unless conservation measures are enforced and planting keeps pace with felling. Two main types of timber are obtained – hardwoods, which are obtained chiefly from broad-leaved trees, and softwoods, which come from coniferous trees. Timber has widespread uses, e.g. as a fuel, in construction, and as a raw material in the production of paper (as wood pulp), turpentine, resins, and cellulose (for paints and rayon).

timber line *See* tree line.

time *See* Greenwich Mean Time, local time, standard time.

time zone *See* standard time.

tin A silvery white metal. It is rarely found in its native state and is derived chiefly from cassiterite (tin oxide, SnO_2). Cassiterite occurs in veins associated with granite in Cornwall in the UK and Bolivia, where it is found in lodes in association with silver. Most of the world's tin ore is obtained from placer deposits, which result from the weathering of tin veins. Tin is a soft weak metal with a low melting point. It is used in the manufacture of tin plate (which is used for food canning); in the production of alloys such as pewter, gunmetal, and bell metal; and in solders. The chief tin-producing countries include Malaysia, Indonesia, Thailand, the USSR, Bolivia, Zaïre, and Australia.

tobacco A plant with large oval-shaped leaves that is best grown in subtropical regions but can also be grown in mid-latitudes as well as the tropics, although the quality and quantity vary according to climatic and soil differences. The leaves are dried and cured in large barns for use in cigarettes, pipe tobacco, cigars, and snuff. The USA is the largest producer with North Carolina, South Carolina, Virginia, and Kentucky as the main states. The other major producing regions are SE Europe, China, India, the USSR, and Brazil.

tombolo A form of coastal spit comprising a ridge of shingle or sand that joins an island to the mainland. Tombolos tend to grow in shallow straits with weak currents and tideless seas, where waves striking the shore at an acute angle cause longshore drift and build up spits. Two tombolos may develop to form a *double tombolo*, which links an island to two points on the mainland and impounds a lagoon in between. Chesil Beach, which joins the Isle of Portland to the mainland in Dorset, is an example of a tombolo; there are also many good examples in the Mediterranean. [Italian]

topography The surface features (i.e. landforms) of an area of land or sea bed. The term was formerly used for the description of both the physical and human features of an area.

A *topographic map* is a map that represents the form of the Earth's surface. [From Greek *topos*: a place]

topology A form of geometry that is concerned with the position and relationships between points, lines, and areas, but not with the distance between points, the straightness of lines, or the size of areas. Topology is used to look at the relationships between places and lines in a transport network. Examples of *topological maps* include the London Underground and British Rail Intercity maps, which enable routes to be clearly seen. *See also* network.

topsoil The well-developed part of the soil at or near the ground surface (i.e. the A horizon). Topsoil is the vital zone for plant growth. The term is generally used for the zone above the subsoil that is moved during cultivation. *Compare* subsoil. *See also* soil.

tor An upstanding body of rock projecting from a hill or gently undulating land surface. Tors take various forms. Examples in the British Isles include the granite tors on Dartmoor and Bodmin Moor, which consist of roughly rectangular blocks with rounded edges, and the gritstone tors formed on the Millstone Grit in Derbyshire. Some blocks rest precariously on others. †D. L. Linton suggested that tors were the remnants of the basal surface of deep chemical weathering, guided by the joint pattern (the least jointed rocks being the least weathered), such as occurs in the present tropics. The loose overlying material was subsequently removed after a change of climate, such as a peri-

glacial climate in glacial periods. More angular tor forms may be formed by frost shattering in a periglacial climate alone. More recent observations suggest that small tors on the upper slopes of the Millstone Grit scarp in Derbyshire may be the result of erosion by seepage waters under a climate similar to that of the present-day.

tornado A violently rotating storm in which winds whirl around a small area of extremely low pressure. It has a diameter of only 50–100 m, and is characterized by a funnel-shaped cloud. It is caused by a very strong updraught of air (convection), and hence is often associated with · thunderstorms. Tornadoes occur most frequently, and are most violent, across the central plains of the USA, where they are sometimes known as twisters. Although they only last about an hour and travel about 50 km, they do tremendous damage along their narrow trail. The meeting of cool dry air from the plains and warm moist air from the Gulf of Mexico provides the ideal conditions for the development of thunderstorms, and the tornadoes form within this area of intense convection. Two factors give it its great destructive power: winds that are thought to reach speeds of 300 km per hour and pressure that is so low that closed up houses have been known to explode because of the great pressure difference between the inside and out. An *African tornado* forms in a similar way when warm moist air from the Gulf of Guinea meets very dry air from the Sahara along the coast of W Africa. However these are less severe forming a squall, which is usually accompanied by a thunderstorm and torrential rain. [From Spanish *tronada*: thunderstorm]

torrid zone One of the three original Greek divisions of the world based on temperature, the others being the temperate and frigid zones. The torrid zone was that part of the Earth's surface lying between the tropics. The term is no longer used.

tourism The business of providing accommodation and other facilities for people who are travelling through or visiting a locality for pleasure. In some UK towns, such as the coastal resorts, tourism is the main industry with many people being employed by the hotels, entertainment centres, etc. A great deal of valuable foreign currency is earned from tourism by countries such as Tanzania, Spain, Thailand, and Jamaica.

town 1. In general, a compact settlement larger than a village and smaller than a city that is engaged chiefly in nonagricultural occupations. There is no universally agreed definition of a town in terms of population size to distinguish it from a city or village. It is essentially an urban settlement with a business centre that serves its own inhabitants and others who live outside its boundaries.
2. In the USA, a unit of local government that is smaller than a county.

town plan The shape and layout of a town. Analysis of the town plan is one of the oldest aspects of urban geography.

townscape The physical form of the spaces and buildings that together compose the urban landscape. †A more acceptable current definition in urban geography is the whole objective visible features of the urban area, or the total subjective 'image of the city'.

township 1. In the highlands and islands of Scotland, a crofting settlement composed of the individually held croft holdings and the common grazing area. A township may contain up to 50 crofts.
2. In the USA, a local administrative subdivision of a county. The term is also used in the newer states and in parts of Canada for an area of approximately six miles that was delineated in the government survey.
3. In Australia, a village and a possible future town.

4. In South Africa, a planned urban settlement of Blacks or Coloureds.

trace element A chemical element that is essential to organisms, but only in minute quantities. In soils, trace elements such as copper, zinc, boron, and iodine are derived naturally from parent material, or may be introduced into the environment in fertilizers.

trade route A route by land, water, or air that is used for trading purposes. The busiest trade route in the world is that across the N Atlantic between Europe and North America. Railway and road networks provide important overland trade routes, e.g. the Trans-Canada Highway and the Hong Kong –Canton railway. The cargo services between Tenerife in the Canary Islands and London, carrying early tomatoes and other fruit, provide an example of a trade route by air.

trade winds The belts of winds that blow in the tropics from the subtropical high-pressure zones towards the equatorial low-pressure trough. In the N hemisphere the prevailing wind direction is from the northeast and hence the winds are known as the *northeast trades*; in the S hemisphere the winds are predominantly from the southeast and are known as the *southeast trades*. The winds are noted for their consistency, both in force and direction, in many areas, especially over the open oceans. However, they are disrupted in coastal areas, where local temperature and pressure differences become the dominant factors, and in monsoon areas (e.g. SE Asia), where they may be completely overwhelmed. Their positions shift about 5° latitude with the seasons, following the apparent N–S movement of the overhead Sun. It is within these wind belts that tropical cyclones develop; these move away from the equator into subtropical and even mid-latitude areas. [Derived from the nautical phrase 'to blow trade', i.e. on a constant course]

trading estate An area of land set aside for the development of small factories and warehouses. In the UK a number of these estates have been set up with government funds to encourage industrial development in areas of high unemployment, or to offer attractive alternative sites to industries in congested locations. The estates are provided with local services such as gas and water and are located close to urban areas where labour is available, e.g. Treforest in S Wales.

tramontana A cold dry northerly wind experienced in winter in the W Mediterranean area, caused by cold air being drawn in behind a depression. It is fairly common in Corsica where it brings snow and very cold conditions, and it may affect places as far W as the Balearic Islands. [Spanish and Italian]

transcurrent fault *See* tear fault.

transect, urban *See* urban transect.

transform fault †A strike-slip fault that occurs within the Earth's crust orientated at right angles to a mid-ocean ridge or trench. The fault offsets the ridge or trench, sometimes for tens or even hundreds of kilometres. Transform faults occur where the motion between two plates is parallel to the boundary that separates the plates.

transgression (marine transgression) A rise in sea level relative to the land that results in an advance of the coastline inland. †The *Flandrian transgression*, which flooded the English Channel and Irish Sea, was the last major advance of the sea. This was part of the worldwide eustatic rise in sea level following the end of the last ice age. Local transgressions may be caused by movements of the land. *See also* eustasy, isostasy.

transhumance The practice of some pastoral farmers to move their flocks or herds of animals seasonally between two regions with different climatic conditions. The movement is necessary to take advantage of seasonal grazing, such as the summer pastures in mountain regions, and leaves the lowlands clear for crops to be grown. In Spain, for example, there are movements of sheep from the dry Meseta and Mediterranean region in summer to the richer pastures of the higher ground and the wetter north. *See also* saeter.

transition zone |*See* zone in transition.

transit trade The movement of goods along a route or through a particular place such as a port or a country to destinations elsewhere. For example, there is a considerable transit trade of copper across Zaïre and Angola from Zambia to the exporting port of Benguela (Angola).

transpiration The loss of water vapour from plants. Plant tissue conducts water upwards from the roots, so that a continuous flow links the soil to leaf surfaces, from which minute pores (stomata) release water vapour to the atmosphere. †Transpiration maintains the circulation of essential minerals through plants, but excessive water loss causes wilting. The rate of transpiration correlates closely with environmental temperatures (i.e. higher temperatures increase transpiration) and is controlled by the opening and closing of the stomata. Low humidity and moderate air movements also assist the process. *See also* evapotranspiration.

transport 1. (*Geomorphology*) The movement of weathered and eroded earth material by fluvial, glacial, and aeolian processes. Agents of erosion may also be agents of transportation, for example rivers, glaciers, and wind. Transport may occur by dragging (as in bedload), saltation (bouncing), or suspension. Some rock fragments such as flat shale gravels may even float in water. *See* bedload, saltation, suspension.
2. (*Human Geography*) The carriage of goods or people from one place to

another. Means of transport vary according to such factors as the goods to be carried, the distance to be travelled, the type of terrain, and time available. In underdeveloped mountainous regions transport may be by animals or human porters. In more advanced regions a variety of forms of transport exist by road, rail, water, and air. In the USA and W Europe there are dense transport networks.

transport costs The costs incurred in moving goods, including carrying raw materials to the factory and finished goods to the market. †The entrepreneur is faced with *fixed transport costs*, which include the purchase of vehicles and loading and offloading facilities, and *variable transport costs*, which are incurred as a result of movement in a transport system, e.g. the costs of fuel and servicing vehicles. There are also different relative costs according to the type of transport used, e.g. air costs for freight are relatively high.

transverse coast See Atlantic type of coastline.

transverse Mercator projection A map projection constructed in the same way as the Mercator projection but with the cylinder tangential to a meridian instead of the equator. This meridian becomes a straight line down the centre of the map. The projection preserves shape near the central meridian but distortion occurs to the E and W of it. It is used mainly for maps of small areas with a N–S orientation. All UK Ordnance Survey (OS) maps are based on this projection.

transverse valley A valley cutting across a ridge at right angles; this may follow the line of a fault or of superimposed drainage.

trap 1. Any fine-grained dark igneous rock, especially basalt, so named because it often takes the form of steps or terraces. [From Swedish *trapp*: a step]

2. A geological structure in which oil or natural gas accumulates. †The oil migrates under the pressure of overlying rocks until it finds an impermeable barrier. There it displaces any water and floats to the top of the trap. If there is excess gas in the oil, this will bubble out and form a 'gas cap' over the oil pool. Many types of trap are recognized, such as *anticlinal traps* and *salt-dome traps* beneath an arched impermeable sedimentary layer, *fault traps* and *unconformity traps*, or *facies traps* where the lithology becomes less permeable within the same rock stratum.

3. (sediment trap) †A device used for catching and measuring the amount of sediment moving in a river, especially a *bedload trap*, which is a large container sunk into the river bed.

travertine A deposit of calcium carbonate formed around a hot spring. The material is brought to the surface in solution in water that has passed through lime-bearing rocks and is deposited as the water cools and pressure is released when the spring emerges. It often forms terraces and cones. Good examples are found in Yellowstone National Park, USA, around Mammoth Springs. *See* sinter.

treaty port A sea or river port or, sometimes, an inland city opened up to foreign trade as a result of a treaty. Foreign ships could load and unload at the port and traders could live there and own property. For example, after the Opium Wars (1839–42) the UK had access to a number of treaty ports in China, including Canton and Shanghai. These rights were surrendered in 1943.

tree line (timber line) The limit to which trees grow, which is determined chiefly by altitude (i.e. the upper limit of growth on mountain sides) or latitude (e.g. the tree line between coniferous forest and tundra). The altitude at which the tree line occurs decreases generally from the tropics to the high latitudes. Above the tree line harsh climatic conditions such as colder temperatures and stronger winds prevent the establishment of seedling trees and the extension of forests. Trees growing actually at such limits are often stunted and deformed.

trellis drainage (trellised drainage) A rectangular pattern of river channels. It may develop where a slope is crossed at right angles by the strike of alternating hard and soft rock strata. Long streams develop along the soft rock strata, parallel with the strike, and short streams follow the slopes.

tremor (earth tremor) A small earthquake of low intensity, such as one set off by the passage of a train, a landslide, an explosion, or similar occurrence.

trench (ocean deep) A deep trough in the abyssal plain, which often resembles a steep-sided valley or canyon on the land and forms the deepest parts of the ocean. Trenches occur where two plates of the Earth's crust are moving together and one is being pushed down below the other; for example, off the W coast of South America and off the coast of Japan and the Philippines. They are frequently over 7500 m in depth, two of the deepest soundings ever made being over 10 500 m in the Mindanao Trench off the Philippines, and over 10 900 m in the Marianas Trench to the E of Guam. Many trenches lie around the margins of the Pacific Ocean; other notable ones are the Puerto Rico Trench (Atlantic Ocean), the Cayman Trench (Caribbean Sea), the South Sandwich Trench (Southern Ocean), and the Java Trench (Indian Ocean). *See also* plate tectonics.

trend The general alignment of a set of structures such as faults, folds, dykes, sills, etc.

triangulation A method of surveying in which an area is divided up into a network of triangles. One side, the base line, is measured accurately. A theodolite or similar instrument is used to measure the angles of the tri-

angles accurately, from which the lengths of the other sides can be computed. The triangulation provides an accurate framework for subsequent more detailed mapping as it establishes a number of initial control points. *See also* theodolite.

Triassic (Trias) The first geological period of the Mesozoic era, extending from the end of the Permian period, about 225 million years ago, to the beginning of the Jurassic period, about 195 million years ago. Three divisions of Triassic rocks are recognized: the Bunter, Muschelkalk, and Keuper. In the British Isles only the Bunter and Keuper rocks are represented. Continental rocks of this period cover the Midlands and Cheshire plains and most of the N of England, continuing S through Gloucestershire and S Wales into Somerset and Devon. Few fossils are found in the Triassic although the European rocks of this period have yielded the earliest-known dinosaurs. [Named by F. A. von Alberti (1834) who recognized that the European rocks of this period formed three distinct divisions]

tributary A stream, river, or glacier that feeds another larger one.

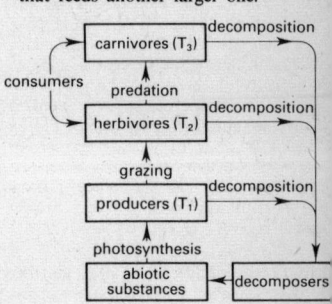

Trophic levels

trophic level †In a natural community, organisms feeding at the same number of steps from the autotrophs (plants) in a food chain are said to be at the same trophic or energy level. The first trophic level consists of the producers, green plants that convert solar energy to food by photosynthesis. On this directly depend the herbivores (primary consumers) of the second trophic level. The primary carnivores (secondary consumers) feeding on the herbivores make up the third level and at the fourth level secondary carnivores (tertiary consumers) eat the primary carnivores. Decomposers such as fungi and bacteria, which break down dead organic material into nutrients usable by the producers, occupy a separate trophic level.

At each trophic level large losses of energy take place so that less is available to the next trophic level. Consequently, successively fewer organisms occupy the higher trophic levels forming a pyramid limited to three, four, or five levels. *See* food chain.

tropical air mass An air mass that has its source region within the subtropical high-pressure areas between 20° and 40°N and S of the equator. If the air mass has originated over an ocean it is classified as *tropical maritime air* (mT), which is warm and moist at source; if it has originated over a continent it is *tropical continental air* (cT), which is hot and dry at source.

tropical climates The group of varying climatic types found within the tropics. Each climatic classification has variations, but it is generally agreed that there are five basic types, all linked by the fact that they have no winter as such (in the Köppen climatic classification, no area has a mean monthly temperature of less than 18°C). The five types are the equatorial climate, with rain all year round; the tropical continental (Sudan) climate, with summer rain and winter drought; the hot desert climate, with little or no rain; the tropical maritime climate; and the tropical monsoon climate.

tropical rainforest *See* equatorial rainforest.

tropopause A plane of discontinuity in the atmosphere between the tropo-

sphere and the stratosphere. It has a variable altitude, but averages 6–8 km over the poles and 16–18 km over the equator. Below it, in the troposphere, temperature decreases with height to about –60°C, while above it, in the stratosphere, temperature increases slightly with height as far as the stratopause, from –60°C to about 0°C. It is now believed that it is a series of discontinuities rather than a single unbroken one.

tropophyte †A plant that alters its physiology to suit changing climatic conditions. For example, savanna trees such as the baobab behave as hygrophytes when rainfall is plentiful, but adopt xerophytic properties, such as losing leaves, during seasonal drought.

troposphere The lowest layer of the Earth's atmosphere, extending from the surface of the Earth upwards to the tropopause. Within the troposphere there is a general decrease of temperature with height of about 6°–8°C per kilometre, but this varies considerably according to place and time. Virtually all weather phenomena (e.g. winds, precipitation, clouds) develop within this layer.

trough 1. A depression between the successive crests of a wave in the sea. **2.** *See* trench.

trough of low pressure A low-pressure system that appears on a weather map as a V-shaped pattern of isobars. Pressure increases towards the point of the V. Troughs are usually associated with unsettled weather with rain and cloud.

truck farming *See* market gardening.

true north The direction along a meridian that leads to the North Pole. It should not be confused with magnetic north, which is not fixed and is a few degrees away from true north.

truncated spur A spur, the end of which has been carved off through erosion by a glacier that formerly occupied the valley. *See also* interlocking spur.

trust territory (trusteeship) A territory supervised or administered by a foreign power, responsible to the United Nations. In 1945 the remaining mandated territories were renamed trust territories and put under the supervision of the United Nations until ready for self-government. Nearly all have now achieved independence or joined independent countries. *See also* mandate.

tsetse fly A small blood-sucking fly found in tropical Africa. It transmits parasites to human beings, which cause sleeping sickness and other diseases. The parasites cause a disease in cattle and wild animals called nagana and cattle rearing is severely limited in infested areas. The fly breeds in lakes and marshes in W, E, and central Africa and some of the breeding grounds are sprayed with pesticides.

tsunami A seismic sea wave generated by an earthquake, explosive volcanic activity, or underwater earth movements. The waves travel through the open ocean, often for considerable distances, reaching heights of about a metre but on reaching shallower water near land their height increases rapidly to tens of metres. They travel at high speeds (about 300–500 knots). Although their occurrence is rare tsunamis can cause great destruction and loss of life through flooding. One of the best known is the tsunami that resulted from the eruption of the volcano Krakatoa in the Sunda Strait between Java and Sumatra (1883), in which waves reached 40 m in height and 36 000 people lost their lives. [Japanese]

tuff A deposit formed from consolidated volcanic ash, usually having a grain size of less than 4 mm. Some tuffs have been laid down under water as they are found interlayered with ordinary sedimentary rocks. There are three main forms of tuff. *Vitric tuff* is composed chiefly of volcanic glass and results from an explosive eruption of liquid lava. *Crystal tuff* is composed of detached crystals and results from lava that has nearly crystallized at the time that it is thrown out in an eruption. *Lithic tuff* is composed chiefly of angular fragments produced by the shattering of solid rock during a volcanic explosion.

tundra A zone that lies between the limits of tree growth in the N hemisphere and the polar ice cap. It extends across the N of Scandinavia, the USSR, and Canada. Temperatures are frequently subzero and the zone is characterized by permafrost (permanently frozen ground). There is little or no daylight in winter and the growing season lasts only two or three months. Lack of available moisture in winter alternates with waterlogged conditions after the thaw. Soils are thin and organic. Few plants can tolerate such conditions and the vegetation thus consists of lichens and mosses that need little soil, low-growing herbs that germinate and flower quickly in the brief summer, and, in the S parts, hardy dwarf shrubs (birch and willow). In the S hemisphere a similar vegetation occurs on high plateaus in lower latitudes. [Lapp]

tundra hummock A small landform shaped like a cushion that is found in regions with a subarctic tundra climate. Tundra hummocks form above permanently frozen ground (permafrost) where the annual depth of thaw is less than 50 cm as a result of soil swelling, frost heaving, and frost cracking. Hummocks vegetated by mosses and lichens are generally inactive and no longer growing.

tungsten A metal obtained from veins of the minerals scheelite and wolframite. The metal has a high melting point and is very strong and hard. It is added to steel for the production of high-speed cutting tools and drills. Tungsten is also used for the filaments in electric light bulbs, as well as in television tubes and X-ray tubes. The chief producing countries are China,

USA, USSR, North Korea, South Korea, Bolivia, and Australia.

turbary A place where peat is dug or the right to dig peat from a piece of land. The fenland adjoining the Norfolk Broads contains many 'turf ponds', shallow peat cuttings that were formed when local people had rights of turbary in the area.

turbidity 1. The opacity of the atmosphere that results from the presence of particles such as dust and smoke.
2. The muddiness of water resulting from suspended sediment. †A *turbidity current* is an underwater flow of sediment or sediment-laden water. It occurs where rivers enter a lake or the sea and on relatively steep submarine slopes (e.g. continental slopes), where earthquakes may trigger it off. Turbidity currents may reach speeds of over 80 km per hour and it has been proposed that the flows could account for the erosion of submarine canyons. *See also* submarine canyon.

turbulence †An irregular disturbed flow of fluid (e.g. water, air) in which the main forward movement of the fluid has superimposed on it a random movement of secondary eddies. Most flows in nature (e.g. rivers and winds) are turbulent.
The erosive potential of a river is directly related to the degree of turbulence. It is important in the transport of sediment; deposition occurs when turbulence is at a minimum. An increase in the velocity of flow, increase in the roughness of the channel bed, and a decrease in the depth of water will cause turbulence.
In the atmosphere turbulence is an important mechanism of mixing in air masses. It occurs, for example, as a result of strong heating of the ground surface causing small eddies to rise. *Compare* laminar flow.

turnpike 1. A barrier that was placed across a road to prevent passage until a toll had been paid. A *turnpike road* was a road along which turnpikes were erected at regular intervals.

Turnpikes were a feature in Britain during the 17th–19th centuries.
2. In the USA, a motorway for the use of which a toll has to be paid.

twilight The reflected light from the Sun that can be seen for a time before sunrise and after sunset. The length of time it lasts depends mainly on latitude and date, which determine the angle that the Sun's path makes with the horizon. In the tropics this is always about 90°, and it is therefore of short duration, but in high latitudes, it can be very acute at certain times of year, making it long. In high latitudes, the Sun may not dip far below the horizon, so it lasts from sunset to sunrise. There are three distinct definitions:
(1) *Astronomical twilight*, which lasts from when the Sun's centre is 18° below the horizon to sunrise in the morning and from sunset to when the Sun's centre is 18° below the horizon in the evening.
(2) *Civil twilight* (lighting-up times), when the angles are 6° respectively.
(3) *Nautical twilight*, when the angles are 12° respectively.

typhoon A tropical cyclone or hurricane occurring in the China Sea and the W and N Pacific Ocean. It brings very high winds (above force 12 on the Beaufort wind scale), torrential rains, and destructive seas to the area. Cyclones are more common in this area than elsewhere in the world, averaging about 36% of the world's total (about 22 per year). *See* cyclone, hurricane. [From Chinese *tai fung*: great wind]

U

ubac The side of a valley that faces away from the equator (i.e. N facing slopes in the N hemisphere; S facing slopes in the S hemisphere). The ubac slope usually remains forested, with pasture and cultivation limited to the opposite (adret) side of the valley. *Compare* adret. *See also* Schattenseite. [French]

ultrabasic rock An igneous rock containing less than 45% silica (by weight), with no free quartz. This type of rock is dark in colour and very dense. An example of an ultrabasic plutonic rock is peridotite from the Lizard Peninsula, Cornwall.

ultraviolet radiation (UV) That part of the Sun's electromagnetic radiation that has wavelengths ranging from 4 to 400 nanometres, i.e. just short of those of visible light. Some ultraviolet rays are absorbed in the ionosphere and by ozone in the atmosphere, but some reach the Earth's surface (the highest intensity being on high mountain tops).

umland See sphere of influence. [German]

unconformity A break within a sequence of rocks representing a period of erosion or nondeposition. The rock strata below the break are likely to have different dips and strikes to the rock strata above. For example, the lower strata may be intensely folded while those above the unconformity are level and unfolded. The significance of an unconformity was first described by the geologist James Hutton in 1788 at Siccar Point, Berwickshire, where Old Red Sandstone rests unconformably on steeply dipping Silurian. See also nonconformity.

undercliff The lower part of a coastal cliff that has experienced landslipping. The undercliff is typically stepped or terraced as a result of successive landslips. Such landslips are common where waves undercut cliffs with weak clay strata near the base. The clays become slippery when saturated and overlying rocks, for example, the chalk in parts of the Isle of Wight and near Lyme Regis, Dorset, tend to slip down the cliff face.

undercutting Erosion of the lower part of a bank or rock face, which eventually results in a rockfall or slumping of the bank or scar. Flow in a river

channel or wave and tide action on the coast are the most common agents of undercutting. Wind-blown sand acting on the base of rock faces in deserts also causes undercutting, which may result in the formation of pedestal rocks.

underdeveloped land An area of land in which the natural resources have not been developed fully. The term may be defined more fully as land that has good prospects for using more capital and skilled labour to utilize more natural resources in order to enable the present population to live at a higher standard of living, or to maintain the present standard despite rapid population growth.

underdevelopment A state of a society in which both the capital and the social structure necessary for development and expansion are lacking. The term is often considered as a euphemism for poverty. The underdeveloped countries of the world are also referred to as the Third-World countries, less-developed countries, and developing countries. See also Third World.

underfit river See misfit river.

underground stream A stream that flows through a cave. Underground streams are commonly found in limestone karst regions where surface streams disappear down swallow holes and continue to flow through underground channels until they reappear as resurgent streams at the edge of the limestone area or where the cave roofs have collapsed. See karst, swallow hole.

underpopulation †A population that is too small to make full use of the resources available to it, or where a higher population could be maintained by the available resources without any reduction in living standards. See also optimum population, overpopulation.

undertow A current of water flowing seaward beneath the waves on the sea shore. It returns seawater brought up the foreshore by the waves. The undertow pulsates and becomes stronger under the wave troughs where it is reinforced by the orbital motion of water in the waves. Such undercurrents can be strong in places, particularly in depressions, where they can cause erosion and be hazardous to bathers.

upland An elevated region, i.e. a hill or mountain area.

upthrow *See* throw.

urban Of or relating to a town or city. Urban settlements are large nucleated settlements in which the majority of the employed inhabitants are engaged in nonagricultural activities. Urban areas may be defined by national governments according to different criteria; for example, size, population density, and type of local government.

urban blight A large cleared area in the inner city that has not yet been redeveloped. Since demolition is a quicker process than rebuilding, the process of urban redevelopment frequently results in the creation of large areas of urban blight.

urban climate The local climate experienced in large urban areas. This differs from the climate of surrounding rural areas as a direct result of the lack of vegetation, the built-up surface, the amount of fuel burnt, etc. Two of the commonest features of this are the heat island effect, caused by the heat-absorbing qualities of the ground surface, and smog, caused by the amount of dust and smoke particles in the air. *See* heat island.

urban conservation area An area within a town or city that has been designated for special protection because of the special architectural or historical interest of its buildings. Buildings in such areas are protected from developments that are permitted elsewhere and grants are available for restoration and other maintenance work on the buildings.

urban field †The area that is economically and socially linked with an urban settlement. The concept of the 'field' is analogous to that of a magnetic field: the degree of attraction of the central place is greatest close to the centre and diminishes with distance. Where more than one centre of attraction exists urban fields may overlap their boundaries. The term is generally regarded as synonymous with hinterland, umland, sphere of influence, and trading area. *See also* gravity model.

urban hierarchy *See* central place hierarchy.

urbanization The process by which an increasing proportion of the population of an area becomes concentrated into the towns and cities. The term is also defined as the level of population concentration in urban areas. The process of urbanization increases both the number and size of towns and cities. It is estimated that two fifths of the world population is now urban and the rate of increase is accelerating, especially in developing countries. Urbanization is now the dominant factor causing changes in the distribution of world population.

urban land value surface †The pattern of land values in a city. This generally reflects accessibility within the city area. Land values reach a peak at the city centre – the most accessible point – and decline away from this. The land values decline least along the major routes that radiate out from the city centre and along the major routes that encircle it; where these routes intersect secondary peaks of high land values occur. *See also* bid rent theory.

urban mesh †The geometrical pattern of the size and spacing of urban settlements. The concept is derived from central place theory – Christaller's theoretical landscape is organized symmetrically in the form of a hexagonal mesh. *See* central place theory.

urban morphology The internal structure of a town or city. It includes the arrangement and layout of buildings, streets, and roads, and the functions of the land and buildings. The study of urban morphology focuses upon the examination of urban structure, the identification of morphological (functional) zones or regions, and the processes that produce their distribution and interrelationships.

urban region 1. The region comprising of a service centre (central place) and its urban field.
2. An internal functional or morphological zone of a town or city.

urban renewal The regeneration of urban areas, especially inner-city areas. Urban renewal has occurred largely as a response to public awareness that the scale of urban problems has now increased to a point where it threatens the functioning of major cities. It is designed to keep people and jobs in the inner-city areas. Urban renewal may occur as spontaneous renewal or comprehensive renewal. Spontaneous renewal occurs when demand for land in the city centre exceeds supply so renewal is profitable for it will bring higher rents. Comprehensive renewal is large-scale renewal involving planned rebuilding of substantial parts of the inner city; for example, the new city centre of Sheffield. The *zone of urban renewal* is that part of a town or city in which renewal occurs.

urban sprawl The spread of urban settlement, both by the simple process of outward growth and by the coalescence of previously separated but closely located urban settlements. The development of motor transport considerably accelerated the rate of this process. As a result of urban sprawl discrete urban settlements have been assimilated into conurbations.

urban structure *See* urban morphology.

urban transect A cross section of a town or city that shows a particular component of its structure. For example, the dominant land-use distribution or the main features of its townscape.

U-shaped valley The typical form of a valley through which a glacier formerly flowed. Erosion of the bed and lower walls of the valley by rock fragments embedded in the moving ice gives a characteristic U-shaped cross section to the valley. The flatness of the valley bottoms is often accentuated by deposits of coarse sediments. *Compare* V-shaped valley.

uvala †A large surface depression, up to about a kilometre across, that is found in limestone areas and formed by carbonation solution. It frequently results from the coalescence of several smaller depressions known as dolines. An uvala is intermediate in size between a doline and a polje. [Serbo-Croat]

V

vadose water Water in the ground that lies above the water table or level of total saturation of the rocks. Vadose water percolates down towards the water table through available joints or fissures in permeable rocks such as limestone. Caves eroded by vadose water show many features typical of open channels on the surface; for example, erosion knicks in the walls marking common water levels and meanders.

vale A broad valley that is often occupied by more than one river and follows the outcrop of relatively less resistant sedimentary rock (e.g. clay vales). Such vales are commonly formed by the erosion of a domed structure of sedimentary rock strata, for example, as in the Weald region of SE England. The more resistant rock strata are left upstanding as cuestas or ridges.

valley A linear depression sculpted by water or ice or formed by movements of the Earth's crust. *See* U-shaped valley, V-shaped valley.

valley glacier A body of ice moving down a valley. Valley glaciers may begin in cirques (known as the alpine type) or they may issue from an ice cap (the outlet type). They mainly occupy pre-existing fluvial valleys, which are modified by the action of the ice over a period of time. *See* glacier. *See also* cirque.

valley wind An anabatic wind that blows up a valley axis during the daytime in mountain areas. The air in contact with the valley sides is heated by conduction from the ground, rises convectionally, and is replaced by air moving up from the valley, thus creating an upslope wind. It is particularly well developed when one side of the valley is heated much more than the other; for example, in valleys that have an E–W trend in mid- and high latitudes. *See* anabatic wind. *See illustration at* mountain wind.

vapour pressure In meteorology, that part of the total atmospheric pressure exerted by the water vapour present in the atmosphere. This increases as the amount of water vapour increases, and reaches a maximum when saturation is reached – the *saturation vapour pressure*. Given saturated air, the vapour pressure is highest when the temperature is highest. Near the Earth's surface it varies from about 5 mb in cold dry arctic air, to about 30 mb in hot humid tropical air. It is measured indirectly from dry- and wet-bulb temperatures and tables are used to obtain a precise value.

vapour trail *See* condensation trail.

varve †An annual layer of sediment laid down in the bottom of a lake in periglacial regions. Varve deposits have a banded appearance caused by alternating layers of different coloured finer and coarser sediments. Each varve consists of a lighter band of coarser clay and silt sediments deposited in spring and summer after the thaw, and a darker-coloured band of finer-grained clay deposited in winter when the freeze-up returns and stream flow slows down. The darker colouring of the winter deposits is caused by the presence of organic material derived from decaying vegetation. As each pair of bands represents one calendar year varves have provided a valuable system for dating. [From Swedish *varv*: layer]

Vauclusian spring A resurgence of an underground stream on the edge of a limestone area where it flows out at the boundary of an impermeable stratum. [Named after the Fontaine de Vaucluse in the lower Rhône valley in SE France]

veering A clockwise change of direction of a wind, e.g. from SW to W to NW. This condition is very common in the UK as the wind veers when a front of a depression passes. *Compare* backing.

vega 1. A fertile irrigated area in the Mediterranean region of Spain that yields only one crop a year compared with the huertas that yield two or more. *See also* huerta.
2. In Cuba, a tobacco field. [Spanish]

vegetable oil Oil obtained from plants, usually from the seed, nut, or fleshy part of the fruit. Among the seeds yielding oil are the soya bean, sunflower, cotton, linseed, maize (corn), and oilseed rape. Oil is also extracted from groundnuts, coconuts, and palm kernels. The fleshy parts of the olive and oil palm fruit also provide valuable vegetable oils. Vegetable oil is used for making margarine, cooking oils, cosmetics, soap, and many other products.

vegetation The total plant life of a particular region. Vegetation worldwide is divided into types or zones, such as equatorial rainforest, which broadly match climatic regions.

vein In geology, a fissure in rock strata containing a thin deposit of crystalline rock formed by extremely hot solutions rising from magma during the final stages of its crystallization. Veins are often the source of economically important minerals, especially copper, tin, lead, and zinc. Very thick veins are called lodes.

veld (veldt) The temperate natural grassland of the plateaus of interior South Africa. The *High Veld* (above 1500 m) is cool and dry and is similar to the Eurasian steppe, although far less extensive. The *Low Veld* (below 900 m) is warmer and has more tree cover. The veld is intensively farmed, for both animals and crops. *See also* grassland. [Afrikaans]

vent An opening or conduit in the surface of the Earth through which volcanic material is ejected. A series of vents may form along a major fracture creating a fissure volcano. Volcanoes also develop with central vents and subsidiary vents on the sides of the cone.

ventifact A stone sculpted by the wind. Wind-blown sand flattens and smooths the surfaces of stones lying on an arid surface as in mid-latitude or tropical deserts or in regions marginal to ice sheets. *See also* dreikanter.

vertical exaggeration The use of a vertical scale on a map, diagram, or model that is greater than the horizontal scale. For example, if the horizontal scale is 1 cm to 50 km (i.e. 1:50 000) and the vertical scale is 1 cm to 10 km (i.e. 1:10 000) then the vertical exaggeration will be 5 times.

vertical heat flux The transfer of heat from the surface of the Earth to the atmosphere. This is achieved by radiation, latent heat, and direct conduction.

village A settlement in a rural area that is larger than a hamlet and smaller than a town. It is essentially a close-knit small community containing a limited number of service functions such as a sub-post office.

viscous Denoting a thick fluid or semimolten rock with a high melting point. Viscous lavas are acid (i.e. silica rich) and do not flow far but solidify quickly, building up a high steep-sided dome. *See also* acid lava.

visibility In meteorology, the greatest distance that an observer can see at a given time. It is usually quoted in kilometres (or metres). Visibility is largely dependent on the presence of particles (e.g. water, ice, dust, and smoke) in the atmosphere.

viticulture The cultivation of the grape vine for the sale of grapes as fruit, as a juice, as sultanas, currants, or raisins, or after fermentation as wine. The main areas of viticulture are France, Germany, Italy, Spain, California (USA), central Chile, and S Australia.

volcanic rock *See* extrusive rock.

volcanism (vulcanism, volcanicity, vulcanicity) The processes responsible for the upward movement of molten rock or magma through the Earth's crust. Volcanism is responsible both for volcanic eruptions and for the intrusion of igneous rocks within the crustal rocks. *See also* igneous rock, extrusive rock, intrusive rock.

volcano The point at which igneous material erupts through the Earth's crust. Volcanoes usually consist of a conical hill or mountain formed from the molten rock or lava, ashes, and rock fragments ejected through the volcanic vent. The exact shape of a volcano depends on many factors, such as the viscosity of the lava and the amount of ash ejected. Volcanoes are described as active, dormant, or extinct depending on the length of time since the last eruption. Extinct volcanoes can take many forms depending on the extent of subsequent erosions. *See* Hawaiian eruption, Peléan eruption.

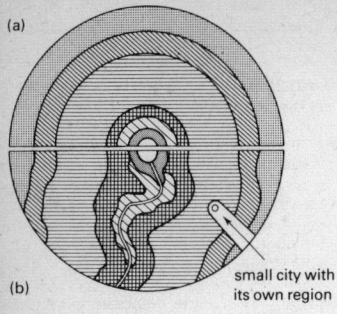

(a)

(b)

small city with its own region

- market gardens
- forestry
- intensive crop rotation
- field grass with emphasis on dairy products
- three-field system
- extensive cattle raising

Von Thünen's model of land use around one central city: (a) isolated state, (b) modified by two variables

von Thünen model †A model of agricultural land use that was devised in 1826 by Heinrich von Thünen, a Prussian landowner. It was designed to explain the factors governing the prices of farm products and the principles by which such price variations are translated into patterns of land use. Von Thünen envisaged a flat featureless plain (isotropic surface) with no variations in soil, fertility, relief, or climate. This area is an 'isolated state' in which one central city is the only source of demand and supply with no external trading. It is inhabited by farmers who supply the city and aim to maximize their profits. There is only one form of transport and transport is possible with equal ease in all directions. Transport costs increase in direct proportion to distance. As a result bulky products or those difficult to transport are grown near the market, while lighter and more easily transported products are grown fur-

ther away. These variations account for variations in the cost of the product at market. This resulted in a land-use pattern of a number of concentric zones for each type of agriculture practised surrounding the market area. Von Thünen modified the model with the addition of two variables: a navigable river and a second market. Other variables such as variations in soil fertility could also be applied.

voralp The lower pastures of an alpine valley that are reached before the main alpine pastures on the higher ground. Many of the valleys in Switzerland have voralps, e.g. the Mattertal. [German]

V-shaped valley A river valley with a V-shaped cross profile. It is the typical form of river-eroded valleys in humid climates along the upper and middle courses of a river, before a broad floodplain has formed. It is common in uplands where it contrasts with U-shaped valleys sculpted by valley glaciers. *Compare* U-shaped valley.

vulcanicity *See* volcanism.

vulcanism *See* volcanism.

W

wadi A normally dry valley in a hot desert or semidesert. It typically has steep walls and a flat bottom. Wadis may experience occasional flash floods after heavy desert thunderstorms, although discharge tends to decrease downstream as the water infiltrates into the sandy bed. *See* flash flood. [Arabic: valley]

Wallace's line A hypothetical line drawn through Indonesia by the 19th-century British zoologist A. R. Wallace. The line runs between Borneo and Sulawesi (formerly Celebes) following the deep Makassar Strait, and further S between the islands of Bali and Lombok. It separates two major zoogeographic regions, Oriental to the

N and W and Australasian to the S and E.

waning slope †A slope at the foot of a hillside that decreases in gradient downslope before reaching a river or floodplain. It is one of the fundamental slope elements in the slope evolution schemes of Walter Penck and Lester King, in which the waning slopes grow as hillslopes retreat back from the valley bottoms. *Compare* waxing slope. *See also* standard hillslope. [Coined in 1942 by Alan Wood]

ward An administrative division of a parish, town, or city, or other area. A ward usually elects its own representatives; for example, to a town or city council. Originally, a ward was a district in the charge of an alderman.

warm front A front separating a retreating cold air mass from an advancing warm air mass, which rises above the cold air. It has a much lower angle (about 1°) than a cold front, and the weather associated with it is seen long before the front reaches the observer. The first indication of its approach is high cirrus cloud, which gradually lowers and thickens to cirrostratus, followed by altostratus, then rain-producing nimbostratus at the front itself, frequently a distance of 1000 km overall. As the front passes temperature and humidity rise, pressure stops falling, and winds veer. The forward speed of warm fronts is very variable but averages about 50 km per hour. *Compare* cold front. *See also* front.

warm sector The region of warmer air of a depression, bounded on its leading edge by the warm front and on its trailing edge by the cold front. It is the active part of the depression and most of the weather associated with a depression occurs within it, especially at the fronts where the warm air is rising. The weather within it generally consists of drizzle and showers falling from a more or less complete cover of stratus or nimbostratus clouds. It is gradually 'pinched out' as the cold front catches up to the warm front, forming an occlusion. *See also* depression.

warm temperate western margin climate *See* Mediterranean climate.

warm wave In the USA, a surge of warm air moving from lower latitudes and experienced in temperate areas, usually in summer. It is caused by air being drawn in front of an eastward moving depression, or to the W of an anticyclone. *Compare* cold wave.

warping The gentle deformation of the Earth's crust over a wide area. A number of causes may be responsible for warping. For example, isostatic adjustment following the melting of an ice cap, which has led to uplift in the Hudson Bay area of Canada. Sedimentation along coasts or in shallow seas leads to the subsidence of these areas. Warping may also result from mountain uplift and faulting.

warren Formerly, an area of rough unimproved land that was kept for breeding game or rabbits. The term is now often used to refer to a collection of sand dunes on the coast.

wash 1. *See* sheet erosion.
2. A mass of fine alluvial material transported and deposited by sheetflow. †The *wash slope* (alluvial toeslope) occurs at the foot of a hillslope where finer material washed down the slope is deposited.
3. An area of sand or mud banks that is regularly washed by the tide or river water, e.g. the Wash on the North Sea coast between Lincolnshire and Norfolk.

waste mantle *See* regolith.

water balance †The budget of incoming and outgoing water in a system. It may be viewed on a global, regional, or local scale. On a global scale this involves the cyclic movement of water between and within the atmosphere and ground surface. On a local scale the water balance of an individual drainage basin may be examined. In assessing the water balance of a drainage basin, it is commonly assumed that river runoff is equal to the precipitation less losses from evaporation and transpiration. This assumes that surface storages (e.g. lakes and glaciers) and subsurface storages (mainly as ground water) remain the same. Over short periods such as an average year this is reasonable, but over long periods and in special circumstances changes in these storages may also have to be taken into account. *See also* hydrological cycle.

water cycle *See* hydrological cycle.

waterfall A steep cliff-like section of a river channel down which water falls vertically. Waterfalls tend to be removed over an extended period of time as rivers achieve a graded state. A waterfall may be created by a number of different processes including rejuvenation, hanging tributary valleys entering a main glaciated valley, or erosion breaching a hard band of rock. †*See also* grade.

water gap A narrow steep-sided valley cut by a stream through a ridge of resistant rock. It is commonly formed during the denudation of an area of domed sedimentary rocks where consequent streams maintain their course as they cut down through more resistant sedimentary strata. Examples of water gaps include Goring Gap, where the River Thames flows between the Berkshire Downs and the Chiltern Hills. *See* consequent stream.

waterhole A natural pool found in semiarid and arid regions formed by a spring or an outcrop of the water table. Waterholes are frequently used by both animals and man. In the savanna grasslands some regularly dry up during the dry season and most dry up from time to time during years of severe drought, with catastrophic effects on the wildlife.

water meadow A low-lying grassy area in humid mid-latitude regions, usually beside a river or stream, that is permanently moist because of a high water table and/or periodic flooding. The associated soil type – meadow soil – produces lush grassland. This is grazed or cropped for hay. Some water meadows are deliberately irrigated to promote early growth.

water power The energy that can be harnessed from running water. Early industrial sites used water power to turn water wheels and drive simple machinery, e.g. the wool industry of East Anglia in the 14th century. Today this form of energy is harnessed as hydroelectric power to produce electricity that can be transmitted to factories a long way from the power station. *See* hydroelectric power.

watershed 1. (divide) †The edge of a drainage basin. The watershed is commonly taken as the line joining the highest points on the boundary of the basin. However, watersheds below the ground surface do not always coincide with the surface watershed and water can 'leak' from one drainage basin into another, especially when there are steeply dipping sedimentary strata. Even surface water may cross the watershed in exceptional cases; for example, in poorly defined basins in flat glaciated areas of the Canadian shield where bogs straddle the watershed.
2. In the USA, a drainage basin.

waterspout A whirling funnel of cloud that extends downwards from the base of a cumulonimbus cloud to the surface of the sea, where water may be picked up from the agitated surface. It is a localized feature and is caused by the development of an extremely intense but small low-pressure area, which behaves in a similar way to the tornado on land.

water table The upper surface of the zone of saturated rocks, i.e. rocks in which all voids are filled with water. Water tables are rarely flat and tend to follow the relief of the land surface but with undulations smoothed out.

They are generally deeper on the crest of a hill than in a valley bottom. The water table rises and falls seasonally. At points where the water table intersects the ground surface water may emerge above ground in the form of springs, marshes, or lakes. *See also* perched water table.

waterway A stretch of water, such as part of a river or a canal, that can be used for the movement of goods or people. The British Waterways Board is responsible for canals in the UK; these are used both for the carriage of goods and as recreational resources. Navigable stretches of rivers such as the Trent and Thames are also examples of waterways.

wave-built terrace A coastal landform that builds up seaward of the wave-cut bench. It is formed by deposition of sediment mainly derived from coastal erosion. *See* wave-cut bench.

wave-cut bench The upper part of a rocky foreshore at the base of a sea cliff. Wave-cut benches are formed by marine erosion and may be enlarged to become wave-cut platforms. *See* wave-cut platform.

wave-cut notch *See* notch.

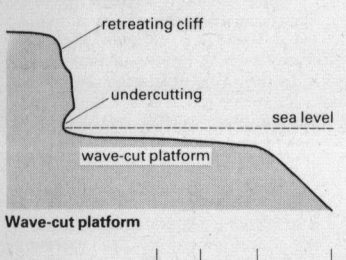

Wave-cut platform

wave-cut platform A gently sloping bare rock platform that extends out from a cliff coastline. Waves wear back the cliffs and extend the platform chiefly along exposed stretches of headland.

wave, ocean An undulation that moves across the surface of the sea caused by the transfer of energy from the wind to the sea. The size of a wave depends on three variables: the speed of the wind, the duration of the wind, and the fetch (i.e. the distance that the wind can blow over the water surface). In the open ocean there are frequently several wave patterns superimposed on each other giving a confused pattern but as they near the coast the waves become more regular and eventually break on the shore. The size is measured by the wavelength (the horizontal distance between successive wave crests) and the wave height (the vertical distance from trough to crest).

waxing slope †A convex slope, which steepens downhill, at the crest of a hillslope. It is the upper slope element in Walter Penck and Lester King's schemes of slope evolution, lying above the free face in the model slope. King imagined that creep was the dominant form of erosion here, but more recent evidence suggests that water rilling and throughflow can also be important. *Compare* waning slope. *See also* standard hillslope. [Coined in 1942 by Alan Wood]

weather The state of the atmosphere, as determined by the meteorological phenomena that are occurring, at any one place and time. The meteorological phenomena include temperature, precipitation, winds, clouds, sunshine, pressure, and visibility.

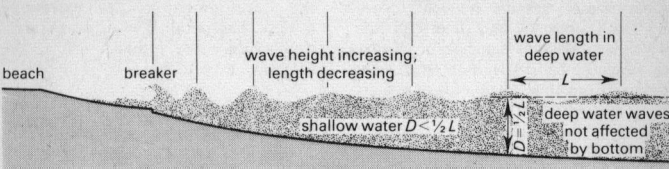

Changes in wave form in shallow water

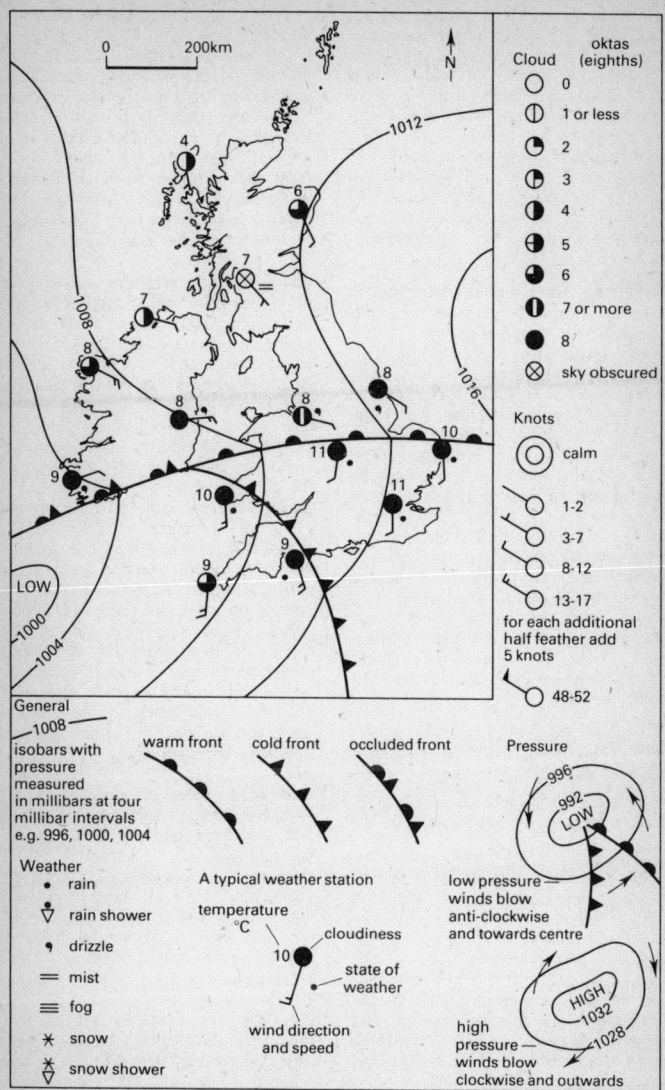

Weather map of the British Isles and weather symbols

weather chart

weather chart (synoptic chart) A map or chart on which is plotted all the data available concerning the weather over the given area at a particular moment of time. The main details plotted concern temperature, pressure (shown as isobars), wind speed and direction, precipitation, and fronts, and these are constantly updated. The chart enables meteorologists and weather forecasters to observe the development of the weather patterns and to anticipate future developments.

weathering The processes that result in the breakdown of rocks and minerals *in situ*. The weathering may result from either physical (mechanical) processes (e.g. freeze–thaw cycles, heating and cooling) or chemical processes (e.g. carbonation, hydration, hydrolysis, solution). *Organic weathering*, which consists of both mechanical and chemical weathering, is caused by plants and animals, particularly when they burrow or release acidic substances onto the rocks. Lichens, for example, cause breakdown as they extract nutrients directly from the rocks. *See* chemical weathering, deep weathering, mechanical weathering, organic weathering. *See also* carbonation, exfoliation, freeze-thaw, frost shattering, hydration, oxidation, solution, †hydrolysis.

weathering front †*See* basal surface of weathering.

weather vane *See* wind vane.

Weber's model †A model of industrial location devised by Alfred Weber in 1909. Weber simplified reality by making a number of assumptions: (1) a homogeneous area in terms of climate and topography; (2) perfect competition with one given price; (3) predetermined raw material, energy, and labour locations; and (4) transport costs increase directly with weight and distance. A *locational triangle* is used to illustrate the model in which the sources of two raw materials (RM₁

and RM₂) and a single market area (A) form the three points of the triangle. The optimum location (OL) at which transport costs, calculated by multiplying the weight of material or product by distance carried, are least lies between these. Weber also demonstrated that the optimum location of an industry might be pulled away from the least-cost transport location in favour of a location with a pool of cheap labour or by the advantages of agglomeration. *See* isodapane.

wedge of high pressure An area of high pressure that separates two depressions and gives a short period of fine weather between the characteristic frontal clouds and rain. It is similar to but narrower than a ridge of high pressure.

weir 1. A structure built across a stream to create a pool for fishing or for diverting flow, e.g. at a mill.
2. †A structure built across a stream in order to measure stream flow. Weirs take various forms but all provide a fixed cross section for flow in such a way that the discharge (i.e. volume of flow per unit time) can be calculated from simply measuring the depth of flow over the weir and applying the *weir equation*, a calibration curve specific to that design of weir. The weir needs to be carefully sited and installed to ensure smooth flow over the structure.

Wentworth scale A method of classifying sedimentary rocks by grain size that was formulated in 1922 by the US geologist C. K. Wentworth, and is now the most common scale used by geologists. The scale ranges, in geometric factors of 2, from particles of clay less than 1/256 mm in diameter to boulders of over 256 mm in diameter.

westerlies The zonal wind belts that flow from the subtropical high-pressure belts around latitudes 35°N and S towards the temperate low-pressure belts around latitudes 65°N and S. In the S hemisphere they have a northwesterly to westerly direction and form a very distinct feature as there are few landmasses to disturb their flow. In the N hemisphere the large landmasses interrupt their flow considerably, but the southwesterly winds are the prevailing winds on the W coasts of Canada and Europe.

wet-bulb temperature The lowest temperature to which an air sample can be cooled at constant pressure, by evaporating water into it (i.e. the temperature of the air sample when it is saturated). It is measured on a *wet-bulb thermometer*, the bulb of which is wrapped in muslin that is kept constantly moist. Evaporation from the muslin ensures that the air around the bulb is kept saturated, and thus the temperature is lowered because of the loss of latent heat through evaporation. If the *dry-bulb temperature* is also known, then the relative humidity of the air can be ascertained by reference to a prepared set of tables.

wet day In the UK, a period of 24 hours, starting at 0900 hours GMT, during which at least 1.0 mm of rain falls. *Compare* rain day.

wet-point site A site at or near a source of water supply such as a stream, lake, or spring, where a settlement was located. For example, the spring-line villages at the foot of chalk scarp slopes (e.g. the Chiltern Hills) in SE England. *See also* defensive site, dry-point site.

wet spell In the UK, a period of 15 consecutive wet days.

whaleback 1. A mound of coarse sand found in deserts that contains generally coarser grains than sand dunes.
2. An elongated rocky hillock similar to a drumlin but carved out of solid rock by a moving ice sheet. Good examples are found in the Canadian Shield.

wheat A cereal grass yielding a grain that can be ground into flour or meal. It is widely cultivated in mid-latitude and subtropical regions. It requires a growing season of at least 90 frost-free days, a rainfall between 300 and 700 mm, and temperatures of 15°C or more in the summer season, with much sunshine and little rain during the ripening period. There are two main varieties, hard and soft. Hard wheats are used for bread flour and pasta and soft wheats for pastries, biscuits, and cake. The two are often blended. The main producers are the USSR, USA, Canada, China, Argentina, W Europe, and Australia.

whirlpool A powerful circular flow of water. In the sea it may result from the meeting of two opposing currents. For example, the Maelstrom off the Lofoten Islands of Norway. In rivers it is found where large lateral differences in the velocity of flow cause horizontal turbulence, for example, near the foot of waterfalls or at the junction of rivers.

whirlwind A small rotating vertical eddy of air, which circulates around an area of low pressure. It is formed by local heating and convectional uprising.

white alkali soil †*See* solonchak.

willy-willy A type of tropical cyclone occurring off the coast of NW Australia. They are fairly infrequent, occurring on average twice a year.

wilting point A level of moisture in the soil at which plants suffer water shortage unless the supply is quickly renewed. It is the point at which water available in the soil cannot replenish that lost by transpiration.

wind The movement of air over the Earth's surface. It is almost totally horizontal, although there is a very small vertical component. The speed and direction are important in meteorology, and are measured at all weather stations. *See also* Beaufort wind scale, general circulation of the atmosphere, planetary winds.

wind break A screen (either natural or artificial) that is erected to check the force of the wind and give shelter to buildings or crops on the lee side, e.g. a line of shrubs or trees. Farm buildings on the polderlands of the Netherlands, for example, are protected by tall hedges and trees. In the Rhône Valley of France protection against the cold mistral wind is provided by tall rows of poplar trees.

wind gap A valley that cuts across a ridge and marks the route of a former river. Wind gaps are a common feature of cuestas in which the original consequent streams have been captured by subsequent streams eroding along vales formed in the weaker rock strata. *See* capture, river.

wind rose A diagram that shows the frequency of winds from each cardinal direction for a particular place and period (usually one year). Radials are drawn from a central point in each of usually 8 or 16 compass directions, each radial being of a length proportionate to the frequency of winds from its direction. This enables prevailing wind patterns to be easily seen. The percentage of days of calm can be indicated in the centre of the diagram. More complex diagrams can be drawn showing variations such as wind speed.

wind shadow An area in the lee of an obstacle that experiences reduced wind speeds. Despite the lower wind speeds there are generally turbulent wind eddies within the wind shadow; for example, eddies in the lee of sand dunes, which maintain the steeper lee slope to the dune.

wind vane (weather vane) An instrument that indicates wind direction. It consists of a freely rotating horizontal arm, usually in the shape of an arrow, mounted over a fixed frame indicating the four main compass points.

windward The side facing the direction from which the wind is blowing. *Compare* lee.

winter The coldest season of the year. In the N hemisphere it is defined as extending from about 22 December, the winter solstice, to about 21 March, the spring equinox; in the S hemisphere it extends from about 21 June to about 21 September. It is popularly assumed to include the months of December, January, and February in the N hemisphere and June, July, and August in the S hemisphere. *See also* season.

winterbourne A stream that only flows during the wetter winter period. It is often found in dry valleys in chalk or limestone.

wool The fibres obtained from the fleeces of some animals, especially sheep. Wool is absorbent, strong, and warm, and can be dyed, spun, and woven to make cloth. The chief wool-exporting countries are Australia, New Zealand, Argentina, and South Africa. Small amounts of high-quality wool are also obtained from the Angora goat (mohair), Kashmir goat (cashmere), llama (alpaca), and vicuna.

working capital That part of a firm's capital that is used for the carrying on of a business and is not invested in buildings or equipment. The amount of working capital required depends on the type of business. For a clothing manufacturer much of the working capital will be in the form of rolls of cloth, whereas for the arable farmer it will consist of seed and fertilizer.

wrench fault *See* tear fault.

X

xenolith A fragment of rock that is contained within an igneous intrusion from which it differs in origin, structure, composition, etc. The rock may be a fragment of the rock through which the magma of the igneous intrusion forced its way. Xenoliths are frequently highly metamorphosed and are found near the top of stocks and batholiths. With increasing depth xenoliths become smaller and eventually disappear, presumably melted fully into the magma. [From Greek *xenos*: foreign]

xerophyte †A plant that can survive prolonged drought. Xerophytes are typically found in deserts and semi-arid climates and include cacti. They show adaptations to such habitats including long or enlarged roots; water-storage organs; and leaves that are thickened and waxy coated, reduced to narrow spines, or deciduous during dry seasons. Xerophytes are typically stout stemmed, thorny, or succulent.

xerosere †A form of sere (complete plant succession) that develops in very dry bare habitats. Xeroseres of rocky environments are called *lithoseres*; those of sandy areas, such as immature dunes, are called *psammoseres*. Xeroseres depend on the pioneer plant communities to enrich the developing soil with humus, and thus initiate succession. *Compare* hydrosere. *See also* sere.

Y

yardang A steep-sided linear rock ridge found in the deserts of central Asia. Yardangs occur in groups separated by corridor-like depressions that have been eroded by wind-blown sand. The wind-blown sand also undercuts the sides of the rock ridges creating a typical overhanging profile. †They are formed of more resistant beds in upended sedimentary strata, especially where the weaker strata can be exploited by the dominant wind.

year The time taken for the Earth to complete one orbit of the Sun. †It is measured in a number of ways. The *sidereal year* is measured with respect to the 'fixed' stars and equals 365.256 36 mean solar days. The *solar year* is the time taken for the Sun to make two successive appearances at the point of Aries. It is 365.242 19 mean solar days. The *calendar year* is regulated using leap years so that its average length is equal to that of the

solar year. To the usual year of 365 days one day is added (between 28 February and 1 March) every fourth year (leap year). This correction is too large, so the leap day is omitted in the century years (1800, 1900, etc.) unless the year is divisible by 400, so that 2000 will be a leap year.

young mountains The fold mountain ranges formed during the Alpine orogeny, the last great period of mountain building.

Z

zambo In South America, a person who is of mixed Negro and American Indian ancestry. [Spanish]

zenith The point in the heavens (i.e. on the celestial sphere) directly above the observer's head. It is opposite to the nadir. *Compare* nadir.

zenithal equal-area projection (azimuthal equal-area projection) A map projection in which the parallels of latitude have been drawn with different radii so that they enclose on the map the same area as they do on the globe. A sheet of paper is regarded as touching the globe at one chosen place and lines of latitude and longitude are projected on to it. This projection is commonly used to show polar areas.

zenithal equidistant projection (azimuthal equidistant projection) A zenithal map projection in which the parallels have been drawn their true distance apart. As a result distances on the map accurately represent the corresponding distances on the globe. For example, in the polar case of this projection distances along meridians of longitude are correct and the directions of all places from the pole are also correct.

zenithal projection (azimuthal projection) A map projection in which all the bearings are true compass directions from the centre of the map. It results from the projection of lines of latitude and longitude onto a plane surface regarded as touching the globe at a chosen point. *See also* zenithal equal-area projection, zenithal equidistant projection.

zeuge (*plural*: zeugen) A desert landform consisting of an upstanding tabular rock undercut around its edges by erosion by wind-blown sand. The cap of the landform is generally more resistant. Zeugen are formed where a hard horizontal rock stratum lies above a softer stratum. Erosion cuts through the harder rock along joints leaving the separate tabular masses. Zeugen may be up to 30 m in height. [German]

zinc A bluish-white brittle metal. It has the property that at a temperature between 100°C and 150°C it can be rolled into sheets and drawn out into wires. The chief ores from which zinc is obtained are sphalerite (zinc blende, ZnS), smithsonite (calamine, $ZnCO_3$), and zincite (ZnO). Metallic zinc is used for making galvanized steel and is a constituent of several alloys, especially brass and nickel-silver. Zinc compounds are used in the preparation of solders and pigments, and in a variety of processes including dyeing. The chief producers are Canada, Australia, the USSR, the USA, Japan, and Mexico.

zonal model of urban land use †*See* concentric model of urban land use.

zonal soil A soil that is well developed and owes its character, especially profile development, mainly to the influence of climate and/or vegetation. Other soil-forming factors are less important. The term applies only to mature free-draining soils (e.g. podzols, chernozems) developed on parent material that is neither very

acid nor very alkaline. †Zonal soils correlate broadly with climatological regions.
Compare azonal soil, intrazonal soil.

zonda 1. A föhn-type wind that blows down from the Andes in Argentina.
2. A hot moist northerly wind that precedes a passing low-pressure system in Argentina and Uruguay.

zone in transition (transition zone) †The second zone in E. W. Burgess's concentric model of urban land use. The zone in transition is contiguous with the central business district (CBD) and is an area of mixed land uses with industry, commerce, and poor housing, which are regularly changing. Its transitional character is the result of invasion by business and light manufacturing industries from the CBD and of physical deterioration and decay. It is not attractive enough to major central functions for them to seek to expand into it and is so close to the city centre area of congestion that it remains largely unattractive to other functions. As a result large portions are abandoned to poor residential uses, which cannot be profitably renewed. *See also* concentric model of urban land use.

zone of urban renewal *See* urban renewal.

zoogeography The scientific study of the geographical distribution of animal life; it is one of the two chief divisions of biogeography. Several zoogeographical regions, separated by natural barriers (such as deserts, mountains, and oceans), broadly define the major groupings of animal species. *See also* Wallace's line.

zoophyte An animal that resembles a plant in its appearance. Zoophytes include corals and sponges.